Inverse Spectral Problems for Differential Operators and Their Applications

ANALYTICAL METHODS AND SPECIAL FUNCTIONS
An International Series of Monographs in Mathematics

Volume 1
Series of Faber Polynomials
P.K. Suetin

Volume 2
Inverse Spectral Problems For Differential Operators and Their Applications
V.A. Yurko

Volume 3
Orthoganal Polynomials in Two Variables
P.K. Suetin

Additional Volumes in Preparation

Bessel Functions and Their Applications
B.G. Korenev

Hypersingular Integrals and Their Applications
S.G. Samko

Fourier Transforms and Approximations
A.M. Sedletskii

This book is part of a series. The publisher will accept continuation orders which may be cancelled at any time and which provide for automatic billing and shipping of each title in the series upon publication. Please write for details.

Inverse Spectral Problems for Differential Operators and Their Applications

V.A. Yurko

University of Saratov
Russia

Gordon and Breach Science Publishers
Australia • Canada • France • Germany • India • Japan • Luxembourg
Malaysia • The Netherlands • Russia • Singapore • Switzerland

Printed in Singapore.

Amsteldijk 166
1st Floor
1079 LH Amsterdam
The Netherlands

British Library Cataloguing in Publication Data

Yurko, V. A.
Inverse spectral problems for linear differential operators and their applications. – (Analytical methods and special functions ; v. 2)
1. Differential operators 2. Spectral theory (Mathematics)
I. Title
515.7′242

ISBN: 90-5699-189-2
ISSN: 1027-0264

CONTENTS

PREFACE

In this monograph the theory of the inverse problems of spectral analysis for ordinary non-self-adjoint differential operators is constructed. Attention is focused on differential equations of an arbitrary order on the half-line and on a finite interval. Inverse problems of spectral analysis consist in recovering operators from their spectral characteristics. Such problems appear frequently in mathematics, mechanics, physics, geophysics, electronics, meteorology and other branches of the natural sciences. Inverse problems also play an important role in solving nonlinear evolution equations of mathematical physics. Interest in this subject has been increasing steadily because of the appearance of new important applications in the natural sciences, and nowadays the inverse problem theory is being developed intensively all over the world. The main goals of this monograph are as follows:

- to highlight spectral characteristics which determine the operators coefficients uniquely;
- to comprehensively investigate analytic, asymptotic and structural properties of spectral characteristics;
- to give a constructive procedure for finding the coefficients of operators;
- to obtain necessary and sufficient conditions for the solvability of the inverse problems;
- to investigate the so-called incomplete inverse problems either when only some part of the operator's coefficients must be found (the rest of them are known a priori) or there is other information about the operator;
- to provide some applications of the theory of the inverse problem which has been constructed.

The book consists of two parts. In Part I we introduce and study the so-called Weyl matrix of the differential operator as the main spectral characteristic. We provide the solution of the inverse problem of recovering differential operators with integrable coefficients from the Weyl matrix. In Part II we study incomplete inverse problems when only some part of the Weyl matrix is given, and there is a priori information about the operator or its spectrum. Such problems often appear in applications. In this direction the method of standard models for solving incomplete inverse problems is suggested. We obtain a constructive solution of a wide class of incomplete inverse problems.

There are results related to applications of the inverse problem theory which has been constructed. In Part II an inverse problem of elasticity is solved when parameters of a beam must be determined from given frequencies of its natural oscillations. This problem relates to incomplete inverse problems, and the solution involves the application of the method of standard models.

The solution of the mixed problem for the nonlinear Boussinesq equation on the half-line is obtained by the inverse problem method. Necessary and sufficient conditions of its solvability are given. The results from Part I of this book for third-order differential operators on the half-line are also applied to this problem.

INTRODUCTION

Inverse problems of spectral analysis consist in recovering differential operators from their spectral characteristics. Such problems often appear in various branches of the natural sciences such as quantum mechanics, electronics, elasticity, geophysics, meteorology and so on. Inverse problems also play an important role in solving nonlinear evolution equations of mathematical physics.

The greatest success in the inverse problem theory was achieved for the Sturm-Liouville operator

$$-y'' + q(x)y. \tag{0.1}$$

The inverse problem for (0.1) was studied by Ambarzumian, Borg, Faddeev, Gasymov, Gel'fand, Krein, Levitan, Levinson, Marchenko, Rofe-Beketov, Sadovnichii and other mathematicians (see [1]–[34]). The first result in this direction belonged to Ambarzumian [1]. He showed that if the eigenvalues of the boundary value problem

$$-y'' + q(x)y = \lambda y, \quad q(x) \in C[0,\pi], \quad y'(0) = y'(\pi) = 0$$

are $\lambda_k + k^2, k \geq 0$, then $q(x) \equiv 0$. However, this result is an exception to the rule, and the specification of the spectrum does not uniquely determine the operator (0.1). Afterwards Borg [2] proved that the specification of two spectra of Sturm-Liouville operators uniquely determines the function $q(x)$. Levinson [3] suggested a different method to prove Borg's results. Tikhonov obtained the uniqueness theorem for the inverse Sturm-Liouville problem on the half-line with a given Weyl function [4].

An important role in the spectral theory of Sturm-Liouville operators was played by the transformation operator. Marchenko [5]–[6] first applied the transformation operator to the solution of the inverse problem. He proved that a Sturm-Liouville operator on the half-line or a finite interval is uniquely determined by specifying the spectral function. Transformation operators were also used in the fundamental paper of Gel'fand and Levitan [10], where they obtained necessary and sufficient conditions along with a method for recovering a Sturm-Liouville operator from its spectral function. Another approach to the inverse problems was suggested by Krein [14]–[15]. Blokh [16] and Rofe-Beketov [17] studied the inverse problem on the line with the given spectral matrix. The inverse scattering theory on the half-line and on the line was constructed in [25]–[34] and other papers. Many works are devoted to the inverse problem theory for partial differential equations. This direction is reflected fairly completely in [35]–[40].

In recent years a new area for applications of the inverse problem theory appeared. In 1967 Gardner, Green, Kruskal and Miura [41] found a remarkable method for solving some important nonlinear equations of mathematical physics connected with the use of the inverse problem theory. This method has been described in [42]–[45] and other works.

In contrast to the case for Sturm-Liouville operators, there are only isolated fragments, not constituting a general picture, in the inverse problem theory for higher-order differential operators of the form

$$ly = y^{(n)} + \sum_{\upsilon=0}^{n-2} p_\upsilon(x)y^{(\upsilon)}. \tag{0.2}$$

Inverse problems for (0.2) in various formulations were studied in [46]–[71] and other works. Fage [72], Leont'ev [73] and Hromov [74] determined that for $n > 2$ transformation operators have a much more complicated structure than for Sturm-Liouville operators which makes it more difficult to use them for solving the inverse problem. However, in the case of analytic coefficients the transformation operators have the same 'triangular' form as for Sturm-Liouville operators (see [46], [52] and [75]). Sakhnovich [47]–[48] and Khachatryan [50]–[51] used a 'triangular' transformation operator to investigate the inverse problem of recovering self-adjoint differential operators on the half-line from the spectral function, as well as the scattering inverse problem. The scattering inverse problem on the line has been treated in various statements in [63]–[69] and other works.

Leibenzon in [53], [54] investigated the inverse problem for (0.2) on a finite interval under the condition of 'separation' of the spectrum. The spectra and 'weight' numbers of certain specially chosen boundary value problems for the differential operators (0.2) appeared as spectral data of the inverse problem. The inverse problem for (0.2) on a finite interval with respect to a system of spectra was investigated under various conditions on the operator in [55]–[59]. Things are more complicated for differential operators on the half-line, since in the non-self-adjoint case the spectrum can have rather 'bad' behavior (see [22], [23]). Moreover, for the differential operator (0.2) on the half-line there is not even a statement of the inverse problem, and it is not clear what the analogue of the discrete data of Leibenzon is, nor what the analogue of the system of the spectra is. The question of the formulation and solution of the inverse problem on a finite interval is also open, when the behavior of the spectrum is arbitrary, since waiving the requirement of 'separation' of the spectra leads to a violation of uniqueness for solution of the inverse problem. So we need to construct the inverse problem theory for the non-self-adjoint differential operator (0.2) when the behavior of the spectrum is arbitrary. In recent years there has been considerable interest in investigating inverse problems for higher-order differential operators because of new applications in elasticity theory and in nonlinear mathematical physics.

In this book we consider the inverse problem of recovering the differential operator (0.2) on the half-line and on a finite interval from the so-called Weyl matrix $\mathfrak{M}(\lambda) = [\mathfrak{M}_{mk}(\lambda)]$, which expresses the spectral properties of the operator most completely; this enables us to construct a general theory of the inverse problem for non-self-adjoint differential operators (0.2) when the behavior of the spectrum is arbitrary. We remark that, in particular, all the inverse problems mentioned above — with the given spectral function, with the given system of spectra, 'incomplete' inverse problems of determining some of the operator's coefficients, with the given traces of solutions of partial differential equations, and others — can be reduced to the inverse problem with the given Weyl functions. For a self-adjoint second-order differential operator the Weyl function introduced here coincides with the classical Weyl function.

In Part I we provide the solution of the inverse problem of recovering the differential operator (0.2) with integrable coefficients from the Weyl matrix

$\mathfrak{M}(\lambda) = [\mathfrak{M}_{mk}(\lambda)]$ on the half-line and on a finite interval. In Chapter 1 we formulate the inverse problem, study the properties of the Weyl functions $\mathfrak{M}_{mk}(\lambda)$ and prove a uniqueness theorem for the solution of the inverse problem. In Chapters 2–3 the solution of the inverse problem on the half-line is obtained. We give a derivation of the main equation of the inverse problem, which is a singular linear equation

$$\tilde{\varphi}(x,\lambda) = \tilde{N}(\lambda)\varphi(x,\lambda) + \frac{1}{2\pi i}\int_{\gamma}\frac{\tilde{H}(x,\lambda,\mu)}{\mu-\lambda}\varphi(x,\mu)d\mu, \quad \lambda\in\gamma, \quad x\geq 0, \tag{0.3}$$

with respect to $\varphi(x,\lambda)$. Here $\varphi(x,\lambda)$ is a vector-function constructed from special solutions of the differential equation $\mathrm{l}y = \lambda y$. The functions $\tilde{\varphi}(x,\lambda), \tilde{N}(\lambda), \tilde{H}(x,\lambda,\mu)$ are constructed from the given differential operator

$$\tilde{\mathrm{l}}y = y^{(n)} + \sum_{\upsilon=0}^{n-2}\tilde{p}_{\upsilon}(x)y^{(\upsilon)},$$

and from the Weyl matrix $\mathfrak{M}(\lambda)$ of the operator (0.2). We give a constructive procedure, as well as necessary and sufficient conditions on the Weyl matrix when the behavior of the spectrum is arbitrary. Further we consider the particular case, namely, differential operators with a simple spectrum. In this case the Weyl matrix is uniquely determined by specifying the so-called spectral data. For self-adjoint second-order differential operators we establish a connection between the main equation of the inverse problem and the Gel'fand-Levitan equation [10].

In Chapter 4 we study the differential operator (0.2) on a finite interval. In this case there are specific difficulties connected with the non-trivial structural properties of the Weyl matrix $\mathfrak{M}(\lambda)$ in the neighborhoods of the points of the spectrum. We provide an algorithm for the solution of the inverse problem, as well as necessary and sufficient conditions of solvability of the inverse problem. A counter example shows that dropping one element of the Weyl matrix violates the uniqueness of the solution of the inverse problem.

In Chapter 5 the inverse problem for self-adjoint higher-order differential operators is investigated. For the self-adjoint case we obtain a procedure for constructing the operator along with necessary and sufficient conditions which are simpler and easier to check than for the non-self-adjoint case.

In Chapter 6 we investigate the inverse problem for differential operators with singular points.

In Part II we study the so-called incomplete inverse problems when only some part of the Weyl matrix is given and there is a priori information about the operator or its spectrum. Such problems often appear in applications. In Chapter 7 we provide the solution of the inverse problem for differential operators with a 'separate' spectrum. In this case to find the operator's coefficients it is sufficient to know $n-1$ Weyl functions. Among other things, the theorems obtained contain the results of Leibenson [54], [55].

In Chapter 8 stability of the solution of the inverse problem is studied for differential operators with a 'separate' spectrum. Here we apply a different method connected with ideas of Levinson [3].

In Chapter 9, we study the inverse problem of recovering $N(1\leq N\leq n-1)$ coefficients of the differential operator (0.2) $\{p_{n-\kappa_j}(x)\}_{j=\overline{1,N}}$ from the given N Weyl functions $\left[\mathfrak{M}_{m_i,\gamma_i}(\lambda)\right]_{j=\overline{1,N}}$ and the known functions $p_k(x) \in L\ (0,T),\ k \neq n-\kappa_j,$ $j=\overline{1,N}$, when the coefficients $p_{n-\kappa_j(x), j=\overline{1,N}}$ are piecewise-analytic functions. So, in contrast to Chapters 7–8, here are a priori conditions on the operator's coefficients. We give a classification of incomplete inverse problems. To solve such inverse problems, the so-called method of standard models is suggested. We obtain the uniqueness theorem and a constructive solution for a wide class of incomplete inverse problems.

In Chapter 10 we apply the method of standard models to solve an inverse problem of elasticity theory when parameters of a beam are to be determined from given frequences of its natural oscillations. This problem can be reduced to the inverse problem of recovering the fourth-order differential operator

$$(h^{\mu}(x)y'')'' = \lambda h(x)y, \quad \mu = 1,2,3$$

from the Weyl function. To solve this problem we use the method described in Chapter 9.

In Chapter 11 we provide the solution of the inverse problem for the operator (0.2) on the half-line with locally integrable coefficients. To solve this problem we introduce the so-called generalized Weyl functions and use connections with an inverse problem for partial differential equations. We also use the Riemann-Fage formula [76] for the solution of the Cauchy problem for higher-order partial differential equations. Note that for $n = 2$ generalized functions for solution of inverse problems were applied by Marchenko [7], [8].

In Chapter 12 an inverse problem for discrete operators of triangular structure is studied. An algorithm for the solution, as well as necessary and sufficient conditions of solvability of this problem are obtained. Applications to difference and differential operators are considered. In Chapter 13 we provide a solution of an inverse problem for integrodifferential operators.

In Appendices I and II there are results related to applications of the inverse problem theory constructed in solving nonlinear evolution equations of mathematical physics.

In Appendix I the mixed problem for the nonlinear Boussinesq equation on the half-line is studied. To solve this problem we use the inverse problem method and results obtained in Part I of this book. An algorithm for the solution, and necessary and sufficient conditions for the solvability of the mixed problem are obtained, and the uniqueness is proved.

In Appendix II using results obtained in Part II of the book we study integrable nonlinear dynamical systems connected with higher-order difference operators by the inverse problem method.

We note that some results related to the topic of this book are in [77]–[91].

Notations:

1. If we consider a differential operator ℓ, then along with ℓ we consider a differential operator $\tilde{\ell}$ of the same form, but with different coefficients. We

agree that if some symbol φ denotes an object relating to $\mathcal{L}$, then $\tilde{\varphi}$ denotes the analogous object realting to $\tilde{\ell}$, and $\hat{\varphi} - \varphi - \tilde{\varphi}$.

2. One and the same symbol C denotes various positive constants in estimates.

3. A matrix A with elemens $a_{ij}, i - \overline{1,r}, j = \overline{1,s}$ will be written in one of the following ways:

$$A = \left[a_{ij}\right]_{i=\overline{1,r},\, j=\overline{1,s}} = \left[a_{i1}, \ldots, a_{is}\right]_{i=\overline{1,r}} = \left[a_{1j}, \ldots, a_{rj}\right]^{T}_{j=\overline{1,s}},$$

where, i is the row index, j is the column index, and T is the symbol for transposition. If A has the maximum rank, we shall write $A \neq 0$.

4. By E we denote the identity matrix of the corresponding dimension or the identity operator on the corresponding space.

5. If for $\lambda \to \lambda_0$

$$F(\lambda) = \sum_{k=-q}^{p} \alpha_k (\lambda - \lambda_0)^k + o((\lambda - \lambda_0)^p),$$

then

$$\left[F(\lambda)\right]^{\langle k \rangle}_{\lambda=\lambda_0} = F_{\langle k \rangle}(\lambda_0) \stackrel{\mathrm{df}}{=} \alpha_k$$

PART I

RECOVERY OF DIFFERENTIAL OPERATORS FROM THE WEYL MATRIX

1. Formulation of the inverse problem. A uniqueness theorem

1.1. We consider a differential equation (DE) and linear forms (LF) $L = (\ell, U)$ of the form

$$\ell y \equiv y^{(n)} + \sum_{\nu=0}^{n-2} p_\nu(x) y^{(\nu)} = \lambda y, \qquad 0 \le x \le T \le \infty, \tag{1.1}$$

$$U_{\xi a}(y) = y^{(\sigma_{\xi a})}(a) + \sum_{\nu=0}^{\sigma_{\xi a}-1} u_{\xi\nu a} y^{(\nu)}(a), \qquad \xi = \overline{1,n}, \tag{1.2}$$

on the half-line ($T = \infty$) or on the finite interval ($T < \infty$). Here $p_\nu(x) \in \mathcal{L}(0,T)$ are complex-valued integrable functions; $a = 0$ for $T = \infty$, and $a = 0, T$ for $T < \infty$; $0 \le \sigma_{\xi a} \le n-1$, $\sigma_{\xi a} \ne \sigma_{\eta a}$ $(\xi \ne \eta)$.

Let $\lambda = \rho^n$. It is known (see [92, p. 53]) that the ρ-plane can be partitioned into sectors S of angle $\frac{\pi}{n}$ $\left(\arg\rho \in \left(\frac{\nu\pi}{n}, \frac{(\nu+1)\pi}{n}\right), \ \nu = \overline{0, 2n-1}\right)$ in which the roots $R_1, \dots, R_n$ of the equation $R^n - 1 = 0$ can be numbered in such a way that

$$\operatorname{Re}(\rho R_1) < \operatorname{Re}(\rho R_2) < \ldots < \operatorname{Re}(\rho R_n), \qquad \rho \in S. \tag{1.3}$$

Let functions $\Phi(x,\lambda) = [\Phi_m(x,\lambda)]_{m=\overline{1,n}}$ be solutions of (1.1) satisfying the conditions $U_{\xi 0}(\Phi_m) = \delta_{\xi m}$, $\xi = \overline{1,m}$ and $U_{\eta T}(\Phi_m) = 0$, $\eta = \overline{1, n-m}$ (for $T < \infty$), $\Phi_m(x,\lambda) = O(\exp(\rho R_m x))$, $x \to \infty$, $\rho \in S$ (for $T = \infty$). Here and in the sequel, $\delta_{\xi m}$ is the Kronecker symbol. Denote $\mathfrak{M}_{mk}(\lambda) = U_{k0}(\Phi_m)$, $k = \overline{m+1, n}$. The functions $\Phi_m(x,\lambda)$ and $\mathfrak{M}_{mk}(\lambda)$ are called the Weyl solutions (WS's) and the Weyl functions (WF's), respectively. The matrix $\mathfrak{M}(\lambda) = [\mathfrak{M}_{mk}(\lambda)]_{m,k=\overline{1,n}}$, $\mathfrak{M}_{mk}(\lambda) = \delta_{mk}$, $k = \overline{1,m}$ is called the Weyl matrix (WM) or the spectrum of L. Thus, $\mathfrak{M}(\lambda) = U_0(\Phi(x,\lambda))$, where $U_a = [U_{\xi a}]^T_{\xi=\overline{1,n}}$. We note that

$$\Phi(x,\lambda) = \mathfrak{M}(\lambda) C(x,\lambda), \tag{1.4}$$

where $C(x,\lambda) = [C_m(x,\lambda)]_{m=\overline{1,n}}$ are the solutions of (1.1) under the conditions $U_{\xi 0}(C_m) = \delta_{\xi m}$, $\xi = \overline{1,n}$.

Formulation of the inverse problem. Given the WM $\mathfrak{M}(\lambda)$, construct the DE and LF $L = (\ell, U)$.

In §1 we study the properties of the WF's and prove the uniqueness theorem of recovering the DE and LF (1.1)–(1.2) on the half-line and on the finite interval from the given WM $\mathfrak{M}(\lambda)$ when the behavior of the spectrum is arbitrary. Below, in §4, we provide a counterexample showing that dropping one element of the WM violates the uniqueness of the solution of the inverse problem (IP).

1.2. Let $\alpha \in (0,T)$, $\rho_\alpha = 2n \max_\nu \|p_\nu\|_{\mathcal{L}(\alpha,T)}$. It is known (see, for example, [92, p. 58]) that in each sector S with the property (1.3) there exists a fundamental system of solutions (FSS) $B_\alpha = \{y_k(x,\rho)\}_{k=\overline{1,n}}$ of the equation (1.1) of the form

$$y_k^{(\nu)}(x,\rho) = (\rho R_k)^\nu \exp(\rho R_k x)(1+O(\rho^{-1})), \qquad |\rho| \to \infty, \qquad x \ge \alpha, \tag{1.5}$$

where for $x \ge \alpha$ and $r_k = k$ the functions $y_k(x,\rho)$ satisfy the equations

$$\begin{aligned} y_k(x,\rho) &= \exp(\rho R_k x) - \int_\alpha^x \sum_{j=1}^{r_k} R_j \exp(\rho R_j(x-t))\, M_t(y_k)\, dt \\ &\quad + \int_x^T \sum_{j=r_k+1}^{n} R_j \exp(\rho R_j(x-t))\, M_t(y_k)\, dt, \end{aligned} \tag{1.6}$$

$$M_t(y_k) = \frac{1}{n}\rho^{1-n} \sum_{\mu=0}^{n-2} p_\mu(t)\, y_k^{(\mu)}(t,\rho).$$

The functions $y_k^{(\nu)}(x,\rho)$, $\nu = \overline{0,n-1}$ are regular for each $x \ge 0$ with respect to $\rho \in S_\alpha = \{\rho : \rho \in S, |\rho| > \rho_\alpha\}$, are continuous for $x \ge 0$ and $\rho \in \bar{S}_\alpha$ and have the estimate

$$|y_k^{(\nu)}(x,\rho)(\rho R_k)^{-\nu} \exp(-\rho R_k x) - 1| \le \rho_\alpha |\rho|^{-1}, \qquad x \ge \alpha, \qquad \rho \in \bar{S}_\alpha.$$

As $|\rho| \to \infty$, $\rho \in S$

$$\det [y_k^{(\nu-1)}(x,\rho)]_{k,j=\overline{1,n}} = \rho^{n(n-1)/2} \det [R_k^{\nu-1}]_{k,\nu=\overline{1,n}} (1+O(\rho^{-1})).$$

Moreover, we require the FSS

$$B_{\alpha m} = \{y_1^0(x,\rho), \ldots, y_m^0(x,\rho), y_{m+1}(x,\rho), \ldots, y_n(x,\rho)\}$$

of DE (1.1), where $y_k(x,\rho) \in B_\alpha$, $k = \overline{m+1,n}$, and the functions $y_k^0(x,\rho)$, $k = \overline{1,m}$ are solutions of (1.6) for $x \ge \alpha$ and $r_k = m$. Furthermore, the functions $y_k^{0(\nu)}(x,\rho)$, $\nu = \overline{0,n-1}$ are continuous for $x \in [0,T]$, $\rho \in \bar{S}_\alpha$, are regular with respect to $\rho \in S_\alpha$ for each $x \in [0,T]$ and satisfy

$$y_k^{0(\nu)}(x,\rho) = O(\rho^\nu \exp(\rho R_m x)), \qquad x \ge \alpha, \qquad |\rho| \to \infty, \qquad \rho \in S.$$

For $T = \infty$ we also introduce the FSS $B_\alpha^0 = \{y_{k0}(x,\rho)\}_{k=\overline{1,n}}$, where the functions $y_{k0}(x,\rho)$ are solutions of (1.6) for $x \geq \alpha$, $r_k = k-1$. The functions $y_{k0}^{(\nu)}(x,\rho)$, $\nu = \overline{0,n-1}$ are continuous for $x \geq 0$, $\rho \in \bar{S}_\alpha$, are regular with respect to $\rho \in S_\alpha$ for each $x \geq 0$ and

$$y_{k0}^{(\nu)}(x,\rho) = (\rho R_k)^\nu \exp(\rho R_k x)(1 + O(\rho^{-1})), \qquad x \geq \alpha, \qquad |\rho| \to \infty,$$

$$\lim_{x\to\infty} y_{k0}^{(\nu)}(x,\rho)(\rho R_k)^{-\nu} \exp(-\rho R_k x) = 1,$$

$$\det [y_{k0}^{(\nu-1)}(x,\rho)]_{k,\nu=\overline{1,n}} \equiv \rho^{n(n-1)/2} \det [R_k^{\nu-1}]_{k,\nu=\overline{1,n}} \,.$$

1.3. Let $\omega_\xi(R) = R^{\sigma_{\xi 0}}$,

$$\Omega(j_1,\dots,j_p) = \det[\omega_{j_\nu}(R_k)]_{\nu,k=\overline{1,p}}, \quad \Omega_\mu(j_1,\dots,j_p) = \det[\omega_{j_\nu}(R_k)]_{\nu=\overline{1,p};k=\overline{1,p+1}\setminus\mu},$$

$$\mu_{mk}^0 = (\Omega(\overline{1,m}))^{-1}\Omega(\overline{1,m-1},k), \quad a_{mk}^0 = (-1)^{m+k}(\Omega(\overline{1,m}))^{-1}\Omega_k(\overline{1,m-1}),$$

and also $\Gamma = \{\lambda : \operatorname{Im}\lambda = 0\}$ and $\Gamma_\pm = \{\lambda : \pm\lambda > 0\}$; Π and $\Pi_{\pm 1}$ are the λ-plane with the cuts Γ and $\Gamma_\pm$ respectively.

Theorem 1.1.

(1) *Let $T < \infty$. Then the WF's $\mathfrak{M}_{mk}(\lambda)$ are meromorphic in λ and*

$$\left.\begin{aligned} &\mathfrak{M}_{mk}(\lambda) = (\Delta_{mm}(\lambda))^{-1}\Delta_{mk}(\lambda), \\ &\Delta_{mk}(\lambda) \overset{df}{=} (-1)^{m+k} \det [U_{\xi T}(C_\nu)]_{\xi=\overline{1,n-m};\nu=\overline{m,n}\setminus k}\,. \end{aligned}\right\} \tag{1.7}$$

(2) *Let $T = \infty$. Then the WF's $\mathfrak{M}_{mk}(\lambda)$ are regular in $\Pi_{(-1)^{n-m}}$ with the exception of an at most countable bounded set Λ'_{mk} of poles. For $(-1)^{n-m}\lambda \geq 0$ the following limits exist and are finite off the bounded sets $\Lambda_{mk}^\pm$:*

$$\mathfrak{M}_{mk}^\pm(\lambda) = \lim_{z\to 0, \operatorname{Re} z>0} \mathfrak{M}_{mk}(\lambda \pm iz).$$

Proof. Let $T = \infty$, $\{y_k(x,\rho)\}_{k=\overline{1,n}}$ is the FSS B_0 of (1.1). We denote

$$\Delta_{mk}^0(\rho) = \det [U_{\xi 0}(y_\nu)]_{\nu=\overline{1,m};\xi=\overline{1,m-1},k}\,.$$

Then

$$\Phi_m(x,\lambda) = \sum_{k=1}^n a_{mk}(\rho) y_k(x,\rho).$$

Using the boundary conditions on $\Phi_m(x,\lambda)$ we obtain

$$a_{mk}(\rho)=0, \qquad k>m,$$

$$\sum_{k=1}^{m} a_{mk}(\rho)U_{\xi 0}(y_k)=\delta_{\xi m}, \qquad \xi=\overline{1,m}.$$

Hence

$$\left.\begin{aligned} &\Phi_m(x,\lambda)=\sum_{k=1}^{m} a_{mk}(\rho)y_k(x,\rho),\\ &a_{mk}(\rho)=(-1)^{m+k}(\Delta^0_{mm}(\rho))^{-1}\det\,[U_{\xi 0}(y_\nu)]_{\xi=\overline{1,m-1};\nu=\overline{1,m}\setminus k}. \end{aligned}\right\} \tag{1.8}$$

Since $\mathfrak{M}_{mk}(\lambda)=U_{k0}(\Phi_m(x,\lambda))$ it follows from (1.8) that

$$\mathfrak{M}_{mk}(\lambda)=(\Delta^0_{mm}(\rho))^{-1}\Delta^0_{mk}(\rho). \tag{1.9}$$

Using the asymptotic properties (1.5) of the functions $y_k^{(\nu)}(x,\rho)$, we have that as $|\rho|\to\infty$, $\rho\in\bar{S}$:

$$\left.\begin{aligned} &a_{mk}(\rho)=\rho^{-\sigma_{m0}}(a^0_{mk}+O(\rho^{-1})),\\ &\Phi_m(x,\lambda)=\rho^{-\sigma_{m0}}\sum_{k=1}^{m}\exp(\rho R_k x)(a^0_{mk}+O(\rho^{-1})), \end{aligned}\right\} \tag{1.10}$$

$$\left.\begin{aligned} &\Delta^0_{mk}(\rho)=\rho^{\sigma_{10}+\dots+\sigma_{m-1,0}+\sigma_{k0}}\Omega(\overline{1,m-1},k)(1+O(\rho^{-1})),\\ &\mathfrak{M}_{mk}(\lambda)=\rho^{\sigma_{k0}-\sigma_{m0}}\mu^0_{mk}(1+O(\rho^{-1})). \end{aligned}\right\} \tag{1.11}$$

Repeating the preceding arguments for the FSS $B_{\alpha m}$ we get that

$$\left.\begin{aligned} &\mathfrak{M}_{mk}(\lambda)=(\Delta^1_{mm}(\rho))^{-1}\Delta^1_{mk}(\rho),\\ &\Delta^1_{mk}(\rho)=\det\,[U_{\xi 0}(y^0_\nu)]_{\nu=\overline{1,m};\xi=\overline{1,m-1},k}. \end{aligned}\right\} \tag{1.12}$$

Let

$$G=\left\{\rho:\arg\in\left(((-1)^{n-m}-1)\frac{\pi}{2n},((-1)^{n-m}+3)\frac{\pi}{2n}\right)\right\}.$$

The domain G consists of two sectors S with the same collection $\{R_\xi\}_{\xi=\overline{1,m}}$. Consequently, the functions $\Delta^1_{mk}(\rho)$ are regular for $\rho\in G$, $|\rho|>\rho_\alpha$ and continuous for $\rho\in\bar{G}$, $|\rho|\geq\rho_\alpha$. The theorem is obtained from this in view of (1.9), (1.11), (1.12), and the arbitrariness of α.

Let $\Lambda_{mk}=\Lambda'_{mk}\cup\Lambda^+_{mk}\cup\Lambda^-_{mk}$, and $\Lambda=\bigcup_{m,k}\Lambda_{mk}$. We say that the spectrum of L has finite multiplicity if for some $p\geq 1$ we have that $\mathfrak{M}(\lambda)=O((\lambda-\lambda_0)^{-p})$, $\lambda\to\lambda_0$, $\lambda_0\in\Lambda$. For example, if $p_\nu(x)\exp(\varepsilon x)\in\mathcal{L}(0,\infty)$, $\varepsilon>0$, then the spectrum of L has finite multiplicity It follows from results of Pavlov [22], [23] that the spectrum can have infinite multiplicity in general.

Let $T < \infty$. Using the boundary conditions on $\Phi_m(x,\lambda)$ we obtain

$$\Phi_m(x,\lambda) = (\Delta_{mm}(\lambda))^{-1} \det [C_\nu(x,\lambda), U_{1T}(C_\nu), \ldots, U_{n-m,T}(C_\nu)]_{\nu=\overline{m,n}}, \qquad (1.13)$$

and consequently the relations (1.7) are valid. Theorem 1.1 is proved.

For $T < \infty$ we denote by $\Lambda_m = \{\lambda_{\ell m}\}_{\ell\geq 1}$ the set of zeroes (with multiplicity) of the entire function $\Delta_{mm}(\lambda)$ and $\Lambda = \bigcup_{m=1}^{n-1} \Lambda_m$. The numbers $\{\lambda_{\ell m}\}$ coincide with eigenvalues of the boundary value problems S_m for the DE (1.1) under the conditions $U_{\xi 0}(y) = U_{\eta T}(y) = 0$, $\xi = \overline{1,m}$, $\eta = \overline{1,n-m}$.

Indeed, let λ_0 be an eigenvalue, and let $\psi(x)$ be an eigenfunction of S_m. Then

$$\psi(x) = \sum_{\mu=1}^{n} \alpha_\mu C_\mu(x,\lambda_0),$$

$$\sum_{\mu=1}^{n} \alpha_\mu U_{\xi 0}(C_\mu(x,\lambda_0)) = 0, \qquad \xi = \overline{1,m},$$

$$\sum_{\mu=1}^{n} \alpha_\mu U_{\eta T}(C_\mu(x,\lambda_0)) = 0, \qquad \eta = \overline{1,n-m}.$$

Since $\psi(x) \not\equiv 0$ this linear algebraic system has only a trivial solution, and consequently $\Delta_{mm}(\lambda_0) = 0$. The inverse assertion can be proved analogously.

It is known (see, for example, [93]) that the following asymptotic formulas are valid

$$\lambda_{\ell m} = (-1)^{n-m} \left(\frac{\pi}{T}\left(\sin\frac{\pi m}{n}\right)^{-1}\left(\ell + \chi_{m0} + O\left(\frac{1}{\ell}\right)\right)\right)^n, \qquad \ell\to\infty. \qquad (1.14)$$

Denote by $G_{\delta,m}$ the λ-plane without circles $|\lambda - \lambda_0| < \delta$, $\lambda_0 \in \Lambda_m$, $G_\delta = \bigcap_{m=1}^{n-1} G_{\delta,m}$. Let

$$s_{mk} = \sigma_{k0} + \sum_{\xi=1}^{m-1} \sigma_{\xi 0} + \sum_{\eta=1}^{n-m} \sigma_{\eta T} - \frac{n(n-1)}{2},$$

$$\begin{aligned}\Delta^1_{mk}(\rho) = \det\, [U_{10}(y_\nu), \ldots, U_{m-1,0}(y_\nu), U_{k0}(y_\nu),& \\ U_{1T}(y_\nu), \ldots, U_{n-m,T}(y_\nu)]_{\nu=\overline{1,n}},&\end{aligned} \qquad (1.15)$$

where $\{y_\nu(x,\rho)\}_{\nu=\overline{1,n}}$ is the FSS B_0 in a sector S with the property (1.3). Then

$$\Phi_m(x,\lambda) = \sum_{k=1}^{n} a_{mk}(\rho) y_k(x,\rho),$$

$$a_{mk}(\rho) = \frac{(-1)^{m+k}}{\Delta^1_{mm}(\rho)} \det \big[U_{10}(y_\nu), \dots, U_{m-1,0}(y_\nu), \\ U_{1T}(y_\nu), \dots, U_{n-m,T}(y_\nu)\big]_{\nu=\overline{1,n}\setminus k}. \tag{1.16}$$

We can write the functions $C_\nu(x,\lambda)$ as follows

$$C_\nu(x,\lambda) = \sum_{\mu=1}^{n} \alpha_{\nu\mu}(\rho) y_\mu(x,\rho).$$

Then

$$\det\left[U_{\xi 0}(C_\nu)\right]_{\xi,\nu=\overline{1,n}} = \det\left[U_{\xi 0}(y_\nu)\right]_{\xi,\nu=\overline{1,n}} \det\left[\alpha_{\nu\mu}\right]_{\nu,\mu=\overline{1,n}},$$

and analogously,

$$\Delta_{mk}(\lambda) = \Delta^1_{mk}(\rho) \det\left[\alpha_{\nu\mu}\right]_{\nu,\mu=\overline{1,n}}.$$

Hence

$$\Delta_{mk}(\lambda) = \Delta^1_{mk}(\rho) \left(\det\left[U_{\xi 0}(y_\nu)\right]_{\xi,\nu=\overline{1,n}}\right)^{-1}.$$

Using (1.15), (1.16) and the asymptotic properties (1.5) of the functions $y_k^{(\nu)}(x,\rho)$ we obtain for $|\lambda| \to \infty$, $\arg((-1)^{n-m}\lambda) = \beta \neq 0$, $\rho \in S$:

$$\left.\begin{aligned} a_{mk}(\rho) &= \rho^{-\sigma_{m0}}(a^0_{mk} + O(\rho^{-1})), & k &= \overline{1,m}, \\ a_{mk}(\rho) &= O(\rho^{-\sigma_{m0}} \exp(\rho(R_m - R_k)T)), & k &= \overline{m+1,n}; \end{aligned}\right\} \tag{1.17}$$

$$\Delta_{mk}(\lambda) = \rho^{s_{mk}} \frac{\Omega(\overline{1,m-1},k)}{\Omega(\overline{1,n})} \det\left[R_\nu^{\sigma_{jT}}\right]_{\nu=\overline{m+1,n};j=\overline{1,n-m}} \\ \times \exp\left(T\rho \sum_{j=m+1}^{n} R_j\right)(1+O(\rho^{-1})); \tag{1.18}$$

$$\left.\begin{aligned} \mathfrak{M}_{mk}(\lambda) &= \rho^{\sigma_{k0}-\sigma_{m0}} \mu^0_{mk}(1+O(\rho^{-1})), \\ \Phi_m(x,\lambda) &= \rho^{-\sigma_{m0}} \sum_{k=1}^{n} \exp(\rho R_k x)(a^0_{mk} + O(\rho^{-1})), \quad x \in [0,T), \end{aligned}\right\} \tag{1.19}$$

and also

$$\left.\begin{aligned} &|\Delta_{mm}(\lambda)| > C|\rho|^{s_{mm}}\left|\exp\left(T\rho\sum_{j=m+1}^{n} R_j\right)\right|, \quad \lambda \in G_{\delta,m},\\ &|\Phi_m^{(\nu)}(x,\lambda)| < C|\rho|^{\nu-\sigma_{mo}}|\exp(\rho R_m x)|, \qquad \lambda \in G_{\delta,m}, \end{aligned}\right\} \tag{1.20}$$

$$\Delta_{mk}(\lambda) = O\left(\rho^{s_{mk}}\exp\left(T\rho\sum_{j=m+1}^{n} R_j\right)\right), \qquad |\lambda|\to\infty. \tag{1.21}$$

1.4. Denote by W_ν the set of functions $f(x)$, $0 < x < T$ such that $f(x)$, $f'(x)$, ..., $f^{(\nu-1)}(x)$ are absolutely continuous and $f^{(k)}(x) \in \mathcal{L}(0,T)$, $k = \overline{0,\nu}$. Let $N \geq 0$ be a fixed integer. We say that $L \in v_N$ if $p_\nu(x) \in W_{\nu+N}$, $\nu = \overline{0,n-2}$. We shall assume below that $L \in v_N$. We define $p_n(x) = 1$, $p_{n-1}(x) = 0$, and $u_{\xi\nu a} = \delta_{\nu,\sigma_{\xi a}}$, $\nu \geq \sigma_{\xi a}$. Let

$$\langle y(x), z(x)\rangle = \langle y(x), z(x)\rangle_\ell = \sum_{\nu,j=0}^{n-1} \mathcal{L}_{\nu j}(x) y^{(\nu)}(x) z^{(j)}(x); \tag{1.22}$$

$$\left.\begin{aligned} &\mathcal{L}_{\nu j}(x) = \sum_{s=j}^{n-\nu-1}(-1)^s C_s^j p_{s+\nu+1}^{(s-j)}(x), \quad \nu + j \leq n-1;\\ &\mathcal{L}_{\nu j}(x) = 0, \qquad \nu + j > n-1. \end{aligned}\right\} \tag{1.23}$$

Here and below $C_s^j = \frac{s!}{j!(s-j)!}$.
We consider the DE and the LF

$$\ell^* z = (-1)^n z^{(n)} + \sum_{\nu=0}^{n-2}(-1)^\nu (p_\nu(x) z)^{(\nu)} = \lambda z, \tag{1.24}$$

$$U_{\xi a}^*(z) = z^{(\sigma_{\xi a}^*)}(a) + \sum_{\nu=0}^{\sigma_{\xi a}^*-1} u_{\xi\nu a}^* z^{(\nu)}(a), \qquad \sigma_{\xi a}^* = n - 1 - \sigma_{n+1-\xi,a}, \tag{1.25}$$

where the LF $U_a^* = \left[(-1)^{n-1-\sigma_{ka}} U_{n-k+1,a}^*\right]_{k=\overline{1,n}}$ are determined from the relation

$$\langle y, z\rangle\Big|_{x=a} = U_a(y) U_a^* = \sum_{k=1}^{n}(-1)^{n-1-\sigma_{ka}} U_{ka}(y) U_{n-k+1,a}^*(z).$$

It is clear that $L^* \in v_N$. Thus, for any sufficiently smooth functions $y(x)$ and $z(x)$

$$\ell y z - y \ell^* z = \frac{d}{dx}\langle y, z\rangle. \tag{1.26}$$

In particular, if the functions $y(x,\lambda)$ and $z(x,\mu)$ are solutions of the DE's $\ell y = \lambda y$ and $\ell^* z = \mu z$, then

$$\frac{d}{dx}\langle y, z\rangle = (\lambda - \mu) yz. \tag{1.27}$$

For definiteness it is assumed here and below that $\sigma_{\xi a} = n - \xi$.

Suppose that the functions $\Phi_m^*(x,\lambda)$, $m = \overline{1,n}$ are solutions of the DE (1.24) under the conditions

$$U_{\xi 0}^*(\Phi_m^*) = \delta_{\xi m}, \qquad \xi = \overline{1,m} \qquad (T \le \infty)$$

$$U_{\eta T}^*(\Phi_m^*) = 0, \qquad \eta = \overline{1,n-m} \qquad (T < \infty)$$

$$\Phi_m^*(x,\lambda) = O(\exp(\rho R_m^* x)), \qquad x \to \infty, \qquad \rho \in S \qquad (T = \infty)$$

$$R_m^* = -R_{n-m+1}.$$

Let $\mathfrak{M}_{mk}^*(\lambda) = U_{k0}^*(\Phi_m^*)$ and

$$\Phi^*(x,\lambda) = [(-1)^{k-1}\Phi_{n-k+1}^*(x,\lambda)]_{k=\overline{1,n}}^T, \qquad \mathfrak{M}^*(\lambda) = U_0^*(\Phi^*(x,\lambda)).$$

We introduce the FSS $C^*(x,\lambda) = [(-1)^{k-1}C_{n-k+1}^*(x,\lambda)]_{k=\overline{1,n}}^T$ of the DE (1.24) under the conditions $U_{\xi 0}^*(C_m^*) = \delta_{\xi m}$, $\xi = \overline{1,n}$. Then

$$\Phi^*(x,\lambda) = C^*(x,\lambda)\mathfrak{M}^*(\lambda). \tag{1.28}$$

The properties of the WF's $\mathfrak{M}_{mk}^*(\lambda)$ are completely analogous to those of the WF's $\mathfrak{M}_{mk}(\lambda)$. For $T < \infty$

$$\mathfrak{M}_{mk}^*(\lambda) = (\Delta_{mm}^*(\lambda))^{-1}\Delta_{mk}^*(\lambda),$$

$$\Delta_{mk}^*(\lambda) = (-1)^{m+k}\det[U_{\xi T}^*(C_\nu^*)]_{\xi=\overline{1,n-m};\nu=\overline{m,n}\setminus k}.$$

For $T = \infty$ the WF's $\mathfrak{M}_{mk}^*(\lambda)$ are regular in $\Pi_{(-1)^m}$ exept for an at most countable set $\Lambda_{mk}^{*'}$ of poles, and for $(-1)^m\lambda \ge 0$ the following limits exist and are finite off the bounded sets $\Lambda_{mk}^{*,\pm}$.

$$\mathfrak{M}_{mk}^{*,\pm}(\lambda) = \lim_{z\to 0, \operatorname{Re} z > 0} \mathfrak{M}_{mk}^*(\lambda \pm iz).$$

Lemma 1.1. $\mathfrak{M}^*(\lambda) = (\mathfrak{M}(\lambda))^1$.

Indeed, it follows from (1.27) for $\lambda = \mu$ that

$$\frac{d}{dx}\langle \Phi_k(x,\lambda), \Phi_j^*(x,\lambda)\rangle = 0.$$

Since for $k + j \le n$

$$\lim_{A\to T} \langle \Phi_k(x,\lambda), \Phi_j^*(x,\lambda)\rangle\big|_{x=A} = 0,$$

$$\begin{aligned}\langle \Phi_k(x,\lambda), \Phi_j^*(x,\lambda)\rangle\big|_{x=0} &= \sum_{\nu=1}^{n}(-1)^{\nu-1}U_{\nu 0}(\Phi_k)U^*_{n-\nu+1,0}(\Phi_j^*)\\ &= \sum_{\nu=1}^{n}(-1)^{\nu-1}\mathfrak{M}_{k\nu}(\lambda)\mathfrak{M}^*_{j,n-\nu+1}(\lambda),\end{aligned}$$

we get

$$\sum_{\nu=1}^{n}(-1)^{\nu-1}\mathfrak{M}_{k\nu}(\lambda)\mathfrak{M}^*_{j,n-\nu+1}(\lambda) = 0, \tag{1.29}$$

i.e., $\mathfrak{M}(\lambda)\mathfrak{M}^*(\lambda) = E$. Lemma 1.1 is proved.

Let $y(x)$ be a sufficiently smooth function. We denote

$$\vec{y}(x) = [y^{(\nu)}(x)]_{\nu=\overline{0,n-1}}.$$

Lemma 1.2. *Suppose that the functions* $y_k(x)$, $k = \overline{1,n-1}$ *are solutions of the DE* (1.1), *and* $z_j(x) = det[y_k^{(\nu)}(x)]_{k=\overline{1,n-1};\nu=\overline{0,n-1}\setminus n-j-1}$. *Then*

$$z_j(x) = \sum_{s=0}^{j}(-1)^s(p_{n-s}(x)z_0(x))^{(j-s)}, \qquad j = \overline{0,n-1}; \tag{1.30}$$

$$\ell^* z_0(x) = \lambda z_0(x), \qquad \det[\vec{y}_1(x),\dots,\vec{y}_{n-1}(x),\vec{y}(x)] = \langle y(x), z_0(x)\rangle. \tag{1.31}$$

Proof. We prove (1.30) by induction. For $j = 0$ (1.30) is obvious. Assume that (1.30) is true for $j = \overline{0,\mu-1}$. Since

$$z'_\mu(x) = z_{\mu+1}(x) + (-1)^\mu p_{n-\mu-1}(x)z_0(x),$$

the use of (1.30) for $j = \mu - 1$ leads to

$$\begin{aligned} z_\mu(x) &= \sum_{s=0}^{\mu-1}(-1)^s(p_{n-s}(x)z_0(x))^{(\mu-s)} + (-1)^\mu p_{n-\mu}(x)z_0(x)\\ &= \sum_{s=0}^{\mu}(-1)^s(p_{n-s}(x)z_0(x))^{(\mu-s)}.\end{aligned}$$

Further, it is obvious that $z'_{n-1}(x) = (\lambda - p_0(x))(-1)^n z_0(x)$. On the other hand, we get from (1.30) that

$$z'_{n-1}(x) = \sum_{s=0}^{n-1}(-1)^s (p_{n-s}(x) z_0(x))^{(n-s)}.$$

Consequently, $\ell^* z_0(x) = \lambda z_0(x)$. Expanding $\det[\vec{y}_1(x), \ldots, \vec{y}_{n-1}(x), \vec{y}(x)]$ by the last column and using (1.30), we get

$$\begin{aligned}
\det[\vec{y}_1(x), \ldots, \vec{y}_{n-1}(x), \vec{y}(x)] &= \sum_{j=0}^{n-1}(-1)^j y^{(n-1-j)}(x) z_j(x) \\
&= \sum_{j=0}^{n-1}(-1)^j y^{(n-1-j)}(x) \sum_{s=0}^{j}(-1)^s (p_{n-s}(x) z_0(x))^{(j-s)} \\
&= \langle y(x), z_0(x) \rangle .
\end{aligned}$$

Lemma 1.2 is proved.

Lemma 1.3.

$$\begin{aligned}
\Phi^*_m(x,\lambda) = \det [\Phi_n^{(s)}(x,\lambda), \ldots, \Phi^{(s)}_{n-m+2}(x,\lambda),& \\
\Phi^{(s)}_{n-m}(x,\lambda), \ldots, \Phi_1^{(s)}(x,\lambda)]_{s=\overline{0,n-2}}.&
\end{aligned} \tag{1.32}$$

Proof. Denote the right-hand side of (1.32) by $y^*_m(x,\lambda)$. It follows from (1.31) that $\ell^* y^*_m(x,\lambda) = \lambda y^*_m(x,\lambda)$ and

$$\begin{aligned}
\det [\vec{\Phi}_n(x,\lambda), \ldots, \vec{\Phi}_{n-m+2}(x,\lambda), \vec{\Phi}_{n-m}(x,\lambda), \ldots, \vec{\Phi}_1(x,\lambda), \vec{y}(x)]\big|_{x=a}& \\
= \langle y(x), y^*_m(x,\lambda)\rangle\big|_{x=a} = \sum_{k=1}^{n}(-1)^{k-1} U_{ka}(y) U^*_{n-k+1,a}(y^*_m).&
\end{aligned} \tag{1.33}$$

In (1.33) we take $y(x) = \Phi_n(x,\lambda), \ldots, y(x) = \Phi_{n-m+1}(x,\lambda)$ successively, and get that $U^*_{\xi 0}(y^*_m) = \delta_{\xi m}$, $\xi = \overline{1,m}$. For $T < \infty$, $a = T$ we take $y(x) = \Phi_1(x,\lambda), \ldots,$ $y(x) = \Phi_{n-m}(x,\lambda)$ successively, and get $U^*_{\eta T}(y^*_m) = 0$, $\eta = \overline{1, n-m}$. For $T = \infty$ from the definition of the functions $y^*_m(x,\lambda)$ and the asymptotic properties of the WS's $\Phi_m^{(s)}(x,\lambda)$ we get that

$$y^*_m(x,\lambda) = O(\exp(\rho R^*_m x)), \qquad x \to \infty, \qquad \rho \in S.$$

Consequently, $y^*_m(x,\lambda) = \Phi^*_m(x,\lambda)$. Lemma 1.3 is proved.

1.5. In this subsection we get the uniqueness theorem for the solution of the inverse problem (IP). Let $C_M(x,\lambda) = [\vec{C}_m(x,\lambda)]^T_{m=\overline{1,n}}$ and $\Phi_M(x,\lambda) = [\vec{\Phi}_m(x,\lambda)]^T_{m=\overline{1,n}}$. Then (1.4) takes the form

$$\Phi_M(x,\lambda) = C_M(x,\lambda)\mathfrak{M}^T(\lambda). \tag{1.34}$$

Since $\det \mathfrak{M}(\lambda) = 1$, (1.34) and the Ostrogradskii–Liouville theorem give us that

$$\det \Phi_M(x,\lambda) = \det C_M(x,\lambda) = (-1)^{n(n-1)/2}. \tag{1.35}$$

Let L, $\tilde{L} \in v_N$. We define the matrix $\mathcal{P}(x,\lambda) = [\mathcal{P}_{jk}(x,\lambda)]_{j,k=\overline{1,n}}$ by the formula $\mathcal{P}(x,\lambda) = \Phi_M(x,\lambda)(\tilde{\Phi}_M(x,\lambda))^{-1}$ or

$$\begin{aligned}
&\mathcal{P}_{jk}(x,\lambda) \\
&= \det [\tilde{\Phi}_\nu^{(n-1)}(x,\lambda), \dots, \tilde{\Phi}_\nu^{(k)}(x,\lambda), \Phi_\nu^{(j-1)}(x,\lambda), \tilde{\Phi}_\nu^{(k-2)}(x,\lambda), \dots, \tilde{\Phi}_\nu(x,\lambda)]_{\nu=\overline{1,n}} \\
&= \sum_{\nu=1}^{n} (-1)^{\nu+k-n-1} \Phi_\nu^{(j-1)}(x,\lambda) \det [\tilde{\Phi}_n^{(s)}(x,\lambda), \dots, \tilde{\Phi}_{\nu-1}^{(s)}(x,\lambda), \\
&\qquad \tilde{\Phi}_{\nu+1}^{(s)}(x,\lambda), \dots, \tilde{\Phi}_1^{(s)}(x,\lambda)]_{s=\overline{0,n-1}\setminus k-1}.
\end{aligned} \tag{1.36}$$

We remark that the idea of using mappings of the solution spaces of DE's for solving the IP is due to Leibenzon [53], [54].

From (1.36) and the asymptotic properties of the WS's $\Phi_m(x,\lambda)$ and $\tilde{\Phi}_m(x,\lambda)$ as $|\lambda| \to \infty$ we get the estimates

$$\left.\begin{aligned}
&|\mathcal{P}_{jk}(x,\lambda)| < C|\rho|^{j-k}, && j,k = \overline{1,n}; \\
&|\mathcal{P}_{1k}(x,\lambda) - \delta_{1k}| < C|\rho|^{-1}, && k = \overline{1,n}.
\end{aligned}\right\} \tag{1.37}$$

($\lambda \in G_\delta$ for $T < \infty$). Let

$$\langle [y_\nu]_{\nu=\overline{0,n-1}}, [z_j]_{j=\overline{0,n-1}} \rangle_\ell \overset{df}{=} \sum_{\nu,j=0}^{n-1} \mathcal{L}_{\nu j}(x) y_\nu z_j.$$

Lemma 1.4. *Let $\vec{y}(x)$ be a sufficiently smooth function. Then*

$$\mathcal{P}(x,\lambda)\vec{\tilde{y}}(x) = \sum_{k=1}^{n} (-1)^{k-1} \langle \vec{y}(x), \tilde{\Phi}^*_{n-k+1}(x,\lambda) \rangle_{\tilde{\ell}} \vec{\Phi}_k(x,\lambda), \tag{1.38}$$

$$\begin{aligned}
&\langle (\mathcal{P}(x,\lambda) - \mathcal{P}(x,\mu)) \vec{\tilde{\Phi}}_k(x,\lambda), \vec{\Phi}_j^*(x,\mu) \rangle_\ell \\
&\qquad = \langle \Phi_k(x,\lambda), \Phi_j^*(x,\mu) \rangle_\ell - \langle \tilde{\Phi}_k(x,\lambda), \tilde{\Phi}_j^*(x,\mu) \rangle_{\tilde{\ell}}.
\end{aligned} \tag{1.39}$$

Proof. Let us use (1.36). We have

$$\mathcal{P}(x,\lambda)\vec{\tilde{y}}(x)=\sum_{k=1}^{n}(-1)^{k-1}\vec{\Phi}_k(x,\lambda)\det[\vec{\tilde{\Phi}}_n(x,\lambda),\dots,$$

$$\vec{\tilde{\Phi}}_{k+1}(x,\lambda),\vec{\tilde{\Phi}}_{k-1}(x,\lambda),\dots,\vec{\tilde{\Phi}}_1(x,\lambda),\vec{\tilde{y}}(x)].$$

From this, using Lemmas 1.2 and 1.3, we get (1.38). Further, since

$$\mathcal{P}(x,\lambda)\vec{\tilde{\Phi}}_k(x,\lambda)=\vec{\Phi}_k(x,\lambda),$$

it follows that

$$\langle\mathcal{P}(x,\lambda)\vec{\tilde{\Phi}}_k(x,\lambda),\vec{\Phi}_j^*(x,\mu)\rangle_\ell=\langle\Phi_k(x,\lambda),\Phi_j^*(x,\mu)\rangle_\ell. \tag{1.40}$$

By (1.38)

$$\langle\mathcal{P}(x,\mu)\vec{\tilde{\Phi}}_k(x,\lambda),\vec{\Phi}_j^*(x,\mu)\rangle_\ell$$

$$=\sum_{s=1}^{n}(-1)^{s-1}\langle\tilde{\Phi}_k(x,\lambda),\tilde{\Phi}_{n-s+1}^*(x,\mu)\rangle_{\tilde{\ell}}\langle\Phi_s(x,\mu),\Phi_j^*(x,\mu)\rangle_\ell.$$

According to (1.27), $\langle\Phi_s(x,\mu),\Phi_j^*(x,\mu)\rangle_\ell$ does not depend on x. Using the conditions on the WS's for $x=0$ and $x=T$ we find that

$$\langle\Phi_s(x,\mu),\Phi_j^*(x,\mu)\rangle_\ell=(-1)^{s-1}\delta_{s,n-j+1}.$$

Thus,

$$\langle\mathcal{P}(x,\mu)\vec{\tilde{\Phi}}_k(x,\lambda),\vec{\Phi}_j^*(x,\mu)\rangle_\ell=\langle\tilde{\Phi}_k(x,\lambda),\tilde{\Phi}_j^*(x,\mu)\rangle_{\tilde{\ell}},$$

which together with (1.40) yields (1.39). Lemma 1.4 is proved.

Theorem 1.2. *If* $\mathfrak{M}(\lambda)=\tilde{\mathfrak{M}}(\lambda)$, *then* $L=\tilde{L}$.

Thus, the specification of the WM $\mathfrak{M}(\lambda)$ determines the DE and LF (1.1), (1.2) uniquely. We remark that the deletion of a single element from the WM leads to non-uniqueness of the solution of the IP.

Proof. We transform the matrix $\mathcal{P}(x,\lambda)$. For this we use (1.34). Under the conditions of the theorem,

$$\begin{aligned}\mathcal{P}(x,\lambda)=\Phi_M(x,\lambda)(\tilde{\Phi}_M(x,\lambda))^{-1}&=C_M(x,\lambda)\mathfrak{M}^T(\lambda)(\tilde{\mathfrak{M}}^T(\lambda))^{-1}(\tilde{C}_M(x,\lambda))^{-1}\\&=C_M(x,\lambda)(\tilde{C}_M(x,\lambda))^{-1}.\end{aligned}$$

In view of (1.35) this leads us to conclude that for each fixed x the matrix-valued function $\mathcal{P}(x,\lambda)$ is an entire analytic function in λ. Using (1.37) and Liouville's theorem ([94, p. 209]), we get that $\mathcal{P}_{11}(x,\lambda) \equiv 1$, $\mathcal{P}_{1k}(x,\lambda) \equiv 0$ for $k = \overline{2,n}$. But then $\Phi_m(x,\lambda) \equiv \tilde{\Phi}_m(x,\lambda)$ for all x, λ and m, and hence $L = \tilde{L}$. Theorem 1.2 is proved.

2. Solution of the inverse problem on the half-line

We consider the DE and LF (1.1)–(1.2) on the half-line ($T = \infty$). In §2 we present a solution of the inverse problem of recovering L from the WM $\mathfrak{M}(\lambda)$, when the behavior of the spectrum is arbitrary. We give a derivation of the main equation of the inverse problem, which is a singular linear integral equation. We obtain necessary and sufficient conditions on the WM and algorithm for the solution of the inverse problem. The main results of §2 are contained in Theorems 2.1 and 2.3.

2.1. We prove some auxiliary assertions.

Lemma 2.1. *The functions*

$$\left.\begin{aligned}&\mathfrak{M}_{mk}(\lambda) - \mathfrak{M}_{m,m+1}(\lambda)\mathfrak{M}_{m+1,k}(\lambda),\\ &\mathfrak{M}^*_{n-m,k}(\lambda) - \mathfrak{M}^*_{n-m,n-m+1}(\lambda)\mathfrak{M}^*_{n-m+1,k}(\lambda),\end{aligned}\right\} \tag{2.0}$$

$$\left.\begin{aligned}&\Phi_m(x,\lambda) - \mathfrak{M}_{m,m+1}(\lambda)\Phi_{m+1}(x,\lambda),\\ &\Phi^*_{n-m}(x,\lambda) - \mathfrak{M}_{m,m+1}(\lambda)\Phi^*_{n-m+1}(x,\lambda),\end{aligned}\right\}$$

are regular for $\lambda \in \Gamma_{(-1)^{n-m}}\setminus\Lambda$.

Proof. Since $\mathfrak{M}_{mk}(\lambda) = \mathfrak{M}^*_{mk}(\lambda) = \delta_{mk}$, $k \le m$, it follows from (1.29) with $k+j = n$ and $k+j = n-1$ that

$$\mathfrak{M}^*_{n-m,n-m+1}(\lambda) = \mathfrak{M}_{m,m+1}(\lambda), \tag{2.1}$$

$$-\mathfrak{M}^*_{n-m-1,n-m+1}(\lambda) = \mathfrak{M}_{m,m+2}(\lambda) - \mathfrak{M}_{m,m+1}(\lambda)\mathfrak{M}_{m+1,m+2}(\lambda),$$

$$-\mathfrak{M}_{n-m-1,n-m+1}(\lambda) = \mathfrak{M}^*_{m,m+2}(\lambda) - \mathfrak{M}^*_{m,m+1}(\lambda)\mathfrak{M}^*_{m+1,m+2}(\lambda).$$

This implies regularity of the functions (2.0) for $\lambda \in \Gamma_{(-1)^{n-m}}\setminus\Lambda$ when $k = m+2$. When $k > m+2$ we interchange the LF's U_{k0} and $U_{m+2,0}$ and repeat the preceding arguments.

The regularity of the functions $\Phi_{n-m}(x,\lambda) - \mathfrak{M}^*_{m,m+1}(\lambda)\Phi_{n-m+1}(x,\lambda)$ for $\lambda \in \Gamma_{(-1)^m}\backslash\Lambda$ is proved by induction. By (1.4) and Lemma 1.1, $C(x,\lambda) = \mathfrak{M}^*(\lambda)\Phi(x,\lambda)$ or

$$C_{n-m}(x,\lambda) = \Phi_{n-m}(x,\lambda) - \sum_{j=0}^{m-1}(-1)^j \mathfrak{M}^*_{m-j,m+1}(\lambda)\Phi_{n-m+j+1}(x,\lambda), \qquad (2.2)$$

$$m = \overline{1, n-1}.$$

It follows from this with $m = 1$ that the function

$$\Phi_{n-1}(x,\lambda) - \mathfrak{M}^*_{12}(\lambda)\Phi_n(x,\lambda)$$

is an entire analytic function in λ. Assume that for $j = 1,\ldots,m-1$ we have proved that the functions $\Phi_{n-j} - \mathfrak{M}^*_{j,j+1}\Phi_{n-j+1}$ are regular for $\lambda \in \Gamma_{(-1)^j}\backslash\Lambda$. Then, using (2.2), we get that the function

$$\Phi_{n-m}(x,\lambda) - \mathfrak{M}^*_{m,m+1}(\lambda)\Phi_{n-m+1}(x,\lambda) - \sum_{j=1}^{[(m-1)/2]} (\mathfrak{M}^*_{m-2j,m+1}(\lambda)$$

$$-\mathfrak{M}^*_{m-2j,m-2j+1}(\lambda)\mathfrak{M}^*_{m-2j+1,m+1}(\lambda))\Phi_{n-m+2j+1}(x,\lambda)$$

is regular for $\lambda \in \Gamma_{(-1)^m}\backslash\Lambda$. Consequently, the function

$$\Phi_{n-m}(x,\lambda) - \mathfrak{M}^*_{m,m+1}(\lambda)\Phi_{n-m+1}(x,\lambda)$$

is regular for $\lambda \in \Gamma_{(-1)^m}\backslash\Lambda$. Thus, in view of (2.1) we have proved the regularity of the functions $\Phi_m(x,\lambda) - \mathfrak{M}_{m,m+1}(\lambda)\Phi_{m+1}(x,\lambda)$ for $\lambda \in \Gamma_{(-1)^{n-m}}\backslash\Lambda$. The regularity of the functions $\Phi^*_{n-m}(x,\lambda) - \mathfrak{M}_{m,m+1}(\lambda)\Phi^*_{n-m+1}(x,\lambda)$ is established similarly. Lemma 2.1 is proved.

We remark that since $L \in v_N$, the asymptotic formula (1.11) can be made more precise; namely,

$$\mathfrak{M}_{mk}(\lambda) = \rho^{m-k}\mu^0_{mk}\left(1 + \sum_{\nu=1}^{n+N-1}\frac{\mu_{mk\nu}}{\rho^\nu} + o\left(\frac{1}{\rho^{n+N-1}}\right)\right), \quad |\rho| \to \infty, \quad \rho \in S.$$

Let $L, \widetilde{L} \in v_N$. In the λ-plane we consider the contour $\gamma = \gamma_{-1} \cup \gamma_0 \cup \gamma_1$ (with a counterclockwise circuit), where γ_0 is a bounded closed contour encircling the set $\Lambda \cup \widetilde{\Lambda} \cup \{0\}$ (i.e., $\Lambda \cup \widetilde{\Lambda} \cup \{0\} \subset \operatorname{int}\gamma_0$), and $\gamma_{\pm 1}$ is the two-sided cut along the arc $\{\lambda : \pm\lambda > 0, \lambda \notin \operatorname{int}\gamma_0\}$. Let $J_\gamma = \{\lambda : \lambda \notin \gamma \cup \operatorname{int}\gamma_0\}$.

Lemma 2.2. *The following relations hold:*

$$\widetilde{\Phi}(x,\lambda)=\Phi(x,\lambda)-\frac{1}{2\pi i}\int\limits_{\gamma}\frac{\langle\widetilde{\Phi}(x,\lambda),\widetilde{\Phi}^*(x,\mu)\rangle_{\widetilde{\ell}}}{\lambda-\mu}\Phi(x,\mu)\,d\mu,\qquad \lambda\in J_\gamma,\tag{2.3}$$

$$\frac{\langle\Phi(x,\lambda),\Phi^*(x,\mu)\rangle_{\ell}}{\lambda-\mu}-\frac{\langle\widetilde{\Phi}(x,\lambda),\widetilde{\Phi}^*(x,\mu)\rangle_{\widetilde{\ell}}}{\lambda-\mu}$$
$$=\frac{1}{2\pi i}\int\limits_{\gamma}\frac{\langle\widetilde{\Phi}(x,\lambda),\widetilde{\Phi}^*(x,\xi)\rangle_{\widetilde{\ell}}}{\lambda-\xi}\frac{\langle\Phi(x,\xi),\Phi^*(x,\mu)\rangle_{\ell}}{\xi-\mu}d\xi,\qquad \lambda,\mu\in J_\gamma.\tag{2.4}$$

In (2.3) (and everywhere below, where necessary) the integral is understood in the principal value sense ([95, p. 27]).

Proof. Let $\gamma_R=(\gamma\cap\{\lambda:|\lambda|\le R\})\cup\{\lambda:|\lambda|=R\}$. Since

$$\frac{1}{\lambda-\mu}\left(\frac{1}{\lambda-\xi}-\frac{1}{\mu-\xi}\right)=\frac{1}{(\lambda-\xi)(\xi-\mu)},\tag{2.5}$$

we have by the Cauchy theorem ([94, p. 166]) that for $\lambda,\ \mu\in J_\gamma\cap\{\xi:|\xi|<R\}$

$$\mathcal{P}_{1k}(x,\lambda)=\frac{1}{2\pi i}\int\limits_{\gamma_R}\frac{\mathcal{P}_{1k}(x,\xi)}{\lambda-\xi}\,d\xi,$$

$$\frac{\mathcal{P}_{jk}(x,\lambda)-\mathcal{P}_{jk}(x,\mu)}{\lambda-\mu}=\frac{1}{2\pi i}\int\limits_{\gamma_R}\frac{\mathcal{P}_{jk}(x,\xi)}{(\lambda-\xi)(\xi-\mu)}\,d\xi.$$

Using (1.37), we get

$$\lim_{R\to\infty}\frac{1}{2\pi i}\int\limits_{|\xi|=R}\frac{\mathcal{P}_{1k}(x,\xi)-\delta_{1k}}{\lambda-\xi}\,d\xi=0,$$

$$\lim_{R\to\infty}\frac{1}{2\pi i}\int\limits_{|\xi|=R}\frac{\mathcal{P}_{jk}(x,\xi)}{(\lambda-\xi)(\xi-\mu)}\,d\xi=0,$$

and hence

$$\left.\begin{aligned}&\mathcal{P}_{1k}(x,\lambda)=\delta_{1k}+\frac{1}{2\pi i}\int\limits_{\gamma}\frac{\mathcal{P}_{1k}(x,\xi)}{(\lambda-\xi)}\,d\xi,\qquad \lambda\in J_\gamma,\\&\frac{\mathcal{P}_{jk}(x,\lambda)-\mathcal{P}_{jk}(x,\mu)}{\lambda-\mu}=\frac{1}{2\pi i}\int\limits_{\gamma}\frac{\mathcal{P}_{jk}(x,\xi)}{(\lambda-\xi)(\xi-\mu)}\,d\xi,\qquad \lambda,\mu\in J_\gamma.\end{aligned}\right\}\tag{2.6}$$

By (1.38) and (2.6),

$$\sum_{k=1}^{n} \mathcal{P}_{1k}(x,\lambda)\widetilde{y}^{(k-1)}(x) = \widetilde{y}(x) + \frac{1}{2\pi i}\int_{\gamma}\sum_{k=1}^{n}\mathcal{P}_{1k}(x,\xi)\widetilde{y}^{(k-1)}(x)\,\frac{d\xi}{\lambda-\xi}$$

$$= \widetilde{y}(x) + \frac{1}{2\pi i}\int_{\gamma}\frac{\langle \widetilde{y}(x), \widetilde{\Phi}^{*}(x,\xi)\rangle_{\widetilde{\ell}}}{\lambda-\xi}\,\Phi(x,\xi)\,d\xi.$$

Setting $\widetilde{y}(x) = \widetilde{\Phi}(x,\lambda)$ here and using the equality

$$\Phi(x,\lambda) = \sum_{k=1}^{n}\mathcal{P}_{1k}(x,\lambda)\widetilde{\Phi}^{(k-1)}(x,\lambda),$$

we get (2.3). Similarly, by (1.38) and (2.6)

$$\frac{\mathcal{P}(x,\lambda)-\mathcal{P}(x,\mu)}{\lambda-\mu}\,\vec{\widetilde{y}}(x) = \frac{1}{2\pi i}\int_{\gamma}\frac{\mathcal{P}(x,\xi)\vec{\widetilde{y}}(x)}{(\lambda-\xi)(\xi-\mu)}\,d\xi$$

$$= \frac{1}{2\pi i}\int_{\gamma}\sum_{s=1}^{n}(-1)^{s-1}\frac{\langle\widetilde{y}(x), \widetilde{\Phi}^{*}_{n-s+1}(x,\xi)\rangle_{\widetilde{\ell}}}{(\lambda-\xi)(\xi-\mu)}\vec{\Phi}_{s}(x,\xi)\,d\xi.$$

From this, by (1.39),

$$\frac{\langle\Phi_k(x,\lambda), \Phi_j^{*}(x,\mu)\rangle_{\ell}}{\lambda-\mu} - \frac{\langle\widetilde{\Phi}_k(x,\lambda), \widetilde{\Phi}_j^{*}(x,\mu)\rangle_{\widetilde{\ell}}}{\lambda-\mu}$$

$$= \langle\frac{\mathcal{P}(x,\lambda)-\mathcal{P}(x,\mu)}{\lambda-\mu}\,\vec{\widetilde{\Phi}}_k(x,\lambda), \vec{\Phi}_j^{*}(x,\mu)\rangle_{\ell}$$

$$= \frac{1}{2\pi i}\int_{\gamma}\frac{\langle\widetilde{\Phi}_k(x,\lambda), \widetilde{\Phi}^{*}(x,\xi)\rangle_{\widetilde{\ell}}}{\lambda-\xi}\,\frac{\langle\Phi(x,\xi), \Phi_j^{*}(x,\mu)\rangle_{\ell}}{\xi-\mu}\,d\xi.$$

Lemma 2.2 is proved.

Let

$$Y = [\delta_{j,k-1}]_{j=\overline{1,n-1};k=\overline{1,n}}; \qquad A_0(\lambda) = \widehat{\mathfrak{M}}(\lambda)\widetilde{\mathfrak{M}}^{-1}(\lambda),$$

$$\widetilde{A}_0(\lambda) = \widehat{\mathfrak{M}}(\lambda)\mathfrak{M}^{-1}(\lambda), \qquad \mathfrak{M}_{\partial}(\lambda) = \operatorname{diag}[\mathfrak{M}_{m,m+1}(\lambda)]_{m=\overline{1,n-1}}.$$

For real λ we define the matrices

$$f(x,\lambda) = [f_k(x,\lambda)]_{k=\overline{2,n}}, \qquad f^{*}(x,\lambda) = [(-1)^{k-1}f^{*}_{n-k+1}(x,\lambda)]^{T}_{k=\overline{1,n-1}}$$

according to the formulas

$$f_k(x,\lambda) = \chi((-1)^{n-k+1}\lambda)\Phi_k(x,\lambda), \qquad f_k^*(x,\lambda) = \chi((-1)^{k-1}\lambda)\Phi_k^*(x,\lambda),$$

where $\chi(\lambda)$ is the Heaviside function. For $\lambda \in \gamma$ let

$$a(\lambda) = \chi_{+1}(\lambda)\chi_{-1}(\lambda)YA_0(\lambda)Y^T, \qquad N(\lambda) = E + \frac{1}{2}a(\lambda),$$

$$\widetilde{a}(\lambda) = \chi_{+1}(\lambda)\chi_{-1}(\lambda)Y\widetilde{A}_0(\lambda)Y^T, \qquad \widetilde{N}(\lambda) = E - \frac{1}{2}\widetilde{a}(\lambda),$$

where $\chi_{\pm 1}(\lambda) = 1$ for $\lambda \in \gamma_0 \cup \gamma_{\pm 1}$, and $\chi_{\pm 1}(\lambda) = 0$ for $\lambda \in \gamma_{\mp 1}$. For $\lambda,\ \mu \in \gamma$ we define the matrices

$$\varphi(x,\lambda) = [\varphi_k(x,\lambda)]_{k=\overline{2,n}}, \qquad g^*(x,\lambda) = [g_k^*(x,\lambda)]^T_{k=\overline{2,n}},$$

$$G^*(x,\lambda) = [G_k^*(x,\lambda)]^T_{k=\overline{1,n}}, \quad r(x,\lambda,\mu) = [r_{kj}(x,\lambda,\mu)]_{k,j=\overline{2,n}},$$

according to the formulas

$$\varphi(x,\lambda) = \begin{cases} Y\Phi(x,\lambda), & \lambda \in \gamma_0 \\ f(x,\lambda), & \lambda \in \gamma_1 \cup \gamma_{-1}, \end{cases}$$

$$g^*(x,\lambda) = \begin{cases} -\Phi^*(x,\lambda)A_0(\lambda)Y^T, & \lambda \in \gamma_0 \\ -f^*(x,\lambda)\widehat{\mathfrak{M}}_\partial(\lambda), & \lambda \in \gamma_1 \cup \gamma_{-1}, \end{cases}$$

$$r(x,\lambda,\mu) = \frac{\langle \varphi(x,\lambda), g^*(x,\mu)\rangle_\ell}{\lambda - \mu}, \qquad G^*(x,\lambda) = g^*(x,\lambda)Y.$$

Similarly, we define the matrices $\widetilde{\varphi}(x,\lambda)$, $\widetilde{g}^*(x,\lambda)$, $\widetilde{G}^*(x,\lambda)$ and $\widetilde{r}(x,\lambda,\mu)$ with $\widetilde{\Phi}$, $\widetilde{f}$, $\widetilde{\Phi}^*$, $\widetilde{f}^*$ and $\widetilde{A}_0$ instead of Φ, f Φ^*, f^* and A_0. Finally, the matrices $\widetilde{\Gamma}(\lambda,\mu) = [\widetilde{\Gamma}_{j\nu}(\lambda,\mu)]_{j,\nu=\overline{1,n}}$ and $\widetilde{A}(\mu) = [\widetilde{A}_{j\nu}(\mu)]_{j,\nu=\overline{1,n}}$, $\mu \in \gamma$ are defined according to the formulas

$$\widetilde{\Gamma}(\lambda,\mu) = -\langle \widetilde{\Phi}(x,\lambda), \widetilde{G}^*(x,\mu)\rangle_{\widetilde{\ell}}\big|_{x=0},$$

$$\widetilde{A}_{j\nu}(\mu) = \delta_{j,\nu-1}\chi_{(-1)^{n-j}}(\mu)\widetilde{\mathfrak{M}}_{j,j+1}(\mu), \qquad \mu \in \gamma_1 \cup \gamma_{-1},$$

$$\widetilde{A}(\mu) = \widetilde{A}_0(\mu), \qquad \mu \in \gamma_0.$$

Since $\widetilde{G}^*(x,\mu) = -\widetilde{\Phi}^*(x,\mu)\widetilde{A}(\mu)$, it follows that

$$\widetilde{\Gamma}(\lambda,\mu) = \langle \widetilde{\Phi}(x,\lambda), \widetilde{\Phi}^*(x,\mu)\rangle_{\widetilde{\ell}|_{x=0}}\widetilde{A}(\mu) = \widetilde{\mathfrak{M}}(\lambda)\widetilde{\mathfrak{M}}^{-1}(\mu)\widetilde{A}(\mu)$$

and hence

$$\left.\begin{aligned}\widetilde{\Gamma}_{j\nu}(\lambda,\mu) &= \delta_{j+1,\nu}\chi_{(-1)^{n-j}}(\mu)\widehat{\mathfrak{M}}_{j,j+1}(\mu), && \nu \le j+1,\\ \widetilde{\Gamma}_{j\nu}(\lambda,\mu) &= \chi_{+1}(\mu)\chi_{-1}(\mu)\widehat{\mathfrak{M}}_{j\nu}(\mu) + \widetilde{\widetilde{\Gamma}}_{j\nu}(\lambda,\mu), && \nu > j+1,\end{aligned}\right\} \tag{2.7}$$

where the functions $\widetilde{\widetilde{\Gamma}}_{j\nu}(\lambda,\mu)$ are constructed from $\mathfrak{M}_{ks}$ and $\widetilde{\mathfrak{M}}_{ks}$ for $s-k<\nu-j$. Further,

$$\begin{aligned}\widetilde{A}_0(\lambda)A_0(\lambda) &= \widehat{\mathfrak{M}}(\lambda)\mathfrak{M}^{-1}(\lambda)\widehat{\mathfrak{M}}(\lambda)\widetilde{\mathfrak{M}}^{-1}(\lambda)\\ &= \widehat{\mathfrak{M}}(\lambda)\widetilde{\mathfrak{M}}^{-1}(\lambda) - \widehat{\mathfrak{M}}(\lambda)\mathfrak{M}^{-1}(\lambda) = A_0(\lambda) - \widetilde{A}_0(\lambda),\end{aligned}$$

i.e.,

$$A_0(\lambda) - \widetilde{A}_0(\lambda) = \widetilde{A}_0(\lambda)A_0(\lambda) \tag{2.8}$$

and hence $a(\lambda) - \widetilde{a}(\lambda) = \widetilde{a}(\lambda)a(\lambda)$. From this we get

$$\widetilde{N}(\lambda)N(\lambda) - \frac{1}{4}\widetilde{a}(\lambda)a(\lambda) = E, \qquad \widetilde{N}(\lambda)a(\lambda) - \widetilde{a}(\lambda)N(\lambda) = 0. \tag{2.9}$$

Theorem 2.1.

$$\widetilde{\varphi}(x,\lambda) = \widetilde{N}(\lambda)\varphi(x,\lambda) + \frac{1}{2\pi i}\int_\gamma \widetilde{r}(x,\lambda,\mu)\varphi(x,\mu)\,d\mu, \qquad \lambda\in\gamma, \tag{2.10}$$

$$\widetilde{N}(\lambda)r(x,\lambda,\mu) - \widetilde{r}(x,\lambda,\mu)N(\mu) + \frac{1}{2\pi i}\int_\gamma \widetilde{r}(x,\lambda,\xi)r(x,\xi,\mu)\,d\xi = 0. \tag{2.11}$$

Equation (2.10) *is the desired main equation of the IP.*

Proof. By (1.4), (1.28) and Lemma 1.1,

$$\begin{aligned}\widetilde{\Phi}^{*(\nu)}(x,\mu)\Phi^{(j)}(x,\mu) &= \widetilde{C}^{*(\nu)}(x,\mu)\widetilde{\mathfrak{M}}^{-1}(\mu)\mathfrak{M}(\mu)C^{(j)}(x,\mu)\\ &= \widetilde{C}^{*(\nu)}(x,\mu)\widetilde{\mathfrak{M}}^{-1}(\mu)\widehat{\mathfrak{M}}(\mu)C^{(j)}(x,\mu) + \widetilde{C}^{*(\nu)}(x,\mu)C^{(j)}(x,\mu)\\ &= \widetilde{\Phi}^{*(\nu)}(x,\mu)\widehat{\mathfrak{M}}(\mu)\mathfrak{M}^{-1}(\mu)\Phi^{(j)}(x,\mu) + \widetilde{C}^{*(\nu)}(x,\mu)C^{(j)}(x,\mu)\\ &= \widetilde{\Phi}^{*(\nu)}(x,\mu)\widetilde{A}_0(\mu)\Phi^{(j)}(x,\mu) + \widetilde{C}^{*(\nu)}(x,\mu)C^{(j)}(x,\mu)\\ &= -\widetilde{g}^{*(\nu)}(x,\mu)\varphi^{(j)}(x,\mu) + \widetilde{C}^{*(\nu)}(x,\mu)C^{(j)}(x,\mu), \qquad \mu\in\gamma_0.\end{aligned}$$

Hence, for $\mu\in\gamma_0$ the function

$$\widetilde{\Phi}^{*(\nu)}(x,\mu)\Phi^{(j)}(x,\mu) + \widetilde{g}^{*(\nu)}(x,\mu)\varphi^{(j)}(x,\mu)$$

is entire in μ. Further, it follows from Lemma 2.1 that the function

$$\widetilde{\Phi}^{*(\nu)}(x,\mu)\Phi^{(j)}(x,\mu) - \widetilde{f}^{*(\nu)}(x,\mu)\widehat{\mathfrak{M}}_\partial(\mu)f^{(j)}(x,\mu)$$

is regular for $\mu \in \Gamma\backslash\Lambda$. This implies that the function

$$\widetilde{\Phi}^{*(\nu)}(x,\mu)\Phi^{(j)}(x,\mu)+\widetilde{g}^{*(\nu)}(x,\mu)\varphi^{(j)}(x,\mu)$$

is regular for $\mu \in \Gamma\backslash\Lambda$. Thus, in view of Cauchy's theorem, we get from (2.3) and (2.4),

$$\widetilde{\Phi}(x,\lambda)=\Phi(x,\lambda)+\frac{1}{2\pi i}\int\limits_{\gamma}\frac{\langle\widetilde{\Phi}(x,\lambda),\widetilde{g}^{*}(x,\mu)\rangle_{\widetilde{\ell}}}{\lambda-\mu}\varphi(x,\mu)\,d\mu;\qquad \lambda\in J_{\gamma},\tag{2.12}$$

$$\frac{\langle\Phi(x,\lambda),\Phi^{*}(x,\mu)\rangle_{\ell}}{\lambda-\mu}-\frac{\langle\widetilde{\Phi}(x,\lambda),\widetilde{\Phi}^{*}(x,\mu)\rangle_{\widetilde{\ell}}}{\lambda-\mu}$$
$$+\frac{1}{2\pi i}\int\limits_{\gamma}\frac{\langle\widetilde{\Phi}(x,\lambda),\widetilde{g}^{*}(x,\xi)\rangle_{\widetilde{\ell}}}{\lambda-\xi}\,\frac{\langle\varphi(x,\xi),\Phi^{*}(x,\mu)\rangle_{\ell}}{\xi-\mu}\,d\xi=0;\tag{2.13}$$
$$\lambda,\mu\in J_{\gamma}.$$

By continuity, (2.12) gives us

$$\widetilde{f}(x,\lambda)=f(x,\lambda)+\frac{1}{2\pi i}\int\limits_{\gamma}\frac{\langle\widetilde{f}(x,\lambda),\widetilde{g}^{*}(x,\mu)\rangle_{\widetilde{\ell}}}{\lambda-\mu}\varphi(x,\mu)\,d\mu,\quad \lambda\in\gamma_{1}\cup\gamma_{-1}.\tag{2.14}$$

Since

$$\langle\widetilde{\Phi}(x,\lambda),\widetilde{g}^{*}(x,\mu)\rangle_{\widetilde{\ell}|_{x=0}}=-\widetilde{\mathfrak{M}}(\lambda)\widetilde{\mathfrak{M}}^{-1}(\mu)\widetilde{A}_{0}(\mu)Y^{T},$$

for $\mu\in\gamma_0$, by (1.27) we get that for $\mu\in\gamma_0$

$$\frac{\langle Y\widetilde{\Phi}(x,\lambda),\widetilde{g}^{*}(x,\mu)\rangle_{\widetilde{\ell}}}{\lambda-\mu}=\frac{Y\widetilde{\mathfrak{M}}(\lambda)\widetilde{\mathfrak{M}}^{-1}(\mu)\widetilde{A}_{0}(\mu)Y^{T}}{\mu-\lambda}+\int\limits_{0}^{x}Y\widetilde{\Phi}(t,\lambda)\widetilde{g}^{*}(t,\mu)\,dt.$$

Consequently, from (2.12) we have by the Sokhotskii formulas [95] that

$$Y\widetilde{\Phi}(x,\lambda)=Y\Phi(x,\lambda)-\frac{1}{2}\widetilde{a}(\lambda)\varphi(x,\lambda)+\frac{1}{2\pi i}\int\limits_{\gamma}\frac{Y\langle\widetilde{\Phi}(x,\lambda),\widetilde{g}^{*}(x,\mu)\rangle_{\widetilde{\ell}}}{\lambda-\mu}\varphi(x,\mu)\,d\mu,$$
$$\lambda\in\gamma_{0},$$

which together with (2.14) yields (2.10).

Further, repeating the preceding arguments, we get from (2.13) that

$$\widetilde{N}(\lambda)\frac{\langle\varphi(x,\lambda),\Phi^{*}(x,\mu)\rangle_{\ell}}{\lambda-\mu}-\frac{\langle\widetilde{\varphi}(x,\lambda),\widetilde{\Phi}^{*}(x,\mu)\rangle_{\widetilde{\ell}}}{\lambda-\mu}$$
$$+\frac{1}{2\pi i}\int\limits_{\gamma}\widetilde{r}(x,\lambda,\xi)\frac{\langle\varphi(x,\xi),\Phi^{*}(x,\mu)\rangle_{\ell}}{\xi-\mu}\,d\xi=0,\quad \lambda\in\gamma,\ \mu\in J_{\gamma}.\tag{2.15}$$

Since for $\xi \in \gamma_0$

$$\frac{\langle \varphi(x,\xi), \Phi^*(x,\mu)\rangle_\ell}{\xi-\mu} = \frac{Y\mathfrak{M}(\xi)\mathfrak{M}^{-1}(\mu)}{\xi-\mu} + \int_0^x \varphi(t,\xi)\Phi^*(t,\mu)\,dt,$$

it follows from (2.15) and the Sokhotskii formula that

$$\widetilde{N}(\lambda)\frac{\langle \varphi(x,\lambda), \Phi^*(x,\mu)\rangle_\ell}{\lambda-\mu} - \frac{\langle \widetilde{\varphi}(x,\lambda), \widetilde{\Phi}^*(x,\mu)\rangle_{\widetilde{\ell}}}{\lambda-\mu} - \frac{1}{2}\widetilde{r}(x,\lambda,\mu)Y$$

$$+\frac{1}{2\pi i}\int_\gamma \widetilde{r}(x,\lambda,\xi)\frac{\langle \varphi(x,\xi), \Phi^*(x,\mu)\rangle_\ell}{\xi-\mu}\,d\xi = 0, \qquad \lambda\in\gamma, \qquad \mu\in\gamma_0.$$

Multiplying both sides of this equality by $-A_0(\mu)Y^T$ from the right and using (2.8), we get

$$\left.\begin{aligned}&\widetilde{N}(\lambda)r(x,\lambda,\mu) - \widetilde{r}(x,\lambda,\mu)(E+a(\mu)) + \frac{1}{2}\widetilde{r}(x,\lambda,\mu)a(\mu)\\ &+\frac{1}{2\pi i}\int_\gamma \widetilde{r}(x,\lambda,\xi)r(x,\xi,\mu)\,d\xi = 0, \qquad \lambda\in\gamma, \qquad \mu\in\gamma_0.\end{aligned}\right. \tag{2.16}$$

By continuity it follows from (2.15) that

$$\widetilde{N}(\lambda)\frac{\langle \varphi(x,\lambda), f^*(x,\mu)\rangle_\ell}{\lambda-\mu} - \frac{\langle \widetilde{\varphi}(x,\lambda), \widetilde{f}^*(x,\mu)\rangle_{\widetilde{\ell}}}{\lambda-\mu}$$

$$+\frac{1}{2\pi i}\int_\gamma \widetilde{r}(x,\lambda,\xi)\frac{\langle \varphi(x,\xi), f^*(x,\mu)\rangle_\ell}{\xi-\mu}\,d\xi = 0, \qquad \lambda\in\gamma, \qquad \mu\in\gamma_1\cup\gamma_{-1}.$$

This and (2.16) yield (2.11). Theorem 2.1 is proved.

We shall assume below for simplicity that L, $\widetilde{L}\in v_N$ are chosen so that

$$\widehat{\mathfrak{M}}_{m,m+1}(\lambda) = O(\rho^{-n-2}), \qquad |\lambda|\to\infty. \tag{2.17}$$

Let us show that

$$\left.\begin{aligned}&|\widetilde{g}_j^{*(\nu)}(x,\mu)| < C|\theta|^{\nu-j-n}|\exp(-\theta R_j x)|,\\ &|\widetilde{g}^{*(\nu)}(x,\mu)\varphi^{(s)}(x,\mu)| < C|\theta|^{\nu+s-2n}, \qquad \mu=\theta^n, \quad \mu\in\gamma_1\cup\gamma_{-1}\end{aligned}\right\} \tag{2.18}$$

Indeed, for $\mu\in\gamma_1\cup\gamma_{-1}$

$$\widetilde{g}_j^{*(\nu)}(x,\mu) = (-1)^{j-1}\chi_{(-1)^{n-j+1}}(\mu)\widetilde{\Phi}^{*(\nu)}_{n-j+2}(x,\mu)\widehat{\mathfrak{M}}_{j-1,j}(\mu),$$

$$|\widetilde{\Phi}^{*(\nu)}_{n-j+2}(x,\mu)| < C|\theta|^{\nu+2-j}|\exp(-\theta R_{j-1}x)|.$$

For $j = n-2\nu$, $\nu = 0,\dots,[\frac{n}{2}]$, we have $\widetilde{g}^*_j(x,\lambda) = 0$ $(\lambda \in \gamma_1)$ and $\operatorname{Re}(\rho(R_j - R_{j-1})) = 0$ $(\lambda \in \gamma_{-1})$, and for $j = n-2\nu-1$, $\nu = 0,\dots,[\frac{n}{2}]$, we have $\widetilde{g}^*_j(x,\lambda) = 0$ $(\lambda \in \gamma_{-1})$ and $\operatorname{Re}(\rho(R_j - R_{j-1})) = 0$ $(\lambda \in \gamma_1)$. This and (2.17) give us the first estimate in (2.18). Since

$$|\varphi_j^{(s)}(x,\mu)| < C\,|\theta|^{j+s-n}\,|\exp(\theta R_j x)|,$$

it follows that

$$|\widetilde{g}^{*(\nu)}(x,\mu)\varphi^{(s)}(x,\mu)| \le \sum_{j=2}^{n} |\widetilde{g}_j^{*(\nu)}(x,\mu)\varphi_j^{(s)}(x,\mu)| < C|\theta|^{\nu+s-2n}.$$

Let

$$\kappa_{\nu s}(x) = \frac{1}{2\pi i}\int_\gamma \widetilde{g}^{*(\nu)}(x,\mu)\varphi^{(s)}(x,\mu)\,d\mu, \qquad \nu+s \le n-1; \tag{2.19}$$

$$\left.\begin{aligned} t_{j\nu}(x) &= -\sum_{\beta=\nu+1}^{j} C_j^\beta C_{\beta-1}^\nu \kappa_{\beta-\nu-1,j-\beta}(x), \qquad j>\nu;\\ t_{i\nu}(x) &= \delta_{j\nu}, \qquad j\le\nu, \qquad j,\nu=\overline{0,n}; \end{aligned}\right\} \tag{2.20}$$

$$\begin{aligned} \xi_\nu(x) \;=\; &\sum_{s=0}^{n-\nu-1}\sum_{j=\nu+1}^{n-s}\Big(C_{j+s}^j C_{j-1}^\nu \widetilde{p}_{j+s}(x)\kappa_{j-\nu-1,s}(x)\\ &+\delta_{s0}(-1)^{j-\nu}\sum_{r=0}^{j-\nu-1} C_{j-\nu-1}^r \widetilde{p}_j^{(j-\nu-1-r)}(x)\kappa_{r0}(x)\Big),\\ &\nu=\overline{0,n-2}; \end{aligned} \tag{2.21}$$

$$\varepsilon_\nu(x) = \xi_\nu(x) - \sum_{j=\nu+1}^{n-2}\varepsilon_j(x)t_{j\nu}(x), \qquad \nu=\overline{0,n-2}. \tag{2.22}$$

The following lemma establishes a connection between the coefficients of the DE's and LF's L and $\widetilde{L}$.

Lemma 2.3.

$$p_\nu(x) = \widetilde{p}_\nu(x) + \varepsilon_\nu(x), \qquad \widetilde{u}_{\xi\nu 0} = \sum_{j=0}^{n-1} u_{\xi j 0} t_{j\nu}(0). \tag{2.23}$$

Proof. Differentiating (2.12) with respect to x and using (1.27), (2.19) and (2.20), we get

$$\sum_{\nu=0}^{n} t_{j\nu}(x)\widetilde{\Phi}^{(\nu)}(x,\lambda) = \Phi^{(j)}(x,\lambda) + \frac{1}{2\pi i}\int_\gamma \frac{\langle \widetilde{\Phi}(x,\lambda), \widetilde{g}^*(x,\mu)\rangle_{\widetilde{\ell}}}{\lambda-\mu}\varphi^{(j)}(x,\mu)\,d\mu, \tag{2.24}$$

$$\lambda \in J_\gamma.$$

Since $\Phi(x,\lambda)$ and $\varphi(x,\lambda)$ are solutions of the DE (1.1), we compute from (2.12) that

$$\begin{aligned} \widetilde{\ell}\widetilde{\Phi}(x,\lambda) &= \ell\Phi(x,\lambda) + \frac{1}{2\pi i}\int_\gamma \frac{\langle \widetilde{\Phi}(x,\lambda), \widetilde{g}^*(x,\mu)\rangle_{\widetilde{\ell}}}{\lambda-\mu}\ell\varphi(x,\mu)\,d\mu \\ &\quad + \frac{1}{2\pi i}\int_\gamma \langle \widetilde{\Phi}(x,\mu), \widetilde{g}^*(x,\mu)\rangle_{\widetilde{\ell}}\varphi(x,\mu)\,d\mu, \qquad \lambda \in J_\gamma. \end{aligned} \tag{2.25}$$

By (2.25), in view of (2.24) and (1.22) we have

$$\widetilde{\ell}\widetilde{\Phi}(x,\lambda) = \sum_{j=0}^{n} p_j(x)\sum_{\nu=0}^{n} t_{j\nu}(x)\widetilde{\Phi}^{(\nu)}(x,\lambda) + \sum_{\nu,j=0}^{n-1} \widetilde{\mathcal{L}}_{\nu j}(x)\widetilde{\Phi}^{(\nu)}(x,\lambda)\kappa_{j0}(x)$$

and hence

$$p_\nu(x) = \widetilde{p}_\nu(x) - \sum_{j=\nu+1}^{n} p_j(x)t_{j\nu}(x) - \sum_{j=0}^{n-1}\widetilde{\mathcal{L}}_{\nu j}(x)\kappa_{j0}(x)$$

or

$$\widehat{p}_\nu(x) = -\sum_{j=\nu+1}^{n} \widehat{p}_j(x)t_{j\nu}(x) - \sum_{j=\nu+1}^{n} \widetilde{p}_j(x)t_{j\nu}(x) - \sum_{j=0}^{n-1}\widetilde{\mathcal{L}}_{\nu j}(x)\kappa_{j0}(x).$$

Using (1.23), (2.20), and (2.21), we get

$$\xi_\nu(x) = \sum_{j=\nu+1}^{n} \widetilde{p}_j(x)t_{j\nu}(x) - \sum_{j=0}^{n-1}\widetilde{\mathcal{L}}_{\nu j}(x)\kappa_{j0}(x)$$

and hence

$$\widehat{p}_\nu(x) = \xi_\nu(x) - \sum_{j=\nu+1}^{n} \widehat{p}_j(x) t_{j\nu}(x).$$

Thus, $\widehat{p}_\nu(x) = \varepsilon_\nu(x)$ and the first relation in (2.23) is proved.

Since $\widetilde{g}^{*(\nu)}(x,\mu)\varphi^{(j)}(x,\mu) = \widetilde{G}^{*(\nu)}(x,\mu)\Phi^{(j)}(x,\mu)$, from (2.24) with $x = 0$ we have

$$\sum_{\nu=0}^{n-1}\left(\sum_{j=0}^{n-1} u_{\xi j0} t_{j\nu}(0)\right)\widetilde{\Phi}^{(\nu)}(0,\lambda)$$

$$= U_{\xi 0}(\Phi(x,\lambda)) + \frac{1}{2\pi i}\int_\gamma \frac{\langle\widetilde{\Phi}(x,\lambda), \widetilde{G}^*(x,\mu)\rangle_{\widetilde{\ell}}}{\lambda-\mu}\bigg|_{x=0} U_{\xi 0}(\Phi(x,\mu))\,d\mu,$$

or

$$\widetilde{\widetilde{U}}_{\xi 0}(\widetilde{\Phi}(x,\lambda)) = U_{\xi 0}(\Phi(x,\lambda)) + \frac{1}{2\pi i}\int_\gamma \frac{\widetilde{\Gamma}(\lambda,\mu)}{\mu-\lambda} U_{\xi 0}(\Phi(x,\mu))\,d\mu,$$

$$\widetilde{\widetilde{U}}_{\xi 0}(y) \stackrel{df}{=} \sum_{\nu=0}^{n-1}\left(\sum_{j=0}^{n-1} u_{\xi j0} t_{j\nu}(0)\right) y^{(\nu)}(0).$$

According to (2.7), $\widetilde{\Gamma}_{j\nu}(\lambda,\mu) = 0$ $(j \ge \nu)$, and hence

$$\widetilde{\widetilde{U}}_{\xi 0}(\widetilde{\Phi}_m(x,\lambda)) = \delta_{\xi m}, \qquad \xi \le m.$$

From this and from the conditions on the WS's $\widetilde{\Phi}_m(x,\lambda)$ we obtain

$$\sum_{\nu=0}^{n-\xi-1} \check{u}_{\xi\nu 0}\widetilde{\Phi}_m^{(\nu)}(0,\lambda) \equiv 0, \qquad 1 \le \xi \le m \le n,$$

where

$$\check{u}_{\xi\nu 0} = \widetilde{u}_{\xi\nu 0} - \sum_{j=0}^{n-1} u_{\xi j0} t_{j\nu}(0).$$

Consequently, $\check{u}_{\xi\nu 0} = 0$, i.e., the second relation in (2.23) is valid. Lemma 2.3 is proved.

We note that we have also obtained the formula

$$\widetilde{U}_{\xi 0}(\widetilde{\Phi}_m(x,\lambda)) = U_{\xi 0}(\Phi_m(x,\lambda)) + \frac{1}{2\pi i}\int_\gamma \sum_{j=m+1}^{n} \frac{\widetilde{\Gamma}_{mj}(\lambda,\mu)}{\mu-\lambda} U_{\xi 0}(\Phi_j(x,\mu))\,d\mu. \quad (2.26)$$

Let

$$\gamma'' = \{\lambda : \lambda \in \gamma_1 \cup \gamma_{-1}, \inf|\lambda - \mu| \geq \delta_0, \mu \in \gamma_0\}, \qquad \delta_0 > 0,$$

and let $\gamma' = \gamma \backslash \gamma''$. Thus, $\gamma = \gamma' \cup \gamma''$.

Lemma 2.4.

$$|\widetilde{r}_{kj}(x,\lambda,\mu)| < \frac{C_x\,|\exp((\rho R_k - \theta R_j)x)|}{|\rho|^{n-k}|\theta|^{n+j}(|\rho-\theta|+1)} \tag{2.27}$$

for $\mu \in \gamma''$, $\lambda \in \gamma$ *or for* $\mu \in \gamma$, $\lambda \in \gamma''$ *and*

$$|\widetilde{r}_{kj}^{(\nu+1)}(x,\lambda,\mu)| < \frac{C_x\,|\exp((\rho R_k - \theta R_j)x)|}{|\rho|^{n-k}|\theta|^{n+j}}(|\rho|+|\theta|)^{\nu} \tag{2.28}$$

for λ, $\mu \in \gamma$, $\nu = \overline{0, n-1}$.

Proof. Suppose that $\mu \in \gamma''$, $\lambda \in \gamma$ or $\mu \in \gamma$, $\lambda \in \gamma''$. Let $|\rho - \theta| \leq \varepsilon_0$, where $\varepsilon_0 > 0$ is a sufficiently small fixed number. Then either λ, $\mu \in \gamma_1$ or λ, $\mu \in \gamma_{-1}$, and hence

$$\widetilde{g}_j^*(x,\mu) = (-1)^{j-1}\chi_{(-1)^{n-j+1}}(\mu)\widetilde{\Phi}^*_{n-j+2}(x,\mu)\widehat{\mathfrak{M}}_{j-1,j}(\mu).$$

For $k = j-1$, either $\varphi_k(x,\lambda) \equiv 0$ or $\widetilde{g}_j^*(x,\mu) \equiv 0$, i.e., $\widetilde{r}_{kj}(x,\lambda,\mu) \equiv 0$. For $k \neq j-1$, we have in view of (1.27) and the equalities

$$\widetilde{r}_{kj}(0,\lambda,\mu) = 0 \quad (k \geq j), \qquad \lim_{x\to\infty} \widetilde{r}_{kj}(x,\lambda,\mu) = 0 \qquad (k \leq j-2)$$

that

$$\widetilde{r}_{kj}(x,\lambda,\mu) = \int_a^x \widetilde{\varphi}_k(t,\lambda)\widetilde{g}_j^*(t,\mu)\,dt, \qquad \begin{cases} a = 0, & (k \geq j) \\ a = \infty, & (k \leq j-2). \end{cases}$$

The estimate (2.27) follows from this by virtue of (2.18) and

$$|\widetilde{\varphi}_k^{(s)}(x,\lambda)| < C\,|\rho|^{k+s-n}\,|\exp(\rho R_k x)|. \tag{2.29}$$

Suppose now that $|\rho - \theta| \geq \varepsilon_0$. Using (1.22), (2.18), and (2.29) we get

$$|\widetilde{r}_{kj}(x,\lambda,\mu)| \leq \left|\frac{\langle\widetilde{\varphi}_k(x,\lambda), \widetilde{g}_j^*(x,\mu)\rangle_{\widetilde{\ell}}}{\lambda-\mu}\right| \leq \frac{C_x\,|\exp((\rho R_k - \theta R_j)x)|}{|\rho|^{n-k}|\theta|^{n+j}|\lambda-\mu|}\sum_{i=0}^{n-1}|\rho|^{n-i-1}|\theta|^i.$$

Let us show that for $\lambda \in \gamma$, $\mu \in \gamma''$ or $\lambda \in \gamma''$, $\mu \in \gamma$

$$\frac{1}{|\lambda-\mu|}\sum_{i=0}^{n-1}|\rho|^{n-i-1}|\theta|^i \leq \frac{C}{|\rho-\theta|+1}, \qquad |\rho-\theta| \geq \varepsilon_0.$$

Indeed, if $\lambda \in \gamma_0$, $\mu \in \gamma''$ or $\lambda \in \gamma''$, $\mu \in \gamma_0$, then this estimate is obvious. If λ, $\mu \in \gamma_1$ or λ, $\mu \in \gamma_{-1}$, then $|\lambda - \mu| = ||\lambda| - |\mu||$, $|\rho - \theta| = ||\rho| - |\theta||$, and hence

$$\frac{1}{|\lambda - \mu|} \sum_{i=0}^{n-1} |\rho|^{n-i-1} |\theta|^i \le \frac{1}{|\rho - \theta|} \le \frac{C_0}{|\rho - \theta| + 1}, \qquad C_0 = \frac{\varepsilon_0 + 1}{\varepsilon_0}.$$

But if $\lambda \in \gamma_1$, $\mu \in \gamma_{-1}$ or $\lambda \in \gamma_{-1}$, $\mu \in \gamma_1$, then $|\lambda - \mu| = |\lambda| + |\mu|$ and hence

$$\frac{1}{|\lambda - \mu|} \sum_{i=0}^{n-1} |\rho|^{n-i-1} |\theta|^i \le \frac{2(n+1)}{|\rho| + |\theta|} \le \frac{C}{|\rho - \theta| + 1}.$$

Thus, (2.27) is proved.

Futher, by (1.27),

$$\tilde{r}_{kj}^{(\nu+1)}(x, \lambda, \mu) = \frac{d^\nu}{dx^\nu} (\tilde{\varphi}_k(x, \lambda) \tilde{g}_j^*(x, \mu)), \qquad \lambda, \mu \in \gamma.$$

From this, by (2.18) and (2.29), we obtain (2.28). Lemma 2.4 is proved.

We note that since $\langle \tilde{\varphi}(x, \lambda), \tilde{g}^*(x, \mu) \rangle_{\tilde{\ell}|_{x=0}} = -Y \widetilde{\mathfrak{M}}(\lambda) \widetilde{\mathfrak{M}}^{-1}(\mu) \tilde{A}_0(\mu) Y^T$ for $\lambda, \mu \in \gamma_0$ we have in view of (1.27) that for λ, $\mu \in \gamma_0$

$$\tilde{r}(x, \lambda, \mu) = \frac{Y \widetilde{\mathfrak{M}}(\lambda) \widetilde{\mathfrak{M}}^{-1}(\mu) \tilde{A}_0(\mu) Y^T}{\mu - \lambda} + \int_0^x \tilde{\varphi}(t, \lambda) \tilde{g}^*(t, \mu) \, dt. \tag{2.30}$$

Suppose for definiteness that $\arg \rho \in (0, 2\pi/n)$. Denote

$$\Omega(x, \lambda) = \operatorname{diag} [\rho^{k-n} \exp(\rho R_k x)]_{k=\overline{2,n}},$$

$$\varphi^+(x, \lambda) = \Omega^{-1}(x, \lambda) \varphi(x, \lambda), \qquad r^+(x, \lambda, \mu) = \Omega^{-1}(x, \lambda) r(x, \lambda, \mu) \Omega(x, \mu),$$

$$a^+(x, \lambda) = \Omega^{-1}(x, \lambda) a(\lambda) \Omega(x, \lambda), \qquad N^+(x, \lambda) = \Omega^{-1}(x, \lambda) N(\lambda) \Omega(x, \lambda).$$

We define the matrices $\tilde{\varphi}^+(x, \lambda)$, $\tilde{r}^+(x, \lambda, \mu)$, $\tilde{a}^+(x, \lambda)$ and $\tilde{N}^+(x, \lambda)$ similarly. It follows from (2.29), (2.30), and Lemma 2.4 that

$$\left.\begin{aligned}
&|\tilde{\varphi}^{+(\nu)}(x, \lambda)| < C|\rho|^\nu, && \lambda \in \gamma, \quad \nu = \overline{0, n-1} \\
&|\tilde{r}_{kj}^+(x, \lambda, \mu)| < \frac{C_x}{|\theta|^{2n}(|\rho - \theta| + 1)}; && \lambda \in \gamma, \mu \in \gamma'' \text{ or } \lambda \in \gamma'', \mu \in \gamma \\
&|\tilde{r}_{kj}^{+(\nu+1)}(x, \lambda, \mu)| < C_x |\theta|^{-2n} (|\rho| + |\theta|)^\nu, && \lambda, \mu \in \gamma, \ \nu = \overline{0, n-1}
\end{aligned}\right\} \tag{2.31}$$

and the functions $\tilde{r}_{kj}^+(x, \lambda, \mu)$ are continuous for λ, $\mu \in \gamma_1$ and λ, $\mu \in \gamma_{-1}$, while for λ, $\mu \in \gamma_0$

$$\widetilde{r}^+(x,\lambda,\mu)=\frac{\widetilde{a}^+(x,\lambda)}{\mu-\lambda}+\widetilde{H}^+(x,\lambda,\mu),$$

where $\widetilde{H}^+(x,\lambda,\mu)$ is a continuous function. The functions $r^+(x,\lambda,\mu)$ and $\varphi^+(x,\lambda)$ have analogous properties. It follows from (2.9) and Theorem 2.1 that the following theorem is valid

Theorem 2.2.

$$\widetilde{\varphi}^+(x,\lambda)=\widetilde{N}^+(x,\lambda)\varphi^+(x,\lambda)+\frac{1}{2\pi i}\int\limits_\gamma \widetilde{r}^+(x,\lambda,\mu)\varphi^+(x,\mu)\,d\mu,\qquad \lambda\in\gamma, \tag{2.32}$$

$$\begin{aligned}&\widetilde{N}^+(x,\lambda)r^+(x,\lambda,\mu)-\widetilde{r}^+(x,\lambda,\mu)N^+(x,\mu)\\ &\qquad+\frac{1}{2\pi i}\int\limits_\gamma \widetilde{r}^+(x,\lambda,\xi)r^+(x,\xi,\mu)\,d\xi=0,\qquad \lambda,\mu\in\gamma;\end{aligned} \tag{2.33}$$

$$\left.\begin{aligned}&\widetilde{N}^+(x,\lambda)N^+(x,\lambda)-\tfrac{1}{4}\widetilde{a}^+(x,\lambda)a^+(x,\lambda)=E,\\ &\widetilde{N}^+(x,\lambda)a^+(x,\lambda)-\widetilde{a}^+(x,\lambda)N^+(x,\lambda)=0,\end{aligned}\right\} \tag{2.34}$$

We introduce the Banach space $B=\mathcal{L}_2^{n-1}(\gamma')\oplus\mathcal{L}_\infty^{n-1}(\gamma'')$ of vector-valued functions $z(\lambda)=[z_j(\lambda)]_{j=\overline{1,n-1}}$, $\lambda\in\gamma$ with the norm

$$\|z\|_B=\sum_{j=1}^{n-1}\left(\|z_j\|_{\mathcal{L}_2(\gamma')}+\|z_j\|_{\mathcal{L}_\infty(\gamma'')}\right).$$

For fixed $x\geq 0$ we consider on B the linear operators

$$\begin{aligned}&\widetilde{A}z(\lambda)=\widetilde{N}^+(x,\lambda)z(\lambda)+\frac{1}{2\pi i}\int\limits_\gamma \widetilde{r}^+(x,\lambda,\mu)z(\mu)\,d\mu,\qquad \lambda\in\gamma,\\ &Az(\lambda)=N^+(x,\lambda)z(\lambda)-\frac{1}{2\pi i}\int\limits_\gamma r^+(x,\lambda,\mu)z(\mu)\,d\mu,\qquad \lambda\in\gamma.\end{aligned} \tag{2.35}$$

Lemma 2.5. *For a fixed $x\geq 0$, the operators A and $\widetilde{A}$ are bounded linear operators on B, and $\widetilde{A}A=A\widetilde{A}=E$.*

Proof. The boundedness of A and $\widetilde{A}$ is obvious. Using the formula for interchanging the order of integration in a singular integral ([95, p. 60]), we get that

$$\begin{aligned}&\frac{1}{2\pi i}\int\limits_\gamma \widetilde{r}^+(x,\lambda,\xi)\,d\xi\,\frac{1}{2\pi i}\int\limits_\gamma r^+(x,\xi,\mu)z(\mu)\,d\mu\\ &\quad=\frac{1}{4}\widetilde{a}^+(x,\lambda)a^+(x,\lambda)+\frac{1}{2\pi i}\int\limits_\gamma\left(\frac{1}{2\pi i}\int\limits_\gamma \widetilde{r}^+(x,\lambda,\xi)r^+(x,\xi,\mu)\,d\xi\right)z(\mu)\,d\mu.\end{aligned}$$

It then follows from (2.33)–(2.35) that

$$\begin{aligned}\widetilde{A}Az(\lambda) &= (\widetilde{N}^+(x,\lambda)N^+(x,\lambda) - \frac{1}{4}\widetilde{a}^+(x,\lambda)a^+(x,\lambda))z(\lambda) \\ &-\frac{1}{2\pi i}\int\limits_\gamma \Big(\widetilde{N}^+(x,\lambda)r^+(x,\lambda,\mu) - \widetilde{r}^+(x,\lambda,\mu)N^+(x,\lambda) \\ &+\frac{1}{2\pi i}\int\limits_\gamma \widetilde{r}^+(x,\lambda,\xi)r^+(x,\xi,\mu)\,d\xi\Big) z(\mu)\,d\mu = z(\lambda),\end{aligned}$$

i.e., $\widetilde{A}A = E$. Similarly, $A\widetilde{A} = E$. Lemma 2.5 is proved.

Corollary 2.1. *For $x \geq 0$ the main equation (2.10) of the IP has a unique solution in the class $\Omega^{-1}(x,\lambda)\varphi(x,\lambda) \in B$ and $\sup\limits_x \|\Omega^{-1}(x,\lambda)\varphi(x,\lambda)\|_B < \infty$.*

2.2. Denote by M the set of matrices $\mathfrak{M}(\lambda) = [\mathfrak{M}_{mk}(\lambda)]_{m,k=\overline{1,n}}$ such that

(1) $\mathfrak{M}_{mk}(\lambda) = \delta_{mk}$, $m \geq k$ and $\mathfrak{M}_{mk}(\lambda) = O(\rho^{m-k})$, $|\lambda| \to \infty$, $m < k$.

(2) the functions $\mathfrak{M}_{mk}(\lambda)$ are regular in $\Pi_{(-1)^{n-m}}$ with the exception of an at most countable bounded set Λ'_{mk} of poles, and are continuous in $\overline{\Pi}_{(-1)^{n-m}}$ with the exception of bounded sets Λ_{mk};

(3) the functions $\mathfrak{M}_{mk}(\lambda) - \mathfrak{M}_{m,m+1}(\lambda)\mathfrak{M}_{m+1,k}(\lambda)$ are regular for $\lambda \in \Gamma_{(-1)^{n-m}} \setminus \Lambda$, $\Lambda = \bigcup\limits_{m,k} \Lambda_{mk}$ (in general the set Λ is different for each matrix $\mathfrak{M}(\lambda)$).

Theorem 2.3. *A matrix $\mathfrak{M}(\lambda) \in M$ is the WM for $L \in v_N$ if and only if the following conditions hold:*

(1) *(asymptotics) there exists $\widetilde{L} \in v_N$ such that $\widehat{\mathfrak{M}}_{m,m+1}(\lambda) = O(\rho^{-n-2})$, $|\lambda| \to \infty$;*

(2) *(condition P) for $x \geq 0$ equation (2.10) has a unique solution in the class $\Omega^{-1}(x,\lambda)\varphi(x,\lambda) \in B$, and $\sup\limits_x \|\Omega^{-1}(x,\lambda)\varphi(x,\lambda)\|_B < \infty$;*

(3) *$\varepsilon_\nu(x) \in W_{\nu+N}$, $\nu = \overline{0, n-2}$, where the functions $\varepsilon_\nu(x)$ are defined by (2.19)–(2.22).*

Under these conditions the DE and LF $L = (\ell, U)$ are constructed according to (2.23).

It can be shown by a counterexample that conditions (2) and (3) in Theorem 2.3 are essential.

The necessity part of Theorem 2.3 was proved above in 2.1. We now prove the sufficiency. Let $\varphi(x,\lambda)$ be a solution of (2.10).

Lemma 2.6. *The functions* $\varphi^{(\nu)}(x,\lambda)$, $\nu = \overline{0,n-1}$ *are absolutely continuous in* x *on each finite interval, and for fixed* $x \geq 0$

$$\Omega^{-1}(x,\lambda)\varphi^{(\nu)}(x,\lambda)\rho^{-\nu} \in B, \qquad \nu = \overline{0,n}.$$

Proof. We first construct the inverse operator for $x = 0$. Let

$$B(\lambda) = \operatorname{diag}\,[\chi((-1)^{n-k+1}\lambda)]_{k=\overline{2,n}}$$

for $\lambda \in \gamma_1 \cup \gamma_{-1}$ and $B(\lambda) = E$ for $\lambda \in \gamma_0$. Then $\widetilde{\varphi}(x,\lambda) = B(\lambda)Y\widetilde{\Phi}(x,\lambda)$. By (1.27),

$$\widetilde{r}(x,\lambda,\mu) = \widetilde{r}^0(\lambda,\mu) + \widetilde{r}^1(x,\lambda,\mu),$$

where

$$\widetilde{r}^1(x,\lambda,\mu) = \int_0^x \widetilde{\varphi}(t,\lambda)\widetilde{g}^*(t,\mu)\,dt, \qquad \widetilde{r}^0(\lambda,\mu) = \widetilde{r}(0,\lambda,\mu) = \frac{\widetilde{\kappa}(\lambda,\mu)}{\mu-\lambda},$$

$$\widetilde{\kappa}(\lambda,\mu) = B(\lambda)Y\widetilde{\mathfrak{M}}(\lambda)\widetilde{\mathfrak{M}}^*(\mu)\widetilde{A}(\mu)Y^T, \qquad \widetilde{\kappa}(\lambda,\lambda) = \widetilde{a}(\lambda).$$

Let

$$r^0(\lambda,\mu) = \frac{\kappa(\lambda,\mu)}{\mu-\lambda}, \qquad \kappa(\lambda,\mu) = B(\lambda)Y\mathfrak{M}(\lambda)\mathfrak{M}^*(\mu)A(\mu)Y^T,$$

$$\mathfrak{M}^*(\lambda) = \mathfrak{M}^{-1}(\lambda), \qquad \kappa(\lambda,\lambda) = a(\lambda).$$

Using (2.5) and the analytic properties of $\mathfrak{M}(\lambda)$ and $\widetilde{\mathfrak{M}}(\lambda)$ we get

$$\frac{\mathfrak{M}(\lambda)\mathfrak{M}^*(\mu)}{\lambda-\mu} - \frac{\widetilde{\mathfrak{M}}(\lambda)\widetilde{\mathfrak{M}}^*(\mu)}{\lambda-\mu} = \frac{1}{2\pi i}\int_\gamma \frac{\widetilde{\mathfrak{M}}(\lambda)\widetilde{\mathfrak{M}}^*(\xi)\mathfrak{M}(\xi)\mathfrak{M}^*(\mu)}{(\lambda-\xi)(\xi-\mu)}\,d\xi.$$

Repeating the arguments in the proof of Theorem 2.1, we find that

$$\widetilde{N}(\lambda)r^0(\lambda,\mu) - \widetilde{r}^0(\lambda,\mu)N(\mu) + \frac{1}{2\pi i}\int_\gamma \widetilde{r}^0(\lambda,\xi)r^0(\xi,\mu)\,d\xi = 0,$$

and, analogously,

$$N(\lambda)\widetilde{r}^0(\lambda,\mu) - r^0(\lambda,\mu)\widetilde{N}(\mu) + \frac{1}{2\pi i}\int_\gamma r^0(\lambda,\xi)\widetilde{r}^0(\xi,\mu)\,d\xi = 0.$$

Let us now consider the operators

$$A_0y(\lambda) = N(\lambda)y(\lambda) - \frac{1}{2\pi i}\int_\gamma r^0(\lambda,\mu)y(\mu)\,d\mu,$$

$$\widetilde{A}_0 y(\lambda) = \widetilde{N}(\lambda) y(\lambda) + \frac{1}{2\pi i} \int\limits_{\gamma} \widetilde{r}^0(\lambda, \mu) y(\mu)\, d\mu.$$

As in the proof of Lemma 2.5, we see that A_0 and $\widetilde{A}_0$ are bounded linear operators on B, and $A_0 \widetilde{A}_0 = \widetilde{A}_0 A_0 = E$.

We regularize equation (2.10). Let $\gamma^- = \{\lambda : \lambda \in \gamma_1 \cup \gamma_{-1}, |\lambda| \geq R\}$ and $\gamma^+ = \gamma \backslash \gamma^-$, where R is a sufficiently large real number. Consider the operator

$$Qy(\lambda) = \begin{cases} y(\lambda), & \lambda \in \gamma^-, \\ N(\lambda) y(\lambda) - \dfrac{1}{2\pi i} \displaystyle\int\limits_{\gamma} r^0(\lambda, \mu) y(\mu)\, d\mu, & \lambda \in \gamma^+. \end{cases}$$

Clearly Q is a bounded linear operator on B. We choose R such that Q^{-1} exists and is bounded. This is possible, because Q "differs little" from A_0, when R is large. Applying Q to both sides of (2.10), we get

$$\widetilde{\widetilde{\varphi}}(x, \lambda) = \varphi(x, \lambda) + \frac{1}{2\pi i} \int\limits_{\gamma} \widetilde{\widetilde{r}}(x, \lambda, \mu) \varphi(x, \mu)\, d\mu,$$

where

$$\widetilde{\widetilde{\varphi}}(x, \lambda) = \begin{cases} N(\lambda) \widetilde{\varphi}(x, \lambda) - \dfrac{1}{2\pi i} \displaystyle\int\limits_{\gamma^+} r^0(\lambda, \mu) \widetilde{\varphi}(x, \mu)\, d\mu, & \lambda \in \gamma^+, \\ \widetilde{\varphi}(x, \lambda), & \lambda \in \gamma^-, \end{cases}$$

$$\widetilde{\widetilde{r}}(x, \lambda, \mu) = \begin{cases} \widetilde{r}(x, \lambda, \mu), & \lambda \in \gamma^-, \\ N(\lambda) \widetilde{r}(x, \lambda, \mu) - \dfrac{1}{2\pi i} \displaystyle\int\limits_{\gamma^+} r^0(\lambda, \xi) \widetilde{r}(x, \xi, \mu)\, d\xi, & \lambda \in \gamma^+, \mu \in \gamma^-, \\ N(\lambda) \widetilde{r}^1(x, \lambda, \mu) - \dfrac{1}{2\pi i} \displaystyle\int\limits_{\gamma^+} r^0(\lambda, \xi) \widetilde{r}^1(x, \xi, \mu)\, d\xi & \\ \quad + \dfrac{1}{2\pi i} \displaystyle\int\limits_{\gamma^-} r^0(\lambda, \xi) \widetilde{r}^0(\xi, \mu)\, d\xi, & \lambda, \mu \in \gamma^+. \end{cases}$$

We make the substitution

$$\widetilde{\widetilde{\varphi}}^+(x, \lambda) = \Omega^{-1}(x, \lambda) \widetilde{\widetilde{\varphi}}(x, \lambda), \qquad \varphi^+(x, \lambda) = \Omega^{-1}(x, \lambda) \varphi(x, \lambda),$$

$$\widetilde{\widetilde{r}}^+(x, \lambda, \mu) = \Omega^{-1}(x, \lambda) \widetilde{\widetilde{r}}(x, \lambda, \mu) \Omega(x, \mu).$$

Then

$$\tilde{\tilde{\varphi}}^{+}(x,\lambda) = \varphi^{+}(x,\lambda) + \frac{1}{2\pi i}\int\limits_{\gamma} \tilde{\tilde{r}}^{+}(x,\lambda,\mu)\varphi^{+}(x,\mu)\,d\mu, \tag{2.36}$$

and

$$\rho^{-\nu}\tilde{\tilde{\varphi}}^{+(\nu)}(x,\lambda) \in B, \qquad \nu = \overline{0,n},$$

$$|\tilde{\tilde{r}}^{+}(x,\lambda,\mu)| \le C_x|\theta|^{-2n}(|\rho-\theta|+1)^{-1},$$

$$|\tilde{\tilde{r}}^{+(\nu+1)}_{kj}(x,\lambda,\mu)| \le C_x|\theta|^{-2n}(|\rho|+|\theta|)^{\nu}, \qquad \nu = \overline{0,n-1}.$$

The operator

$$\tilde{\tilde{A}}y(\lambda) = y(\lambda) + \frac{1}{2\pi i}\int\limits_{\gamma} \tilde{\tilde{r}}^{+}(x,\lambda,\mu)y(\mu)\,d\mu, \qquad B \to B$$

is a bounded linear Fredholm operator for each $x \ge 0$. Let us show that $\tilde{\tilde{A}}^{-1}$ exists and is bounded. To do this it suffices to prove that the homogeneous equation

$$y^{+}(\lambda) + \frac{1}{2\pi i}\int\limits_{\gamma} \tilde{\tilde{r}}^{+}(x,\lambda,\mu)y^{+}(\mu)\,d\mu = 0, \qquad y^{+}(\lambda) \in B$$

has only the zero solution. Denote $y(x) = \Omega(x,\lambda)y^{+}(\lambda)$. Then

$$y(\lambda) + \frac{1}{2\pi i}\int\limits_{\gamma} \tilde{\tilde{r}}(x,\lambda,\mu)y(\mu)\,d\mu = 0.$$

We set

$$\tilde{z}(\lambda) = \tilde{N}(\lambda)y(\lambda) + \frac{1}{2\pi i}\int\limits_{\gamma} \tilde{r}(x,\lambda,\mu)y(\mu)\,d\mu.$$

Applying Q we get

$$Q\tilde{z}(\lambda) = y(\lambda) + \frac{1}{2\pi i}\int\limits_{\gamma} \tilde{\tilde{r}}(x,\lambda,\mu)y(\mu)\,d\mu = 0.$$

Since Q is invertible, $\tilde{z}(\lambda) = 0$. Then

$$\tilde{N}(\lambda)y(\lambda) + \frac{1}{2\pi i}\int\limits_{\gamma} \tilde{r}(x,\lambda,\mu)y(\mu)\,d\mu = 0, \qquad \Omega^{-1}(x,\lambda)y(\lambda) \in B.$$

This and condition P in Theorem 2.3 imply that $y(\lambda) = 0$ and hence $y^{+}(\lambda) = 0$. Thus, the operator $\tilde{\tilde{A}}^{-1}$ exists and is bounded. Further, using the Fredholm equation (2.36), we obtain the result by a well-known method (see, for example, [12, p. 39]) that the functions $\varphi^{+(\nu)}(x,\lambda)$, $\nu=\overline{0,n-1}$ are absolutely continuous in x and that $\rho^{-\nu}\varphi^{|(\nu)}(x,\lambda)\in B$, $\nu=\overline{0,n}$ for fixed $x\geq 0$. Lemma 2.6 is proved.

We construct the functions $\Phi(x,\lambda)=[\Phi_m(x,\lambda)]_{m=\overline{1,n}}$ by the formula

$$\Phi(x,\lambda)=\widetilde{\Phi}(x,\lambda)-\frac{1}{2\pi i}\int\limits_{\gamma}\frac{\langle\widetilde{\Phi}(x,\lambda),\widetilde{g}^*(x,\mu)\rangle_{\widetilde{\ell}}}{\lambda-\mu}\varphi(x,\mu)\,d\mu,\qquad \lambda\in J_\gamma,\tag{2.37}$$

and also the DE and LF $L=(\ell,U)$ by (2.23), where $\varepsilon_\nu(x)$ and $t_{j\nu}(x)$ are defined by (2.19)–(2.22). It is clear that $L\in v_N$.

Lemma 2.7.

$$\ell\varphi(x,\lambda)=\lambda\varphi(x,\lambda),\quad \lambda\in\gamma;\qquad \ell\Phi(x,\lambda)=\lambda\Phi(x,\lambda),\quad \lambda\in J_\gamma.$$

Proof. Differentiating (2.10) and (2.37) with respect to x and using (1.27), (2.19), and (2.20), we get

$$\sum_{\nu=0}^{n}t_{j\nu}(x)\widetilde{\varphi}^{(\nu)}(x,\lambda)=\widetilde{N}(\lambda)\varphi^{(j)}(x,\lambda)+\frac{1}{2\pi i}\int\limits_{\gamma}\widetilde{r}(x,\lambda,\mu)\varphi^{(j)}(x,\mu)\,d\mu,\tag{2.38}$$
$$\lambda\in\gamma,$$

$$\sum_{\nu=0}^{n}t_{j\nu}(x)\widetilde{\Phi}^{(\nu)}(x,\lambda)=\Phi^{(j)}(x,\lambda)+\frac{1}{2\pi i}\int\limits_{\gamma}\frac{\langle\widetilde{\Phi}(x,\lambda),\widetilde{g}^*(x,\mu)\rangle_{\widetilde{\ell}}}{\lambda-\mu}\varphi^{(j)}(x,\mu)\,d\mu,\tag{2.39}$$
$$\lambda\in J_\gamma.$$

Let us now show that

$$\begin{aligned}\widetilde{\ell}\widetilde{\varphi}(x,\lambda)&=\widetilde{N}(\lambda)\ell\varphi(x,\lambda)+\frac{1}{2\pi i}\int\limits_{\gamma}\widetilde{r}(x,\lambda,\mu)\ell\varphi(x,\mu)\,d\mu\\&\quad+\frac{1}{2\pi i}\int\limits_{\gamma}\langle\widetilde{\varphi}(x,\lambda),\widetilde{g}^*(x,\mu)\rangle_{\widetilde{\ell}}\varphi(x,\mu)\,d\mu,\qquad\lambda\in\gamma,\end{aligned}\tag{2.40}$$

$$\begin{aligned}\widetilde{\ell}\widetilde{\Phi}(x,\lambda)&=\ell\Phi(x,\lambda)+\frac{1}{2\pi i}\int\limits_{\gamma}\frac{\langle\widetilde{\Phi}(x,\lambda),\widetilde{g}^*(x,\mu)\rangle_{\widetilde{\ell}}}{\lambda-\mu}\ell\varphi(x,\mu)\,d\mu\\&\quad+\frac{1}{2\pi i}\int\limits_{\gamma}\langle\widetilde{\Phi}(x,\lambda),\widetilde{g}^*(x,\mu)\rangle_{\widetilde{\ell}}\varphi(x,\mu)\,d\mu,\qquad\lambda\in J_\gamma.\end{aligned}\tag{2.41}$$

Indeed, by (2.19)–(2.23),

$$\widetilde{p}_\nu(x) = p_\nu(x) + \sum_{j=\nu+1}^{n} p_j(x)t_{j\nu}(x) + \sum_{j=0}^{n-1} \widetilde{\mathcal{L}}_{\nu j}(x)\kappa_{j0}(x).$$

From this and (1.22) and (2.38) we get

$$\widetilde{N}(\lambda)\ell\varphi(x,\lambda) + \frac{1}{2\pi i}\int_\gamma \widetilde{r}(x,\lambda,\mu)\ell\varphi(x,\mu)\,d\mu + \frac{1}{2\pi i}\int_\gamma \langle\widetilde{\varphi}(x,\lambda),\widetilde{g}^*(x,\mu)\rangle_{\widetilde{\ell}}\varphi(x,\mu)\,d\mu$$

$$= \sum_{j=0}^{n} p_j(x)\sum_{\nu=0}^{n} t_{j\nu}(x)\widetilde{\varphi}^{(\nu)}(x,\lambda) + \sum_{\nu,j=0}^{n-1} \widetilde{\mathcal{L}}_{\nu j}(x)\kappa_{j0}(x)\widetilde{\varphi}^{(\nu)}(x,\lambda)$$

$$= \sum_{\nu=0}^{n} \widetilde{p}_\nu(x)\widetilde{\varphi}^{(\nu)}(x,\lambda) = \widetilde{\ell}\widetilde{\varphi}(x,\lambda).$$

An analogous proof works for (2.41).

By (2.40),

$$\lambda\widetilde{\varphi}(x,\lambda) = \widetilde{N}(\lambda)\ell\varphi(x,\lambda) + \frac{1}{2\pi i}\int_\gamma \widetilde{r}(x,\lambda,\mu)\ell\varphi(x,\mu)\,d\mu$$

$$+\frac{1}{2\pi i}\int_\gamma (\lambda-\mu)\widetilde{r}(x,\lambda,\mu)\varphi(x,\mu)\,d\mu$$

or

$$\widetilde{N}(\lambda)(\ell\varphi(x,\lambda) - \lambda\varphi(x,\lambda)) + \frac{1}{2\pi i}\int_\gamma \widetilde{r}(x,\lambda,\mu)(\ell\varphi(x,\lambda) - \mu\varphi(x,\mu))\,d\mu = 0.$$

Denote

$$y(x,\lambda) \overset{df}{=} \ell\varphi(x,\lambda) - \lambda\varphi(x,\lambda), \qquad y(x,\lambda) = [y_k(x,\lambda)]_{k=\overline{2,n}}.$$

Then

$$\widetilde{N}(\lambda)y(x,\lambda) + \frac{1}{2\pi i}\int_\gamma \widetilde{r}(x,\lambda,\mu)y(x,\mu)\,d\mu = 0, \qquad \lambda\in\gamma. \tag{$*$}$$

By Lemma 2.6, $\frac{1}{\lambda}\Omega^{-1}(x,\lambda)y(x,\lambda) \in B$. We have from $(*)$ that

$$y_k(x,\lambda) = -\frac{1}{2\pi i}\int_\gamma \sum_{j=2}^{n} \widetilde{r}_{kj}(x,\lambda,\mu)y_j(x,\mu)\,d\mu, \qquad \lambda\in\gamma''.$$

From this, by Lemma 2.4,

$$|y_k(x,\lambda)| < C_x\, |\rho|^{k-n}\, |\exp(\rho R_k x)|, \qquad \lambda \in \gamma''.$$

Hence $\Omega^{-1}(x,\lambda)y(x,\lambda) \in B$. Then, by condition P in Theorem 2.3, the homogenous equation $(*)$ has only the zero solution, i.e., $\ell\varphi(x,\lambda) = \lambda\varphi(x,\lambda)$, $\lambda \in \gamma$.

Further, from (2.41),

$$\begin{aligned}\lambda\widetilde{\Phi}(x,\lambda) &= \ell\Phi(x,\lambda) + \frac{1}{2\pi i}\int\limits_{\gamma} \frac{\langle\widetilde{\Phi}(x,\lambda), \widetilde{g}^*(x,\mu)\rangle_{\widetilde{\ell}}}{\lambda-\mu}\, \mu\varphi(x,\mu)\, d\mu \\ &\quad + \frac{1}{2\pi i}\int\limits_{\gamma} \langle\widetilde{\Phi}(x,\lambda), \widetilde{g}^*(x,\mu)\rangle_{\widetilde{\ell}}\varphi(x,\mu)\, d\mu\end{aligned}$$

or

$$\ell\Phi(x,\lambda) = \lambda\left(\widetilde{\Phi}(x,\lambda) - \frac{1}{2\pi i}\int\limits_{\gamma} \frac{\langle\widetilde{\Phi}(x,\lambda), \widetilde{g}^*(x,\mu)\rangle_{\widetilde{\ell}}}{\lambda-\mu}\, \varphi(x,\mu)\, d\mu\right),$$

and hence $\ell\Phi(x,\lambda) = \lambda\Phi(x,\lambda)$ by (2.37). Lemma 2.7 is proved.

Lemma 2.8. *The functions $\Phi(x,\lambda) = [\Phi_m(x,\lambda)]_{m=\overline{1,n}}$ are WS's for L.*

Proof. It follows from (2.37) that for a fixed x the functions $\Phi^{(\nu)}(x,\lambda)$, $\nu = \overline{0,n}$ are regular for $\lambda \in J_\gamma$. Moreover, it follows from (2.10) and (2.37), as in the proof of Theorem 2.1, that

$$\lim_{z\to\lambda,\, z\in J_\gamma} \Phi_m^{(\nu)}(x,z) = \varphi_m^{(\nu)}(x,\lambda), \qquad \lambda \in \gamma_0 \cup \gamma_{(-1)^{n-m+1}}, \qquad m = \overline{2,n}.$$

Since $\widetilde{g}^{*(\nu)}(x,\mu)\varphi^{(j)}(x,\mu) = \widetilde{G}^{*(\nu)}(x,\mu)\Phi^{(j)}(x,\mu)$, from (2.39) we get

$$\begin{aligned}&\sum_{j=0}^{n-1} u_{\xi j0} \sum_{\nu=0}^{n} t_{j\nu}(0)\widetilde{\Phi}^{(\nu)}(0,\lambda) \\ &\quad = U_{\xi 0}(\Phi(x,\lambda)) + \frac{1}{2\pi i}\int\limits_{\gamma} \frac{\langle\widetilde{\Phi}(x,\lambda), \widetilde{G}^*(x,\mu)\rangle_{\widetilde{\ell}}}{\lambda-\mu}\, U_{\xi 0}(\Phi(x,\mu))\, d\mu\end{aligned}$$

or

$$\widetilde{U}_{\xi 0}(\widetilde{\Phi}(x,\lambda)) = U_{\xi 0}(\Phi(x,\lambda)) + \frac{1}{2\pi i}\int\limits_{\gamma} \frac{\widetilde{\Gamma}(\lambda,\mu)}{\mu-\lambda}\, U_{\xi 0}(\Phi(x,\mu))\, d\mu.$$

From this, by (2.7),

$$\widetilde{U}_{\xi 0}(\widetilde{\Phi}_m(x,\lambda)) = U_{\xi 0}(\Phi_m(x,\lambda)) + \frac{1}{2\pi i}\int_{\gamma} \sum_{j=m+1}^{n} \frac{\widetilde{\Gamma}_{mj}(\lambda,\mu)}{\mu-\lambda} U_{\xi 0}(\Phi_j(x,\mu))\, d\mu. \quad (2.42)$$

Taking $\xi = n, n-1, \dots, 1$ successively in (2.42), and using the fact that $\widetilde{U}_{\xi 0}(\widetilde{\Phi}_m) = \delta_{\xi m}$, $\xi = \overline{1,m}$, we find that

$$U_{\xi 0}(\Phi_m(x,\lambda)) = \delta_{\xi m}, \qquad \xi = \overline{1,m}.$$

Further, using the estimates $\|\varphi^{+}(x,\lambda)\|_B < C$ and (2.18) we get

$$\int_{\gamma} |\widetilde{g}^{*(\nu)}(x,\mu)\varphi(x,\mu)|\, |d\mu| < C\exp(ax), \qquad a>0, \qquad \nu = \overline{0,n-1}. \qquad (2.43)$$

In the λ-plane we consider the domain

$$G_{\varepsilon}^{0} = \{\lambda : \arg(\pm\lambda) \notin (-\varepsilon,\varepsilon), \inf|\lambda-\mu| \geq \varepsilon, \mu\in\gamma_0\}, \qquad \varepsilon>0.$$

It follows from (2.37) in view of (1.22) and (2.43) that

$$|\Phi_m(x,\lambda)| < C\,|\rho|^{m-n}\exp(ax)\,|\exp(\rho R_m x)|, \qquad x\geq 0, \qquad \lambda\in G_{\varepsilon}^{0}.$$

Denote by $\Phi^{0}(x,\lambda) = [\Phi_m^{0}(x,\lambda)]_{m=\overline{1,n}}$ the WS's for L, and let $\psi_m(x,\lambda) = \Phi_m(x,\lambda) - \Phi_m^{0}(x,\lambda)$. Then the functions $\psi_m(x,\lambda)$ are regular for $\lambda\in J_{\gamma}$ and

$$|\psi_m(x,\lambda)| < C|\rho|^{m-n}\exp(ax)|\exp(\rho R_m x)|, \quad x\geq 0, \quad \lambda\in G_{\varepsilon}^{0}.$$

Moreover, $\ell\psi_m(x,\lambda) = \lambda\psi_m(x,\lambda)$ and $U_{\xi 0}(\psi_m(x,\lambda)) = 0$, $\xi = \overline{1,m}$.

We show that $\psi_m(x,\lambda)\equiv 0$. Indeed, since the functions $[\Phi_m^{0}(x,\lambda)]_{m=\overline{1,n}}$ form an FSS of the DE $\ell y = \lambda y$, it follows that

$$\psi_m(x,\lambda) = \sum_{j=1}^{n} \alpha_{jm}(\lambda)\Phi_j^{0}(x,\lambda).$$

Applying the LF's $U_{10},\dots,U_{m0}$ successively we find that $\alpha_{jm}(\lambda) = 0$, $j = \overline{1,m}$. Thus,

$$\psi_m(x,\lambda) = \sum_{j=m+1}^{n} \alpha_{jm}(\lambda)\Phi_j^{0}(x,\lambda),$$

and the functions $\alpha_{jm}(\lambda)$ are regular for $\lambda\in J_{\gamma}$. Assume that for some s $(m+1\leq s\leq n)$ we have $\alpha_{sm}(\lambda)\not\equiv 0$ and $\alpha_{jm}(\lambda)\equiv 0$, $j=\overline{s+1,n}$. Choose a $\lambda^{*}\in \bar{G}_{\varepsilon}^{0}$ such that $\alpha_{sm}(\lambda^{*})\neq 0$ and $\operatorname{Re}(\rho^{*}(R_s - R_{s-1})) > a$. Then

$$\Phi_s^{0}(x,\lambda^{*}) = \frac{1}{\alpha_{sm}(\lambda^{*})}\left(\psi_m(x,\lambda^{*}) - \sum_{j=m+1}^{s-1} \alpha_{jm}(\lambda^{*})\Phi_j^{0}(x,\lambda^{*})\right),$$

and hence

$$|\Phi_s^0(x,\lambda^*)| < C^* \exp(ax)\, |\exp(\rho^* R_{s-1}x)|, \qquad x > 0.$$

On the other hand, it follows from (1.10) that

$$|\Phi_s^0(x,\lambda^*)| > C_1^* \,|\exp(\rho^* R_s x)|, \qquad x > 0.$$

This contradiction proves that $\alpha_{jm}(\lambda) \equiv 0$, $j = \overline{m+1,n}$, i.e. $\psi_m(x,\lambda) \equiv 0$. Consequently, $\Phi_m(x,\lambda) \equiv \Phi_m^0(x,\lambda)$. Lemma 2.8 is proved.

Lemma 2.9. *The matrix $\mathfrak{M}(\lambda)$ is the WM for L.*

Proof. Let $\mathfrak{M}_{m\xi}^0(\lambda) = U_{\xi 0}(\Phi_m)$ be the WF's for L, and let $\check{\mathfrak{M}}_{m\xi}(\lambda) = \mathfrak{M}_{m\xi}(\lambda) - \mathfrak{M}_{m\xi}^0(\lambda)$. According to (2.26),

$$\tilde{U}_{\xi 0}(\tilde{\Phi}_m(x,\lambda)) = U_{\xi 0}(\Phi_m(x,\lambda)) + \frac{1}{2\pi i}\int_\gamma \sum_{j=m+1}^{n} \frac{\tilde{\Gamma}_{mj}^0(\lambda,\mu)}{\mu-\lambda} U_{\xi 0}(\Phi_j(x,\mu))\,d\mu, \quad (2.44)$$

where $\tilde{\Gamma}^0(\lambda,\mu)$ has the same form as $\tilde{\Gamma}(\lambda,\mu)$ but with $\mathfrak{M}_{m\xi}^0$ instead of $\mathfrak{M}_{m\xi}$. Comparing (2.42) and (2.44), we get

$$\frac{1}{2\pi i}\int_\gamma (\tilde{\Gamma}_{m\xi}(\lambda,\mu) - \tilde{\Gamma}_{m\xi}^0(\lambda,\mu))\frac{d\mu}{\mu-\lambda} + \frac{1}{2\pi i}\int_\gamma \sum_{j=m+1}^{\xi-1} (\tilde{\Gamma}_{mj}(\lambda,\mu)$$
$$-\tilde{\Gamma}_{mj}^0(\lambda,\mu))\mathfrak{M}_{j\xi}^0(\mu)\frac{d\mu}{\mu-\lambda} = 0, \qquad \xi > m, \qquad \lambda \in J_\gamma. \quad (2.45)$$

Let us show by induction that $\check{\mathfrak{M}}_{m\xi}(\lambda) \equiv 0$, $\xi > m$. For $\xi = m+1$ we have from (2.45) in view of (2.7) that

$$\frac{1}{2\pi i}\int_\gamma \chi_{(-1)^{n-m}}(\mu)\check{\mathfrak{M}}_{m,m+1}(\mu)\frac{d\mu}{\mu-\lambda} = 0, \qquad \lambda \in J_\gamma$$

or

$$\frac{1}{2\pi i}\int_{\gamma_0\cup\gamma_{(-1)^{n-m}}} \frac{\check{\mathfrak{M}}_{m,m+1}(\mu)}{\mu-\lambda}\,d\mu = 0, \qquad \lambda \in J_\gamma$$

and hence $\check{\mathfrak{M}}_{m,m+1}(\lambda) \equiv 0$.

Assume that $\check{\mathfrak{M}}_{mj}(\lambda) \equiv 0$, $j = \overline{m+1,\xi-1}$. Then, by (2.7),

$$\tilde{\Gamma}_{mj}(\lambda,\mu) = \tilde{\Gamma}_{mj}^0(\lambda,\mu), \qquad j = \overline{m+1,\xi-1}; \qquad \tilde{\tilde{\Gamma}}_{m\xi}(\lambda,\mu) = \tilde{\tilde{\Gamma}}_{m\xi}^0(\lambda,\mu).$$

Thus, by (2.45),

$$\int\limits_{\gamma_0} \frac{\overset{\vee}{\mathfrak{M}}_{m\xi}(\mu)}{\mu-\lambda}\,d\mu = 0, \qquad \lambda \in J_\gamma. \tag{2.46}$$

The function $\overset{\vee}{\mathfrak{M}}_{n\xi}(\lambda)$ is regular for $\lambda \in J_\gamma$ and $\lambda \in \gamma_{(-1)^{n-m+1}}$. Further, the functions $\mathfrak{M}_{m\xi}(\lambda) - \mathfrak{M}_{m,m+1}(\lambda)\mathfrak{M}_{m+1,\xi}(\lambda)$ and $\mathfrak{M}^0_{m\xi}(\lambda) - \mathfrak{M}^0_{m,m+1}(\lambda)\mathfrak{M}^0_{m+1,\xi}(\lambda)$ are regular for $\lambda \in \gamma_{(-1)^{n-m}}$. Since $\overset{\vee}{\mathfrak{M}}_{mj}(\lambda) \equiv 0$, $j = \overline{m+1,\xi-1}$, the function $\overset{\vee}{\mathfrak{M}}_{m\xi}(\lambda)$ is regular for $\lambda \in \gamma_{(-1)^{n-m}}$. Thus, the function $\overset{\vee}{\mathfrak{M}}_{m\xi}(\lambda)$ is regular for $\lambda \notin \operatorname{int}\gamma_0$ and $\overset{\vee}{\mathfrak{M}}_{m\xi}(\lambda) = O(\rho^{-1})$, $|\lambda| \to \infty$. Then it follows from (2.46) that $\overset{\vee}{\mathfrak{M}}_{m\xi}(\lambda) \equiv 0$. Lemma 2.9 is proved.

Thus, the proof of Theorem 2.3 is complete.

Remark. The method described in §2 also allows one to solve the inverse problem for differential operators with non-integrable coefficients. Indeed, suppose a matrix $\mathfrak{M}(\lambda) \in M$ satisfies the conditions 1 and 2 of Theorem 2.3. Then according to the above-mentioned procedure we can construct a DE and LF (1.1)–(1.2), and in general the coefficients $p_k(x)$ will be non-integrable (see 3.4).

3. Differential operators with a simple spectrum

We consider the DE and LF (1.1)–(1.2) on the half-line ($T = \infty$). If the spectrum of L has finite multiplicity, then the main equation obtained in §2 can be contracted to the set $\Gamma \cup \Lambda$. For convenience we confine ourselves here to the case of a simple spectrum (see Definition 3.1). For differential operators with a simple spectrum the main equation can be transformed to the form (3.16)–(3.18), and the WM is uniquely determined from the so-called spectral data (see Definition 3.2). In particular, if only the discrete spectrum is perturbed, then the main equation of the inverse problem is a linear algebraic system (3.24). For $n = 2$ we obtain the Gel'fand–Levitan equation from the main equation using a Fourier transform.

3.1.

Definition 3.1. We will say that L has a simple spectrum if for each $\lambda_0 \in \Lambda' \overset{df}{=} \Lambda\backslash\{0\} \subset \Pi$ there exist the finite limits

$$\left.\begin{aligned} \mathfrak{M}_{<-1>}(\lambda_0) &= \lim_{\lambda\to\lambda_0} (\lambda-\lambda_0)\mathfrak{M}(\lambda), \\ \mathfrak{M}^*_{<-1>}(\lambda_0) &= \lim_{\lambda\to\lambda_0} (\lambda-\lambda_0)\mathfrak{M}^*(\lambda), \end{aligned}\right\} \tag{3.1}$$

and

$$\mathfrak{M}_{mk}(\lambda) = O(\rho^{m-k}), \qquad \lambda \to 0. \tag{3.2}$$

Lemma 3.1. *If L has a simple spectrum, then Λ is a finite set, and*

$$\mathfrak{M}(\lambda) = E + \frac{1}{2\pi i}\int_{-\infty}^{\infty} \frac{M^0(\mu)}{\mu - \lambda}\, d\mu + \sum_{\lambda_0 \in \Lambda'} \frac{Q(\lambda_0)}{\lambda - \lambda_0}, \qquad \operatorname{Im}\lambda \neq 0, \tag{3.3}$$

where

$$M^0(\lambda) = \mathfrak{M}^+(\lambda) - \mathfrak{M}^-(\lambda),$$

$$Q(\lambda_0) = \begin{cases} \frac{1}{2}\mathfrak{M}_{<-1>}(\lambda_0), & \lambda_0 \in \Lambda'_{\mathrm{Re}} \overset{df}{=} \Lambda' \cap \Gamma, \\ \mathfrak{M}_{<-1>}(\lambda_0), & \lambda_0 \in \Lambda'_{\mathrm{Im}} \overset{df}{=} \Lambda' \backslash \Lambda'_{\mathrm{Re}}, \end{cases}$$

$$M^0(\lambda) = [M^0_{mk}(\lambda)]_{m,k=\overline{1,n}}, \qquad Q(\lambda_0) = [Q_{mk}(\lambda_0)]_{m,k=\overline{1,n}}.$$

Proof. The set Λ is bounded and, by virtue of (3.1) and (3.2) has no limit points. Hence Λ is a finite set. Since $\mathfrak{M}(\lambda) = E + O(\rho^{-1})$, Cauchy's theorem gives us that

$$\mathfrak{M}(\lambda) = E + \frac{1}{2\pi i}\int_{\gamma} \frac{\mathfrak{M}(\mu)}{\lambda - \mu}\, d\mu, \qquad \lambda \in J_\gamma, \qquad \Lambda \subset \operatorname{int}\gamma_0. \tag{3.4}$$

For each $\lambda_0 \in \Lambda'_{\mathrm{Re}}$ on the upper (lower) side of the cut we take a half-circumference $\gamma(\lambda_0) : |\mu - \lambda_0| = \delta$, $\operatorname{Im}\mu > 0$ ($\operatorname{Im}\mu < 0$) and choose $\delta > 0$ such that $\operatorname{int}\gamma(\lambda_0)$ do not intersect each other, and do not contain points of the set Λ'_{Im}. Let $\mathcal{L} = \mathcal{L}_\delta \cup \left(\bigcup_{\lambda_0 \in \Lambda'_{\mathrm{Re}}} \gamma(\lambda_0) \right)$ be a contour (with counterclockwise circuit), where $\mathcal{L}_\delta$ is a two-sided cut $(-\infty, \infty)$ without δ-neighbourhoods of the points $\lambda_0 \in \Lambda'_{\mathrm{Re}}$. Let Γ_δ be the real axis without δ-neighbourhoods of the points $\lambda_0 \in \Lambda'_{\mathrm{Re}}$.

Contracting the contour γ in (3.4) to the real axis and using (3.1) and (3.2) we obtain

$$\mathfrak{M}(\lambda) = E + \frac{1}{2\pi i}\int_{\mathcal{L}} \frac{\mathfrak{M}(\mu)}{\lambda - \mu}\, d\mu + \sum_{\lambda_0 \in \Lambda'_{\mathrm{Im}}} \frac{\mathfrak{M}_{<-1>}(\lambda_0)}{\lambda - \lambda_0}. \tag{3.5}$$

Since

$$\frac{1}{2\pi i}\int_{\mathcal{L}_\delta} \frac{\mathfrak{M}(\mu)}{\lambda - \mu}\, d\mu = \frac{1}{2\pi i}\int_{\Gamma_\delta} \frac{M^0(\mu)}{\mu - \lambda}\, d\mu,$$

$$\frac{1}{2\pi i}\int_{\gamma(\lambda_0)} \frac{\mathfrak{M}(\mu)}{\lambda - \mu}\, d\mu = \frac{1}{2}\frac{\mathfrak{M}_{<-1>}(\lambda_0)}{\lambda - \lambda_0} + o(1), \qquad \delta \to 0,$$

For $\delta \to 0$ (3.5) gives us (3.4). Lemma 3.1 is proved.

Thus the WM $\mathfrak{M}(\lambda)$ is uniquely determined from

$$\{M^0(\lambda)\}_{\lambda\in\Gamma}, \qquad \{\lambda_0, \mathfrak{M}_{<-1>}(\lambda_0)\}_{\lambda_0\in\Lambda'}.$$

But actually to construct $\mathfrak{M}(\lambda)$ we need less.

Definition 3.2. Suppose L has a simple spectrum. The set

$$\mathfrak{M}' = (\{M(\lambda)\}_{\lambda\in\Gamma}, \{\lambda_0, \mathfrak{M}_{<-1>}(\lambda_0)\}_{\lambda_0\in\Lambda'}),$$

where

$$M(\lambda) = \operatorname{diag}[M_m(\lambda)]_{m=\overline{1,n-1}}, \qquad M_m(\lambda) = \mathfrak{M}^+_{m,m+1}(\lambda) - \mathfrak{M}^-_{m,m+1}(\lambda)$$

is called the spectral data of L.

The specification of the spectral data uniquely determines the WM $\mathfrak{M}(\lambda)$. Indeed, by virtue of Lemma 2.1, the functions $\mathfrak{M}_{mk}(\lambda) - \mathfrak{M}_{m,m+1}(\lambda)\mathfrak{M}_{m+1,k}(\lambda)$ are regular for $\lambda \in \Gamma_{(-1)^{n-m}}\backslash\Lambda$, and hence $M^0_{mk}(\lambda) = M_m(\lambda)\mathfrak{M}_{m+1,k}(\lambda)$. It then follows from (3.3) that

$$\mathfrak{M}_{mk}(\lambda) = \frac{1}{2\pi i}\int_{-\infty}^{\infty} M_m(\mu)\mathfrak{M}_{m+1,k}(\mu)\frac{d\mu}{\lambda-\mu} + \sum_{\lambda_0\in\Lambda'}\frac{Q_{mk}(\lambda_0)}{\lambda-\lambda_0}, \qquad k > m. \tag{3.6}$$

Thus, the specification of the spectral data uniquely determines L. To construct L from $\mathfrak{M}'$ we can construct $\mathfrak{M}(\lambda)$ by the recurrence formulae (3.6), and then use the method described in §2. But if there is more precise asymptotics of $\mathfrak{M}(\lambda)$ in the neighbourhoods of the points of Λ', we can construct the main equation on the set $\Gamma\cup\Lambda$ and solve the IP directly from the spectral data $\mathfrak{M}'$.

Suppose that for each $\lambda_0 \in \Lambda'$ the following asymptotics are valid for $\lambda \to \lambda_0$:

$$\left.\begin{aligned} \mathfrak{M}(\lambda) &= \frac{\mathfrak{M}_{<-1>}(\lambda_0)}{\lambda-\lambda_0} + \mathfrak{M}_{<0>}(\lambda_0) + (\lambda-\lambda_0)\mathfrak{M}_{<1>}(\lambda_0) + o(\lambda-\lambda_0),\\ \mathfrak{M}^*(\lambda) &= \frac{\mathfrak{M}^*_{<-1>}(\lambda_0)}{\lambda-\lambda_0} + \mathfrak{M}^*_{<0>}(\lambda_0) + (\lambda-\lambda_0)\mathfrak{M}^*_{<1>}(\lambda_0) + o(\lambda-\lambda_0). \end{aligned}\right\} \tag{3.7}$$

We will say that $L \in v'_N$ if $L \in v_N$ and (3.2) and (3.7) hold.

Let $L \in v'_N$, $\lambda_0 \in \Lambda'$. Denote $\mathfrak{N}(\lambda_0) = \mathfrak{M}_{<-1>}(\lambda_0)(\mathfrak{M}_{<0>}(\lambda_0))^{-1}$. Let us show that

$$\left.\begin{aligned} &\mathfrak{N}(\lambda_0)\mathfrak{N}(\lambda_0) = 0,\\ &\Phi_{<-1>}(x,\lambda_0) = \mathfrak{N}(\lambda_0)\Phi_{<0>}(x,\lambda_0),\\ &\Phi^*_{<-1>}(x,\lambda_0) = -\Phi^*_{<0>}(x,\lambda_0)\mathfrak{N}(\lambda_0). \end{aligned}\right\} \tag{3.8}$$

Indeed, since $\mathfrak{M}^*(\lambda)\mathfrak{M}(\lambda) = E$, we get

$$\mathfrak{M}^*_{<-1>}(\lambda_0)\mathfrak{M}_{<-1>}(\lambda_0) = 0,$$

$$\mathfrak{M}^*_{<-1>}(\lambda_0)\mathfrak{M}_{<0>}(\lambda_0) + \mathfrak{M}^*_{<0>}(\lambda_0)\mathfrak{M}_{<-1>}(\lambda_0) = 0.$$

This implies that $\mathfrak{N}(\lambda_0) = -(\mathfrak{M}^*_{<0>}(\lambda_0))^{-1}\mathfrak{M}^*_{<-1>}(\lambda_0)$, and consequently

$$\mathfrak{N}(\lambda_0)\mathfrak{N}(\lambda_0) = -(\mathfrak{M}^*_{<0>}(\lambda_0))^{-1}\mathfrak{M}^*_{<-1>}(\lambda_0)\mathfrak{M}_{<-1>}(\lambda_0)(\mathfrak{M}_{<0>}(\lambda_0))^{-1} = 0.$$

Further, it follows from (1.4), (1.28) and Lemma 1.1 that $C(x,\lambda) = \mathfrak{M}^*(\lambda)\Phi(x,\lambda)$, $C^*(x,\lambda) = \Phi^*(x,\lambda)\mathfrak{M}(\lambda)$. Then

$$\mathfrak{M}^*_{<0>}(\lambda_0)\Phi_{<-1>}(x,\lambda_0) + \mathfrak{M}^*_{<-1>}(\lambda_0)\Phi_{<0>}(x,\lambda_0) = 0,$$

$$\Phi^*_{<-1>}(x,\lambda_0)\mathfrak{M}_{<0>}(\lambda_0) + \Phi^*_{<0>}(x,\lambda_0)\mathfrak{M}_{<-1>}(\lambda_0) = 0.$$

Hence

$$\Phi_{<-1>}(x,\lambda_0) = \mathfrak{N}(\lambda_0)\Phi_{<0>}(x,\lambda_0), \qquad \Phi^*_{<-1>}(x,\lambda_0) = -\Phi^*_{<0>}(x,\lambda_0)\mathfrak{N}(\lambda_0).$$

3.2. Let $L, \widetilde{L} \in v'_N$. For simplicity, we assume that $\widehat{M}(\lambda) = O(\rho^{-n-2})$, $|\lambda| \to \infty$. Denote $J = \Lambda' \cup \widetilde{\Lambda}'$, $J_0 = J \cap \Gamma$. For $\lambda_0 \in J$ we define the matrices

$$\widetilde{D}_s(x,\lambda,\lambda_0) = -\left[\frac{\langle\widetilde{\Phi}(x,\lambda), \widetilde{\Phi}^*(x,\mu)\rangle_{\widetilde{\ell}}}{\lambda-\mu}\right]^{<s>}_{|\mu=\lambda_0}, \qquad s=-1,0,$$

$$\widetilde{d}_s(x,\lambda,\lambda_0) = -\left[\frac{\langle\widetilde{f}(x,\lambda), \widetilde{\Phi}^*(x,\mu)\rangle_{\widetilde{\ell}}}{\lambda-\mu}\right]^{<s>}_{|\mu=\lambda_0}, \qquad s=-1,0.$$

Lemma 3.2. *Let $\lambda_0 \in J$. The functions*

$$\widetilde{\Phi}^{*(\nu)}(x,\mu)\Phi^{(j)}(x,\mu) - \left(\frac{\widetilde{\Phi}^{*(\nu)}(x,\lambda_0)}{(\mu-\lambda_0)^2} + \frac{\widetilde{\Phi}^{*(\nu)}(x,\lambda_0)}{\mu-\lambda_0}\right)\hat{\mathfrak{N}}(\lambda_0)\Phi^{(j)}_{<0>}(x,\lambda_0),$$

$$\frac{\langle\widetilde{\Phi}(x,\lambda), \widetilde{\Phi}^*(x,\mu)\rangle_{\widetilde{\ell}}}{\lambda-\mu}\Phi(x,\mu) - \left(\frac{\widetilde{D}_{-1}(x,\lambda,\lambda_0)}{(\mu-\lambda_0)^2} + \frac{\widetilde{D}_0(x,\lambda,\lambda_0)}{\mu-\lambda_0}\right)\hat{\mathfrak{N}}(\lambda_0)\Phi_{<0>}(x,\lambda_0)$$

are continuous for $\mu = \lambda_0$.

Indeed, the functions

$$\widetilde{\Phi}^{*(\nu)}(x,\mu)\Phi^{(j)}(x,\mu)-(\mu-\lambda_0)^{-2}\widetilde{\Phi}^{*(\nu)}_{<-1>}(x,\lambda_0)\Phi^{(j)}_{<-1>}(x,\lambda_0)$$
$$-(\mu-\lambda_0)^{-1}\left(\widetilde{\Phi}^{*(\nu)}_{<-1>}(x,\lambda_0)\Phi^{(j)}_{<0>}(x,\lambda_0)+\widetilde{\Phi}^{*(\nu)}_{<0>}(x,\lambda_0)\Phi^{(j)}_{<-1>}(x,\lambda_0)\right),$$

$$\frac{\langle\widetilde{\Phi}(x,\lambda),\widetilde{\Phi}^*(x,\mu)\rangle_{\widetilde{\ell}}}{\lambda-\mu}\Phi(x,\mu)+\frac{\widetilde{D}_{-1}(x,\lambda,\lambda_0)\Phi_{<-1>}(x,\lambda_0)}{(\mu-\lambda_0)^2}$$
$$+(\mu-\lambda_0)^{-1}\left(\widetilde{D}_{-1}(x,\lambda,\lambda_0)\Phi_{<0>}(x,\lambda_0)+\widetilde{D}_0(x,\lambda,\lambda_0)\Phi_{<-1>}(x,\lambda_0)\right),$$

are continuous for $\mu=\lambda_0$, *and*

$$\widetilde{D}_{-1}(x,\lambda,\lambda_0)\Phi_{<-1>}(x,\lambda_0)=-\frac{\langle\widetilde{\Phi}(x,\lambda),\widetilde{\Phi}^*_{<-1>}(x,\lambda_0)\rangle_{\widetilde{\ell}}}{\lambda-\lambda_0}\Phi_{<-1>}(x,\lambda_0),$$

$$\widetilde{D}_{-1}(x,\lambda,\lambda_0)\Phi_{<0>}(x,\lambda_0)+\widetilde{D}_0(x,\lambda,\lambda_0)\Phi_{<-1>}(x,\lambda_0)$$
$$=\frac{\langle\widetilde{\Phi}(x,\lambda),\widetilde{\Phi}^*_{<-1>}(x,\lambda_0)\rangle_{\widetilde{\ell}}}{\lambda-\lambda_0}\Phi_{<0>}(x,\lambda_0)-\frac{\langle\widetilde{\Phi}(x,\lambda),\widetilde{\Phi}^*_{<0>}(x,\lambda_0)\rangle_{\widetilde{\ell}}}{\lambda-\lambda_0}$$
$$\times\,\Phi_{<-1>}(x,\lambda_0)-\frac{\langle\widetilde{\Phi}(x,\lambda),\widetilde{\Phi}^*_{<-1>}(x,\lambda_0)\rangle_{\widetilde{\ell}}}{(\lambda-\lambda_0)^2}\Phi_{<-1>}(x,\lambda_0).$$

Using (3.8), we obtain

$$\widetilde{\Phi}^{*(\nu)}_{<-1>}(x,\lambda_0)\Phi^{(j)}_{<-1>}(x,\lambda_0)=-\widetilde{\Phi}^{*(\nu)}_{<0>}(x,\lambda_0)\widehat{\mathfrak{N}}(\lambda_0)\mathfrak{N}(\lambda_0)\Phi^{(j)}_{<0>}(x,\lambda_0)$$
$$=\widetilde{\Phi}^{*(\nu)}_{<-1>}(x,\lambda_0)\widehat{\mathfrak{N}}(\lambda_0)\Phi^{(j)}_{<0>}(x,\lambda_0),$$

$$\widetilde{\Phi}^{*(\nu)}_{<-1>}(x,\lambda_0)\Phi^{(j)}_{<0>}(x,\lambda_0)+\widetilde{\Phi}^{*(\nu)}_{<0>}(x,\lambda_0)\Phi^{(j)}_{<-1>}(x,\lambda_0)$$
$$=\widetilde{\Phi}^{*(\nu)}_{<0>}(x,\lambda_0)\widehat{\mathfrak{N}}(\lambda_0)\Phi^{(j)}_{<0>}(x,\lambda_0).$$

This implies the assertion of the lemma.

Denote $\varphi_{<0>}(x,\lambda_0)=Y\Phi_{<0>}(x,\lambda_0)$, $\lambda_0\in J$; $Q_1(\lambda_0)=\mathfrak{N}(\lambda_0)$ for $\lambda_0\in J\backslash J_0$ and $Q_1(\lambda_0)=\frac{1}{2}\mathfrak{N}(\lambda_0)$ for $\lambda_0\in J_0$.

Lemma 3.3. *The following relations hold*

$$\begin{aligned}\kappa_{\nu s}(x) &= \frac{1}{2\pi i}\int_{-\infty}^{\infty}\Big\{\widetilde{f}^{*(\nu)}(x,\mu)\widehat{M}(\mu)f^{(s)}(x,\mu)\\ &\quad+\sum_{\lambda_0\in J_0}\frac{(\pm\widetilde{\Phi}^{*(\nu)}(x,\lambda_0)\widetilde{\mathfrak{N}}(\lambda_0)\mathfrak{N}(\lambda_0)Y^T)}{(\mu-\lambda_0)^2}\varphi^{(s)}_{<0>}(x,\lambda_0)\Big\}\,d\mu\\ &\quad-\sum_{\lambda_0\in J}(\widetilde{\Phi}^{*(\nu)}_{<0>}(x,\lambda_0)Q_1(\lambda_0)Y^T)\varphi^{(s)}_{<0>}(x,\lambda_0),\qquad \nu+s\le n-1\end{aligned}\tag{3.9}$$

$$\begin{aligned}\widetilde{\Phi}(x,\lambda) &= \Phi(x,\lambda)+\frac{1}{2\pi i}\int_{-\infty}^{\infty}\Big\{\frac{\langle\widetilde{\Phi}(x,\lambda),\widetilde{f}^*(x,\mu)\rangle_{\widetilde{\ell}}}{\lambda-\mu}\widehat{M}(\mu)f(x,\mu)\\ &\quad+\sum_{\lambda_0\in J_0}(\pm\widetilde{D}_{-1}(x,\lambda,\lambda_0)\mathfrak{N}(\lambda_0)Y^T)\frac{\varphi_{<0>}(x,\lambda_0)}{(\mu-\lambda_0)^2}\Big\}\,d\mu\\ &\quad+\sum_{\lambda_0\in J}(\widetilde{D}_0(x,\lambda,\lambda_0)\widehat{Q}_1(\lambda_0)Y^T)\varphi_{<0>}(x,\lambda_0),\end{aligned}\tag{3.10}$$

where we write + (-) when λ_0 lies on the upper (lower) side of the cut.

To prove (3.9) we use (2.19). It was shown above, in the proof of Theorem 2.1, that for $\mu\in\gamma_0$ the function

$$\widetilde{\Phi}^{*(\nu)}(x,\mu)\Phi^{(s)}(x,\mu)+\widetilde{g}^{*(\nu)}(x,\mu)\varphi^{(s)}(x,\mu)$$

is entire in μ. Since for $\mu\in\gamma_1\cup\gamma_{-1}$

$$\widetilde{g}^{*(\nu)}(x,\mu)\varphi^{(s)}(x,\mu)=-\widetilde{f}^{*(\nu)}(x,\mu)\mathfrak{M}_\partial(\mu)f^{(s)}(x,\mu),$$

it follows from (2.19) that

$$\begin{aligned}\kappa_{\nu s}(x) &= -\frac{1}{2\pi i}\int_{\gamma_1\cup\gamma_{-1}}\widetilde{f}^{*(\nu)}(x,\mu)\mathfrak{M}_\partial(\mu)f^{(s)}(x,\mu)\,d\mu\\ &\quad-\frac{1}{2\pi i}\int_{\gamma_0}\widetilde{\Phi}^{*(\nu)}(x,\mu)\Phi^{(s)}(x,\mu)\,d\mu.\end{aligned}$$

We contract the contour γ_0 to the real axis through the poles $\lambda_0\in J\backslash J_0$. It follows from (3.2), (1.4), (1.28) and Lemma 1.1 that $\widetilde{\Phi}^{*(\nu)}(x,\lambda)\Phi^{(s)}(x,\lambda)=O(\rho^{1-n})$, $\lambda\to 0$. Since the function

$$\widetilde{\Phi}^{*(\nu)}(x,\mu)\Phi^{(s)}(x,\mu)-\widetilde{f}^{*(\nu)}(x,\mu)\mathfrak{M}_\partial(\mu)f^{(s)}(x,\mu)$$

is regular for $\mu \in \Gamma\backslash\Lambda$, we get for a small $\delta > 0$, in view of Lemma 3.2

$$\begin{aligned}\kappa_{\nu s}(x) &= -\frac{1}{2\pi i}\int\limits_{\mathcal{L}_\delta} \widetilde{f}^{*(\nu)}(x,\mu)\mathfrak{M}_\partial(\mu)f^{(s)}(x,\mu)\,d\mu \\ &\quad - \sum_{\lambda_0\in J_0}\frac{1}{2\pi i}\int\limits_{\gamma(\lambda_0)} \widetilde{\Phi}^{*(\nu)}(x,\mu)\Phi^{(s)}(x,\mu)\,d\mu \\ &\quad - \sum_{\lambda_0\in J\backslash J_0} \widetilde{\Phi}^{*(\nu)}_{<0>}(x,\lambda_0)\mathfrak{N}(\lambda_0)Y^T\varphi^{(s)}_{<0>}(x,\lambda_0),\end{aligned} \tag{3.11}$$

where $\mathcal{L}_\delta$ is a two-sided cut $(-\infty,\infty)$ without δ-neighbourhoods of the points $\lambda_0 \in J_0$, and $\gamma(\lambda_0)$ is a half-circumference $|\mu-\lambda_0| = \delta$, $\operatorname{Im}\mu > 0$ ($\operatorname{Im}\mu < 0$) for $\lambda_0 \in J_0$ lying on the upper (lower) side of the cut. Using Lemma 3.2, we compute

$$\begin{aligned}-\frac{1}{2\pi i}\int\limits_{\gamma(\lambda_0)} \widetilde{\Phi}^{*(\nu)}(x,\mu)\Phi^{(s)}(x,\mu)\,d\mu &= -\frac{1}{2}\widetilde{\Phi}^{*(\nu)}_{<0>}(x,\lambda_0)\mathfrak{N}(\lambda_0)\Phi^{(s)}_{<0>}(x,\lambda_0) \\ &\quad \pm\frac{1}{\pi i\delta}\widetilde{\Phi}^{*(\nu)}_{<0>}(x,\lambda_0)\widetilde{\mathfrak{N}}(\lambda_0)\mathfrak{N}(\lambda_0)\Phi^{(s)}_{<0>}(x,\lambda_0) + o(1), \qquad \delta\to 0.\end{aligned} \tag{3.12}$$

Since

$$\begin{aligned}-\frac{1}{2\pi i}&\int\limits_{\mathcal{L}_\delta} \widetilde{f}^{*(\nu)}(x,\mu)\mathfrak{M}_\partial(\mu)f^{(s)}(x,\mu)\,d\mu \\ &= \frac{1}{2\pi i}\int\limits_{\Gamma_\delta} \widetilde{f}^{*(\nu)}(x,\mu)\widehat{M}(\mu)f^{(s)}(x,\mu)\,d\mu,\end{aligned} \tag{3.13}$$

$$\left(\int\limits_{-\infty}^{\lambda_0-\delta} + \int\limits_{\lambda_0+\delta}\right)\frac{d\mu}{(\mu-\lambda_0)^2} = \frac{2}{\delta}, \tag{3.14}$$

where

$$\Gamma_\delta = \Gamma\backslash\Big(\bigcup_{\lambda_0\in J_0}(\lambda_0-\delta,\lambda_0+\delta)\Big)$$

is the real axis without δ-neighbourhoods of the points $\lambda_0 \in J_0$, we obtain (3.9) from (3.11). The relation (3.10) is proved analogously.

Lemma 3.4. *Let* $z_0 \in \Gamma$, $\kappa = (z_0 - \delta, z_0 + \delta)$. *Denote*

$$T(\lambda) = \frac{1}{2\pi i} \int_{\kappa} \left(\frac{t(\mu)}{\lambda - \mu} - \frac{t_{-2}}{(\lambda - z_0)(\mu - z_0)^2} \right) d\mu, \qquad \lambda \notin \kappa,$$

$$t(\mu) = \frac{t_{-2}}{(\mu - z_0)^2} + \frac{t_{-1}}{\mu - z_0} + t_0 + t_1(\mu),$$

where $t_1(\mu) \in C[z_0 - \delta, z_0 + \delta]$, $t_1(z_0) = 0$. *Then*

$$T(\lambda) = \frac{T^{\pm}_{<-2>}(z_0)}{(\lambda - z_0)^2} + \frac{T^{\pm}_{<-1>}(z_0)}{\lambda - z_0} + T^{\pm}_{<0>}(z_0) + o(1), \qquad \lambda \to z_0, \qquad \pm \operatorname{Im} \lambda > 0,$$

where

$$T^{\pm}_{<0>}(z_0) = \frac{t_{-1}}{\pi i \delta} \mp \frac{t_0}{2} + o(1), \qquad T^{\pm}_{<-1>}(z_0) = \frac{t_{-2}}{\pi i \delta} \mp \frac{t_{-1}}{2} + o(1),$$

$$T^{\pm}_{<-2>}(z_0) = \mp \frac{t_{-2}}{2} + o(1), \qquad \delta \to 0.$$

Proof. Denote

$$\mathcal{F}(\lambda) = \frac{1}{2\pi i} \int_{\kappa} \frac{d\mu}{\mu - \lambda}.$$

Using (2.5), we get

$$\frac{1}{2\pi i} \int_{\kappa} \frac{d\mu}{(\lambda - \mu)(\mu - z_0)} = \frac{1}{\lambda - z_0} \frac{1}{2\pi i} \int_{\kappa} \left(\frac{1}{\lambda - \mu} - \frac{1}{z_0 - \mu} \right) d\mu = -\frac{1}{\lambda - z_0} \mathcal{F}(\lambda).$$

Since

$$\frac{1}{(\lambda - \mu)(\mu - z_0)^2} - \frac{1}{(\lambda - z_0)(\mu - z_0)^2} = \frac{1}{(\lambda - z_0)} \frac{1}{(\lambda - \mu)(\mu - z_0)},$$

it follows that

$$T(\lambda) = -\left(\frac{t_{-2}}{(\lambda - z_0)^2} + \frac{t_{-1}}{\lambda - z_0} + t_0 \right) \mathcal{F}(\lambda) + \frac{1}{2\pi i} \int_{\kappa} \frac{t_1(\mu)}{\lambda - \mu} d\mu. \tag{3.15}$$

We have, by the Sokhotskii formulas, that $\mathcal{F}^{\pm}_{<0>}(z_0) = \pm \frac{1}{2}$. Since

$$\mathcal{F}'(\lambda) = \frac{1}{2\pi i} \int_{\kappa} \frac{d\mu}{(\mu - \lambda)^2} = \frac{\delta}{\pi i} \frac{1}{(z_0 - \lambda)^2 - \delta^2},$$

it follows that $\mathcal{F}^{\pm}_{<1>}(z_0) = -(\pi i \delta)^{-1}$. Analogously we calculate $\mathcal{F}^{\pm}_{<2>}(z_0) = 0$. From this and (3.15) we obtain the assertion of Lemma 3.4.

Denote

$$Y' = [\delta_{j,k-1}]_{j=\overline{1,n-2},k=\overline{1,n-1}}, \qquad Y_0 = [\delta_{j,k-1}]_{j,k=\overline{1,n-1}}, \qquad \chi(x,\lambda) = Y' f(x,\lambda),$$

$$V_1(z_0) = Y_0(\widehat{M}_{<-1>}(z_0) + \widehat{M}_{<0>}(z_0) Y \mathfrak{N}(z_0) Y^T), \qquad z_0 \in J_0,$$

$$V_2(z_0) = Y_0(\widehat{M}_{<0>}(z_0) + \widehat{M}_{<1>}(z_0) Y \mathfrak{N}(z_0) Y^T), \qquad z_0 \in J_0,$$

$$V_3(z_0) = Y_0 \widehat{M}_{<-1>}(z_0) Y'^T, \qquad z_0 \in J_0;$$

$$V_k(z_0) = 0, \qquad z_0 \in J \setminus J_0, \qquad k = \overline{1,3}.$$

Theorem 3.1.

$$\begin{aligned} \widetilde{f}(x,\lambda) \;&=\; f(x,\lambda) + \frac{1}{2\pi i} \int_{-\infty}^{\infty} \left\{ \frac{\langle \widetilde{f}(x,\lambda), \widetilde{f}^*(x,\mu)\rangle_{\widetilde{\ell}}}{\lambda - \mu} \widehat{M}(\mu) f(x,\mu) \right. \\ &\quad + \left. \sum_{\lambda_0 \in J_0} (\pm \widetilde{d}_{-1}(x,\lambda,\lambda_0) \widehat{\mathfrak{N}}(\lambda_0) Y^T) \frac{\varphi_{<0>}(x,\lambda_0)}{(\mu - \lambda_0)^2} \right\} d\mu \\ &\quad + \sum_{\lambda_0 \in J} (\widetilde{d}_0(x,\lambda,\lambda_0) \widehat{Q}_1(\lambda_0) Y^T) \varphi_{<0>}(x,\lambda_0), \qquad \lambda \in \Gamma, \end{aligned} \tag{3.16}$$

$$\begin{aligned} &\widetilde{\varphi}_{<0>}(x,z_0) \\ &= \varphi_{<0>}(x,z_0) + \frac{1}{2\pi i} \int_{-\infty}^{\infty} \left\{ \left[\frac{\langle Y \widetilde{\Phi}(x,\lambda), \widetilde{f}^*(x,\mu)\rangle_{\widetilde{\ell}}}{\lambda - \mu} \right]^{<0>}_{|\lambda = z_0} \widehat{M}(\mu) f(x,\mu) \right. \\ &\quad + \frac{V_1(z_0) \varphi_{<0>}(x,z_0)}{(\mu - z_0)^2} \\ &\quad + \left. \sum_{\lambda_0 \in J_0} (\pm Y \widetilde{D}_{-1,<0>}(x,z_0,\lambda_0) \widehat{\mathfrak{N}}(\lambda_0) Y^T) \frac{\varphi_{<0>}(x,\lambda_0)}{(\mu - \lambda_0)^2} \right\} d\mu \\ &\quad \pm \frac{1}{2} (V_2(z_0) \varphi_{<0>}(x,z_0) + V_3(z_0) \chi_{<1>}(x,z_0)) \\ &\quad + \sum_{\lambda_0 \in J} (Y \widetilde{D}_{0,<0>}(x,z_0,\lambda_0) \widehat{Q}_1(\lambda_0) Y^T) \varphi_{<0>}(x,\lambda_0), \qquad z_0 \in J, \end{aligned} \tag{3.17}$$

$$\begin{aligned}
&\widetilde{\chi}_{<1>}(x, z_0)\\
&= \chi_{<1>}(x, z_0) + \frac{1}{2\pi i}\int\limits_{-\infty}^{\infty}\Bigg\{\left[\frac{\langle\widetilde{\chi}(x,\lambda), \widetilde{f}^*(x,\mu)\rangle_{\widetilde{\ell}}}{\lambda-\mu}\right]^{<1>}_{|\lambda=z_0}\widehat{M}(\mu)f(x,\mu)\\
&+ \sum_{\lambda_0\in J_0}(\pm Y'\widetilde{d}_{-1,<1>}(x, z_0, \lambda_0)\mathfrak{N}(\lambda_0)Y^T)\,\frac{\varphi_{<0>}(x,\lambda_0)}{(\mu-\lambda_0)^2}\Bigg\}\,d\mu\\
&+ \sum_{\lambda_0\in J}(Y'\widetilde{d}_{0,<1>}(x, z_0, \lambda_0)\widehat{Q}_1(\lambda_0)Y^T)\,\varphi_{<0>}(x,\lambda_0), \qquad z_0\in J_0.
\end{aligned}\tag{3.18}$$

Proof. Let $z_0 \in J_0$. Since $Y_0\widehat{M}(\mu)f(x,\mu) = Y_0\widehat{M}(\mu)Y\Phi(x,\mu)$, it follows from (3.8) that

$$\left.\begin{aligned}
&\left[Y_0\widehat{M}(\mu)f(x,\mu)\right]^{<-1>}_{\mu=z_0} = V_1(z_0)\varphi_{<0>}(x,z_0),\\
&\left[Y_0\widehat{M}(\mu)f(x,\mu)\right]^{<0>}_{\mu=z_0} = V_2(z_0)\varphi_{<0>}(x,z_0) + V_3(z_0)\chi_{<1>}(x,z_0).
\end{aligned}\right\}\tag{3.19}$$

Using (1.27), Lemma 1.1 and the relation $\langle\widetilde{\Phi}(x,\lambda), \widetilde{\Phi}^*(x,\mu)\rangle|_{x=0} = \widetilde{\mathfrak{M}}(\lambda)\widetilde{\mathfrak{M}}^*(\mu)$, we obtain

$$\frac{\langle\widetilde{\Phi}(x,\lambda), \widetilde{\Phi}^*(x,\mu)\rangle}{\lambda-\mu} = \frac{E}{\lambda-\mu} + \varepsilon_0(x,\lambda,\mu),$$

$$\varepsilon_0(x,\lambda,\mu) = \frac{(\widetilde{\mathfrak{M}}(\lambda) - \widetilde{\mathfrak{M}}(\mu))\mathfrak{M}^*(\mu)}{\lambda-\mu} + \int\limits_0^x \widetilde{\Phi}(t,\lambda)\widetilde{\Phi}^*(t,\mu)\,dt.$$

This implies that

$$\frac{\langle Y\widetilde{\Phi}(x,\lambda), \widetilde{f}^*(x,\mu)\rangle_{\widetilde{\ell}}}{\lambda-\mu}\widehat{M}(\mu)f(x,\mu) = \frac{Y_0\widehat{M}(\mu)f(x,\mu)}{\lambda-\mu} + \varepsilon(x,\lambda,\mu),\tag{3.20}$$

where

$$\varepsilon(x,\lambda,\mu) = Y\varepsilon_0(x,\lambda,\mu)B^*(\mu)\widehat{M}(\mu)f(x,\mu),$$

$$B^*(\mu) = [\delta_{jk}\chi((-1)^{n-k}\mu)]_{\mu=\overline{1,n},k=\overline{1,n-1}}.$$

Multiply (3.10) by Y at the left and divide the integral along the line $(-\infty,\infty)$ into two integrals along $\kappa=(z_0-\delta,z_0+\delta)$ and $(-\infty,\infty)\backslash\kappa$, $\delta>0$. In the integral at κ we use (3.20). Then, in view of (3.14), (3.19) and Lemma 3.4, we get (3.17). For $z_0\in J\backslash J_0$ relation (3.17) follows clearly from (3.10). Further, from (3.10), by virtue of continuity, we obtain (3.16), and hence (3.18). Theorem 3.1 is proved.

Solution of the inverse problem. Relations (3.16)–(3.18) are the main equations of the IP with respect to $\{f(x,\lambda)\}_{\lambda\in\Gamma}$, $\{\varphi_{<0>}(x,z_0)\}_{z_0\in J}$, $\{\chi_{<1>}(x,z_0)\}_{z_0\in J_0}$. They allow the solution of the IP by recovering the DE and LF $L\in v'_N$ from the given spectral data $\mathfrak{M}'$. For constructing L we need to solve the main equations (3.16)–(3.18) for each fixed $x\ge 0$ and then find the DE and LF via (2.23), where the functions $\varepsilon_\nu(x)$ and $t_{j\nu}(x)$ are defined by (3.9) and (2.20)–(2.22).

3.3. Consider a perturbation of the discrete spectrum. Let $L,\ \widetilde{L}\in v'_N$ and

$$\widehat{\mathfrak{M}}_{m,m+1}(\lambda)=\sum_{\lambda_0\in J}\frac{\widehat{Q}_{m,m+1}(\lambda_0)}{\lambda-\lambda_0}, \tag{3.21}$$

i.e. $\widehat{M}(\lambda)\equiv 0$. Denote

$$\widetilde{P}(x,\lambda,\lambda_0)=\widetilde{D}_0(x,\lambda,\lambda_0)\widehat{Q}_1(\lambda_0)Y^T,\qquad \widetilde{G}(x,z_0,\lambda_0)=[Y\widetilde{P}(x,\lambda,\lambda_0)]^{<0>}\big|_{\lambda=z_0},$$

$$\widetilde{g}^*(x,\lambda_0)=-\widetilde{\Phi}^*_{<0>}(x,\lambda_0)\widehat{Q}_1(\lambda_0)Y^T,\qquad \varphi(x)=[\varphi_{<0>}(x,\lambda_0)]_{\lambda_0\in J},$$

$$\widetilde{g}^*(x)=[\widetilde{g}^*(x,\lambda_0)]^T_{\lambda_0\in J},\qquad \widetilde{G}(x)=[\widetilde{G}(x,z_0,\lambda_0)]_{z_0,\lambda_0\in J}.$$

Theorem 3.2.

$$\kappa_{\nu s}(x)=\widetilde{g}^{*(\nu)}(x)\varphi^{(s)}(x), \tag{3.22}$$

$$\widetilde{\Phi}(x,\lambda)=\Phi(x,\lambda)+\sum_{\lambda_0\in J}\widetilde{P}(x,\lambda,\lambda_0)\varphi_{<0>}(x,\lambda_0), \tag{3.23}$$

$$\widetilde{\varphi}(x)=(E+\widetilde{G}(x))\varphi(x). \tag{3.24}$$

Proof. Use (3.11). Since $\widehat{M}(\mu)\equiv 0$, from (3.12) and (3.13) we have

$$\begin{aligned}\kappa_{\nu s}(x) &= -\sum_{\lambda_0\in J}\widetilde{\Phi}^{*(\nu)}_{<0>}(x,\lambda_0)\widehat{Q}_1(\lambda_0)Y^T\varphi^{(s)}_{<0>}(x,\lambda_0)\\ &\quad+\sum_{\lambda_0\in J_0}\left(\pm\frac{1}{\pi i\delta}\widetilde{\Phi}^{*(\nu)}_{<0>}(x,\lambda_0)\widetilde{\mathfrak{N}}(\lambda_0)\widehat{\mathfrak{N}}(\lambda_0)\Phi^{(s)}_{<0>}(x,\lambda_0)\right)+o(1),\quad \delta\to 0.\end{aligned}$$

Since $\kappa_{\nu s}(x)$ does not depend on δ, we obtain (3.22). An analogous proof works for (3.23). We note that (3.22) and (3.23) are the particular case of (3.9) and (3.10) when the integrals in (3.9) and (3.10) are equal to zero. Further, multiplying (3.23) by Y at the left, we get

$$\widetilde{\varphi}_{<0>}(x, z_0) = \varphi_{<0>}(x, z_0) + \sum_{\lambda_0 \in J} \widetilde{G}(x, z_0, \lambda_0)\varphi_{<0>}(x, z_0),$$

i.e. (3.24) is valid.

Equation (3.24) is the main equation of the IP. For each fixed $x \geq 0$, (3.24) is a linear algebraic system with respect to $\varphi(x)$ and $\det[E + \widetilde{G}(x)] \neq 0$. To solve the IP we must find $\varphi(x)$ from (3.24) and then construct the DE and LF L via (2.23) where the functions $\varepsilon_\nu(x)$, $t_{j\nu}(x)$ are defined by (3.22) and (2.20)–(2.22).

3.4. Consider a differential operator $L = (\ell, U) \in v_N$ of the form

$$\ell y \equiv -y'' + q(x)y = \lambda y = \rho^2 y, \qquad q(x) \in W_N, \qquad \operatorname{Im} \rho \geq 0, \tag{3.25}$$

$$U(y) = y'(0) - hy(0), \tag{3.26}$$

where $q(x)$ is a complex-valued function, and h is a complex number. Here we reformulate the results obtained above for Sturm–Liouville differential operators (3.25) and (3.26) and establish a connection between the main equation of the IP and the Gel'fand–Levitan equation [10]. Some difference in the notations is pointed out below.

Let M be a set of functions $\mathfrak{M}(\lambda)$ which are regular in Π_{+1} with the exception of an at most countable bounded set of poles, and are continuous in $\bar{\Pi}_{+1}$ with the exception of a bounded set Λ (in general, the set Λ is different for each matrix $\mathfrak{M}(\lambda)$), and $\mathfrak{M}(\lambda) = O(\rho^{-1})$, $|\lambda| \to \infty$.

Let $\varphi(x, \lambda)$ be a solution of (3.25) under the conditions $\varphi(0, \lambda) = 1$, $\varphi'(0, \lambda) = h$. It is known (see, for example, [9]) that the following representation is valid

$$\varphi(x, \lambda) = \cos \rho x + \int_0^x K(x, t) \cos \rho t \, dt. \tag{3.27}$$

Let $e(x, \rho)$ be the Jost solution of (3.25) under the condition

$$\lim_{x \to \infty} e(x, \rho) \exp(-i\rho x) = 1.$$

Denote

$$\Phi(x, \lambda) = e(x, \rho)(e'(0, \rho) - he(0, \rho))^{-1}, \qquad \mathfrak{M}(\lambda) = \Phi(0, \lambda).$$

The functions $\Phi(x, \lambda)$ and $\mathfrak{M}(\lambda)$ are the WS and the WF for L respectively. It is clear that $\mathfrak{M}(\lambda) \in M$. It follows from Theorem 1.2 that the specification of the WF

$\mathfrak{M}(\lambda)$ determines L uniquely. Note that $\Phi(x,\lambda) = S(x,\lambda) + \mathfrak{M}(\lambda)\varphi(x,\lambda)$, where $S(x,\lambda)$ is a solution of (3.25) under the conditions $S(0,\lambda) = 0$, $S'(0,\lambda) = 1$. Denote

$$M(\lambda) = \mathfrak{M}^-(\lambda) - \mathfrak{M}^+(\lambda), \qquad \mathfrak{M}^\pm(\lambda) = \lim_{z\to\lambda,\, \pm \operatorname{Im} z>0} \mathfrak{M}(z).$$

Let L, $\widetilde{L} \in v_N$. Denote

$$\widetilde{D}(x,\lambda,\mu) = \frac{\langle \widetilde{\varphi}(x,\lambda), \widetilde{\varphi}(x,\mu)\rangle}{\lambda-\mu} = \int_0^x \widetilde{\varphi}(t,\lambda)\widetilde{\varphi}(t,\mu)\,dt,$$

$$\widetilde{P}(x,\lambda,\mu) = \widetilde{D}(x,\lambda,\mu)\widehat{\mathfrak{M}}(\mu),$$

where $\langle y(x), z(x)\rangle = y(x)z'(x) - y'(x)z(x)$. In the λ-plane we consider the contour $\gamma = \gamma_0 \cup \gamma_1$ (with counterclockwise circuit), where γ_0 is a bounded closed contour encircling the set $\Lambda \cup \widetilde{\Lambda} \cup \{0\}$, and γ_1 is a two-sided cut along the arc$\{\lambda : \lambda > 0, \lambda \notin \operatorname{int}\gamma_0\}$. By virtue of Theorem 2.1, we have

$$\widetilde{\varphi}(x,\lambda) = \varphi(x,\lambda) + \frac{1}{2\pi i}\int_\gamma \widetilde{P}(x,\lambda,\mu)\varphi(x,\mu)\,d\mu. \tag{3.28}$$

Equation (3.28) is the main equation of the IP, and for each fixed $x \geq 0$ (3.28) is uniquely solvable. From results of §2 we obtain the following theorem

Theorem 3.3. *For a function $\mathfrak{M}(\lambda) \in M$ to be the WF for $L \in v_N$ of the form (3.25) and (3.26) it is necessary and sufficient that the following conditions hold:*

1) *(asymptotics) there exists an $\widetilde{L} \in v_N$ such that $\widehat{M}(\lambda) = O(\lambda^{-2})$, $\lambda \to \infty$;*

2) *(condition P) for each fixed $x \geq 0$ equation (3.28) has a unique solution $\varphi(x,\lambda) \in C(\gamma)$;*

3) *$\varepsilon(x) \in W_N$, where*

$$\varepsilon(x) = -\frac{1}{\pi i}\int_\gamma \frac{d}{dx}(\widetilde{\varphi}(x,\mu)\varphi(x,\mu))\widehat{\mathfrak{M}}(\mu)\,d\mu. \tag{3.29}$$

Under these conditions the differential operator L is constructed by (3.30):

$$q(x) = \widetilde{q}(x) + \varepsilon(x), \qquad h = \widetilde{h} - \frac{1}{2\pi i}\int_\gamma \widehat{\mathfrak{M}}(\mu)\,d\mu. \tag{3.30}$$

The case of a simple spectrum. Denote by M_0 the set of functions $\mathfrak{M}(\lambda) \in M$ such that

$$\mathfrak{M}(\lambda) = \frac{\mathfrak{M}_{<-1>}(\lambda_0)}{\lambda-\lambda_0} + \mathfrak{M}_{<0>}(\lambda_0) + o(1), \qquad \lambda \to \lambda_0, \qquad \lambda_0 \in \Lambda' = \Lambda\backslash\{0\},$$

$$\mathfrak{M}(\lambda) = O(\rho^{-1}), \qquad \lambda \to 0.$$

We shall say that $L \in v'_N$ if $L \in v_N$ and the WF $\mathfrak{M}(\lambda) \in M_0$. It follows from Lemma 3.1 that for $L \in v'_N$ the set Λ is finite, and

$$\mathfrak{M}(\lambda) = \frac{1}{2\pi i}\int_0^\infty \frac{M(\mu)}{\mu - \lambda}\,d\mu + \sum_{\lambda_0 \in \Lambda'} \frac{Q(\lambda_0)}{\lambda - \lambda_0}, \tag{3.31}$$

where $Q(\lambda_0) = \mathfrak{M}_{<-1>}(\lambda_0)$ for $\lambda_0 \notin (0, \infty)$ and $Q(\lambda_0) = \frac{1}{2}\mathfrak{M}_{<-1>}(\lambda_0)$ for $\lambda_0 > 0$. The set $\mathfrak{M}' = (\{M(\lambda)\}_{\lambda > 0}, \{\lambda_0, \mathfrak{M}_{<-1>}(\lambda_0)\}_{\lambda_0 \in \Lambda'})$ is called the spectral data for L. The specification of $\mathfrak{M}'$ uniquely determines L.

Let $L, \widetilde{L} \in v'_N$, $J = \Lambda' \cup \widetilde{\Lambda}'$. According to Theorem 3.1 and Lemma 3.3, (3.28) and (3.29) become

$$\begin{aligned}\widetilde{\varphi}(x,\lambda) &= \varphi(x,\lambda) + \frac{1}{2\pi i}\int_0^\infty \widetilde{D}(x,\lambda,\mu)\widehat{M}(\mu)\varphi(x,\mu)\,d\mu \\ &\quad + \sum_{\lambda_0 \in J} \widetilde{D}(x,\lambda,\lambda_0)\widehat{Q}(\lambda_0)\varphi(x,\lambda_0),\end{aligned} \tag{3.32}$$

$$\begin{aligned}\varepsilon(x) &= -\frac{1}{\pi i}\int_0^\infty \frac{d}{dx}(\widetilde{\varphi}(x,\mu)\varphi(x,\mu))\widehat{M}(\mu)\,d\mu \\ &\quad -2\sum_{\lambda_0 \in J} \frac{d}{dx}(\widetilde{\varphi}(x,\lambda_0)\varphi(x,\lambda_0))\widehat{Q}(\lambda_0).\end{aligned} \tag{3.33}$$

Connection with the Gel'fand–Levitan equation. Let $q(x)$ and h be real, and let $\sigma(\lambda)$ be a spectral function of the self-adjoint differential operator L. Then $L \in v'_N$, $\Lambda' \in (-\infty, 0)$, the function $\sigma(\lambda)$ is continuously differentiable for $\lambda > 0$, and $\sigma'(\lambda) = \frac{1}{2\pi i}M(\lambda)$. The relations (3.31) and (3.32) become

$$\mathfrak{M}(\lambda) = \int_{-\infty}^\infty \frac{d\sigma(\mu)}{\mu - \lambda}, \tag{3.34}$$

$$\widetilde{\varphi}(x,\lambda) = \varphi(x,\lambda) + \int_{-\infty}^\infty \widetilde{D}(x,\lambda,\mu)\varphi(x,\mu)\,d\widehat{\sigma}(\mu). \tag{3.35}$$

Thus, (3.35) is the main equation of the IP for the self-adjoint case.

For definiteness we suppose that $\widetilde{q}(x) = \widetilde{h} = 0$. Then $\widetilde{\varphi}(x,\lambda) = \cos\rho x$, $\widetilde{\sigma}'(\lambda) = \frac{1}{\pi\rho}$ $(\lambda > 0)$, $\widetilde{\sigma}(\lambda) = 0$ $(\lambda < 0)$. The Gel'fand–Levitan equation has a form

$$K(x,t) + F(x,t) + \int_0^x K(x,\tau)F(t,\tau)\,d\tau = 0, \tag{3.36}$$

where

$$F(x,t)=\int_{-\infty}^{\infty}\cos\sqrt{\mu}x\,\cos\sqrt{\mu}t\,d\widehat{\sigma}(\mu).$$

Multiplying (3.36) by $\cos\sqrt{\lambda}t$ and integrating with respect to t, we get

$$\int_0^x K(x,t)\cos\sqrt{\lambda}t\,dt+\int_0^x\cos\sqrt{\lambda}t\,dt\int_{-\infty}^{\infty}\cos\sqrt{\mu}x\,\cos\sqrt{\mu}t\,d\widehat{\sigma}(\mu)$$

$$+\int_0^x\cos\sqrt{\lambda}t\,dt\int_0^x K(x,\tau)\,d\tau\int_{-\infty}^{\infty}\cos\sqrt{\mu}t\,\cos\sqrt{\mu}\tau\,d\widehat{\sigma}(\mu)=0.$$

From this and (3.27) we obtain (3.35). Thus, the Gel'fand–Levitan equation is the Fourier transform of the main equation of the IP (3.35).

Perturbations of the discrete spectrum. Let $\widetilde{L}\in v_N$, and let J be a finite set of the λ-plane, and $\{a_{\lambda_0}\}_{\lambda_0\in J}$ be complex numbers. Let

$$\mathfrak{M}(\lambda)=\widetilde{\mathfrak{M}}(\lambda)+\sum_{\lambda_0\in J}\frac{a_{\lambda_0}}{\lambda-\lambda_0}. \tag{3.37}$$

In this case, $\widehat{M}(\lambda)=0$. It then follows from (3.28)–(3.30) that

$$\widetilde{\varphi}(x,z_0)=\varphi(x,z_0)+\sum_{\lambda_0\in J}(\widetilde{D}(x,z_0,\lambda_0)a_{\lambda_0}\varphi(x,\lambda_0),\qquad z_0\in J, \tag{3.38}$$

$$\varepsilon(x)=-2\sum_{\lambda_0\in J}a_{\lambda_0}\frac{d}{dx}(\widetilde{\varphi}(x,\lambda_0)\varphi(x,\lambda_0)), \tag{3.39}$$

$$q(x)=\widetilde{q}(x)+\varepsilon(x),\qquad h=\widetilde{h}-\sum_{\lambda_0\in J}a_{\lambda_0}. \tag{3.40}$$

Thus, the main equation of the IP is the linear algebraic system (3.38) with the determinant $\det[E+\widetilde{G}(x)]$, where $\widetilde{G}(x)=[\widetilde{D}(x,z_0,\lambda_0)a_{\lambda_0}]_{z_0,\lambda_0\in J}$. From Theorem 3.3 we have

Theorem 3.4. *Let $\widetilde{L}\in v_N$. For a function $\mathfrak{M}(\lambda)$ of the form (3.37) to be the WF for $L\in v_N$ it is necessary and sufficient that*

$$\det[E+\widetilde{G}(x)]\neq 0,\qquad x\geq 0, \tag{3.41}$$

and $\varepsilon(x)\in W_N$, where $\varepsilon(x)$ is defined via (3.39), and $\{\varphi(x,\lambda_0)\}_{\lambda_0\in J}$ is the solution of (3.38). Under these conditions the differential operator L is constructed by (3.40).

Example. Let $\tilde{q}(x) = \tilde{h} = 0$, i.e. $\tilde{\mathfrak{M}}(\lambda) = \frac{1}{i\rho}$. Consider the function

$$\mathfrak{M}(\lambda) = \tilde{\mathfrak{M}}(\lambda) + \frac{a}{\lambda - \lambda_0},$$

where a and λ_0 are complex numbers. Then the main equation of the IP (3.38) becomes $\tilde{\varphi}(x, \lambda_0) = \mathcal{F}(x)\varphi(x, \lambda_0)$, where

$$\tilde{\varphi}(x, \lambda_0) = \cos \rho_0 x, \qquad \mathcal{F}(x) = 1 + a \int_0^x \cos^2 \rho_0 t \, dt, \qquad \lambda_0 = \rho_0^2.$$

Then (3.41) takes the form:

$$\mathcal{F}(x) \neq 0, \qquad x \geq 0. \tag{3.42}$$

and the function $\varepsilon(x)$ is found by the formula

$$\varepsilon(x) = \frac{2a\rho_0 \sin 2\rho_0 x}{\mathcal{F}(x)} + \frac{2a^2 \cos^4 \rho_0 x}{\mathcal{F}^2(x)}.$$

Case 1. Let $\lambda_0 = 0$. Then $\mathcal{F}(x) = 1 + ax$, and (3.42) is equivalent to the condition

$$a \notin (-\infty, 0). \tag{3.43}$$

If (3.43) is fulfilled then $\mathfrak{M}(\lambda)$ is the WF for L of the form (3.25)–(3.26), and

$$q(x) = \frac{2a^2}{(1 + ax)^2}, \qquad h = -a.$$

If $a < 0$ then the solvability condition is not fulfilled, and the function $\mathfrak{M}(\lambda)$ is not the WF. For the self-adjoint case $a > 0$ the solvability condition is fulfilled automaticaly.

Case 2. Let λ_0 be a real number, and $a > 0$. Then $\mathcal{F}(x) \geq 1$, and (3.42) is fulfilled. But in this case $\varepsilon(x) \notin \mathcal{L}(0, \infty)$, i.e. $\varepsilon(x) \notin W_N$ for any $N \geq 0$.

4. Solution of the inverse problem on a finite interval

We consider the DE and LF $L \in v_N$ of the form (1.1)–(1.2) on a finite interval ($T < \infty$). In §4 we provide a solution of the IP of recovering L from the given WM $\mathfrak{M}(\lambda)$. We use the notations and the results of §1. For IP on a finite interval there are specific difficulties connected with the properties S_1 and S_2 of the WM $\mathfrak{M}(\lambda)$ (see Lemmas 4.1 and 4.2). We obtain necessary and sufficient conditions on the WM, a procedure of constructing coefficients of the DE and LF from the given WM $\mathfrak{M}(\lambda)$, and study the stability problem. The main results of §4 are expressed in Theorems 4.1–4.4. In p. 4.7 there is a counterexample showing that dropping one element of the WM leads to non-uniqueness of the solution of the IP.

4.1. We shall say that $L \in v'_N$ if $L \in v_N$ and the functions $\Delta_{mm}(\lambda)$, $m = \overline{1, n-1}$ have only simple zeros. If $L \in v'_N$ then the WM $\mathfrak{M}(\lambda)$ and $\mathfrak{M}^*(\lambda)$ have only simple poles. For simplicity, in the sequel we shall assume that $L \in v'_N$.

Since $L \in v_N$ the asymptotic formula (1.14) can be made more precise:

$$\lambda_{\ell m} = (-1)^{n-m} \left(\frac{\pi}{T} \left(\sin \frac{\pi m}{n} \right)^{-1} \left(\ell + \sum_{\nu=0}^{n+N-1} \frac{\chi_{m\nu}}{\ell^\nu} + \frac{\kappa_{\ell m}}{\ell^{n+N-1}} \right) \right)^n, \tag{4.1}$$

where $\kappa_{\ell m} = o(1)$ as $\ell \to \infty$. Denote $\beta_{\ell mk} = \operatorname{res}_{\lambda_{\ell m}} \mathfrak{M}_{mk}(\lambda)$.

In virtue of (1.7), we have $\beta_{\ell mk} = (\dot{\Delta}_{mm}(\lambda_{\ell m}))^{-1} \Delta_{mk}(\lambda_{\ell m})$, and for $\ell \to \infty$

$$\left. \begin{aligned} \beta_{\ell mk} &= \ell^{n+m-k-1} \left(\sum_{\nu=0}^{n+N-1} \frac{\chi_{mk\nu}}{\ell^\nu} + \frac{\kappa_{\ell mk}}{\ell^{n+N-1}} \right), \qquad \chi_{mk0} \neq 0, \\ \mathfrak{M}_{m+1,k,<0>} &= \ell^{m+1-k} \left(\sum_{\nu=0}^{n+N-1} \frac{\eta_{mk\nu}}{\ell^\nu} + \frac{\kappa^0_{\ell mk}}{\ell^{n+N-1}} \right), \qquad \eta_{mk0} \neq 0, \end{aligned} \right. \tag{4.2}$$

where $\kappa_{\ell mk} = o(1)$, $\kappa^0_{\ell mk} = o(1)$. It follows from (1.7), (1.20) and (1.21) that

$$|\mathfrak{M}_{mk}(\lambda)| < C|\rho|^{m-k}, \qquad \lambda \in G_\delta. \tag{4.3}$$

(G_δ is the λ-plane without the circles $|\lambda - \lambda_0| < \delta$, $\lambda_0 \in \Lambda$), and hence

$$\mathfrak{M}_{mk}(\lambda) = \sum_{\ell=1}^{\infty} \frac{\beta_{\ell mk}}{\lambda - \lambda_{\ell m}}. \tag{4.4}$$

Thus, the WF $\mathfrak{M}_{mk}(\lambda)$ is uniquely determined by its zeros and residues $\{\lambda_{\ell m}, \beta_{\ell mk}\}_{\ell \geq 1}$. For $\lambda_0 \in \Lambda$ we define the matrix

$$\mathfrak{N}(\lambda_0) = [\mathfrak{N}_{j\nu}(\lambda_0)]_{j,k=\overline{1,n}} \qquad \text{via} \qquad \mathfrak{N}(\lambda_0) = \mathfrak{M}_{<-1>}(\lambda_0)(\mathfrak{M}_{<0>}(\lambda_0))^{-1}.$$

Since $\mathfrak{M}_{mk}(\lambda) = \delta_{mk}$, $m \geq k$, it follows that $\mathfrak{N}_{jk}(\lambda_0) = 0$ for $j \geq k$. The following relations hold

$$\left. \begin{aligned} &\mathfrak{N}(\lambda_0)\mathfrak{N}(\lambda_0) = 0, \\ &\Phi_{<-1>}(x, \lambda_0) = \mathfrak{N}(\lambda_0)\Phi_{<0>}(x, \lambda_0) \\ &\Phi^*_{<-1>}(x, \lambda_0) = -\Phi^*_{<0>}(x, \lambda_0)\mathfrak{N}(\lambda_0) \end{aligned} \right\} \tag{4.5}$$

(see the proof of (3.8)). It follows from $\ell\Phi_m(x, \lambda) = \lambda\Phi_m(x, \lambda)$ and (1.13) that

$$U_{n-m+1,T}(\Phi_m(x,\lambda)) = (-1)^{n-m}(\Delta_{mm}(\lambda))^{-1}\Delta_{m-1,m-1}(\lambda), \tag{4.6}$$

$$\left.\begin{aligned} \ell\Phi_{m,<-1>}(x,\lambda_0) &= \lambda_0\Phi_{m,<-1>}(x,\lambda_0), \\ \ell\Phi_{m,<0>}(x,\lambda_0) &= \lambda_0\Phi_{m,<0>}(x,\lambda_0) + \Phi_{m,<-1>}(x,\lambda_0). \end{aligned}\right\} \tag{4.7}$$

We prove two important properties of the WM. Define $\Lambda_0 = \Lambda_n = \emptyset$.

Lemma 4.1. *(property S_1) If $\lambda_0 \notin \Lambda_m$, then $\mathfrak{M}_{j,m+1}(\lambda_0) = \ldots = \mathfrak{M}_{jn}(\lambda_0) = 0$, $j = \overline{1,m}$. If, moreover, $\lambda_0 \in \Lambda_{\nu+1} \cap \ldots \cap \Lambda_{m-1}$, $\lambda_0 \notin \Lambda_\nu$, $1 \le \nu+1 < m \le n$, then $\mathfrak{M}_{\nu+1,m}(\lambda_0) \neq 0$.*

The first assertion of the lemma will be proved by induction. Since $\lambda_0 \in \Lambda_m$ it follows from (1.13) that $\Phi_{m,<-1>}(x,\lambda_0) = 0$. On the other hand, in view of (4.5), we get

$$\Phi_{m,<-1>}(x,\lambda_0) = \mathfrak{M}_{m,m+1}(\lambda_0)\Phi_{m+1,<0>}(x,\lambda_0) + \ldots + \mathfrak{M}_{mn}(\lambda_0)\Phi_{n,<0>}(x,\lambda_0).$$

Applying here the LF $U_{m+1,0}, \ldots, U_{n0}$, we find successively

$$\mathfrak{M}_{m,m+1}(\lambda_0) = \ldots = \mathfrak{M}_{mn}(\lambda_0) = 0.$$

Suppose that $\mathfrak{M}_{j,m+1}(\lambda_0) = \ldots = \mathfrak{M}_{jn}(\lambda_0) = 0$, $j = \overline{m-s+1,m}$, $s \ge 1$. According to (4.5), we have

$$\begin{aligned} \Phi_{m-s,<-1>}(x,\lambda_0) &= \mathfrak{M}_{m-s,m-s+1}(\lambda_0)\Phi_{m-s+1,<0>}(x,\lambda_0) \\ &\quad + \ldots + \mathfrak{M}_{m-s,n}(\lambda_0)\Phi_{n,<0>}(x,\lambda_0) \end{aligned}$$

or

$$\begin{aligned} &\Phi_{m-s,<-1>}(x,\lambda_0) - \sum_{i=1}^{s} \mathfrak{M}_{m-s,m-s+i}(\lambda_0)\Phi_{m-s+i,<0>}(x,\lambda_0) \\ &\quad = \sum_{i=s+1}^{n-m+s} \mathfrak{M}_{m-s,m-s+i}(\lambda_0)\Phi_{m-s+i,<0>}(x,\lambda_0) \stackrel{df}{=} \psi(x). \end{aligned} \tag{4.8}$$

Since $\Phi_{m,<-1>}(x,\lambda_0) = 0$, (4.7) gives the result that the functions $\Phi_{m-s,<-1>}(x,\lambda_0)$ and $\Phi_{m,<0>}(x,\lambda_0)$ are solutions of the DE $\ell y = \lambda_0 y$. Further, using (4.5) and the assumption of induction, we obtain

$$\sum_{i=1}^{s-1} \mathfrak{N}_{m-s,m-s+i}(\lambda_0)\Phi_{m-s+i,<-1>}(x,\lambda_0)$$

$$= \sum_{i=1}^{s-1} \mathfrak{N}_{m-s,m-s+i}(\lambda_0) \sum_{\nu=m-s+i+1}^{m} \mathfrak{N}_{m-s+i,\nu}(\lambda_0)\Phi_{\nu,<0>}(x,\lambda_0)$$

$$= \sum_{\nu=m-s+2}^{m} \Phi_{\nu,<0>}(x,\lambda_0) \sum_{i=1}^{\nu-m+s-1} \mathfrak{N}_{m-s,m-s+i}(\lambda_0)\mathfrak{N}_{m-s+i,\nu}(\lambda_0) = 0,$$

and consequently, the function $\sum\limits_{i=1}^{s-1} \mathfrak{N}_{m-s,m-s+i}(\lambda_0)\Phi_{m-s+i,<0>}(x,\lambda_0)$ is a solution of the DE $\ell y = \lambda_0 y$. This and (4.8) implies that $\ell\psi(x) = \lambda_0\psi(x)$. Using (4.8) again, we compute $U_{\xi 0}(\psi) = U_{\eta T}(\psi) = 0$, $\xi = \overline{1,m}$, $\eta = \overline{1,n-m}$. Since λ_0 is not an eigenvalue of S_m, we conclude that $\psi(x) \equiv 0$. Applying the LF $U_{m+1,0}, \ldots, U_{n0}$ to (4.8), we find successively $\mathfrak{N}_{m-s,k}(\lambda_0) = 0$, $k = \overline{m+1,n}$.

Let us go on to the second assertion of the lemma. Since $\Delta_{\nu\nu}(\lambda_0) \neq 0$, $\Delta_{ss}(\lambda_0) = 0$, $s = \overline{\nu+1,m-1}$, it follows from (4.6) that

$$U_{n-s+1,T}(\Phi_{s,<0>}(x,\lambda_0)) \neq 0, \qquad s = \overline{\nu+2,m-1}; \qquad \Phi_{\nu+1,<-1>}(x,\lambda_0) \not\equiv 0.$$

Suppose that $\mathfrak{N}_{\nu+1,m}(\lambda_0) = 0$. Then

$$\Phi_{\nu+1,<-1>}(x,\lambda_0) = \mathfrak{N}_{\nu+1,\nu+2}(\lambda_0)\Phi_{\nu+2,<0>}(x,\lambda_0) + \ldots + \mathfrak{N}_{\nu+1,m-1}(\lambda_0)\Phi_{m-1,<0>}(x,\lambda_0).$$

Applying the LF $U_{n-m+2,T}, \ldots, U_{n-\nu-1,T}$ successively, we obtain $\mathfrak{N}_{\nu+1,m-1}(\lambda_0) = \ldots = \mathfrak{N}_{\nu+1,\nu+2}(\lambda_0) = 0$, i.e. $\Phi_{\nu+1,<-1>}(x,\lambda_0) \equiv 0$. Lemma 4.1 is proved.

Denote $A_s(\lambda_0) = [\mathfrak{N}_{j\nu}(\lambda_0)]_{j=\overline{1,n-s};\nu=\overline{n-s,n}}$, $s = \overline{1,n-1}$.

Lemma 4.2. *(property S_2)*

$$\operatorname{rank} A_s(\lambda_0) \le 1, \qquad s = \overline{1,n-1}.$$

We will prove the lemma by induction. Let us show that $\operatorname{rank} A_1(\lambda_0) \le 1$. Indeed, if $\Delta_{n-2,n-2}(\lambda_0) = \Delta_{n-1,n-1}(\lambda_0) = 0$, then from (4.6) we have $U_{2,T}(\Phi_{n-1,<-1>}(x,\lambda_0)) = 0$, $U_{2,T}(\Phi_{n-1,<0>}(x,\lambda_0)) \neq 0$. Applying the LF $U_{2,T}$ to the equality $\Phi_{<-1>}(x,\lambda_0) = \mathfrak{N}(\lambda_0)\Phi_{<0>}(x,\lambda_0)$, we get

$$\mathfrak{N}_{j,n-1}(\lambda_0)U_{2,T}(\Phi_{n-1,<0>}(x,\lambda_0)) + \mathfrak{N}_{jn}(\lambda_0)U_{2,T}(\Phi_{n,<0>}(x,\lambda_0)) = 0, \quad j = \overline{1,n-1}$$

and hence $\operatorname{rank} A_1(\lambda_0) \le 1$. If $\Delta_{n-1,n-1}(\lambda_0) \neq 0$ or $\Delta_{n-2,n-2}(\lambda_0) \neq 0$ then by Lemma 4.1, $\mathfrak{N}_{jn}(\lambda_0) = 0$, $j = \overline{1,n-1}$, i.e. $\operatorname{rank} A_1(\lambda_0) \le 1$.

Suppose that the relations $\operatorname{rank} A_k(\lambda_0) \leq 1$, $k = \overline{1, s-1}$ have been proved. If $\Delta_{n-s-1,n-s-1}(\lambda_0) = \Delta_{n-s,n-s}(\lambda_0) = 0$, it follows from (4.6) that

$$U_{s+1,T}(\Phi_{n-s,<-1>}(x,\lambda_0)) = 0, \qquad U_{s+1,T}(\Phi_{n-s,<0>}(x,\lambda_0)) \neq 0,$$

and hence

$$\sum_{k=n-s}^{n} \mathfrak{N}_{jk}(\lambda_0) U_{s+1,T}(\Phi_{k,<0>}(x,\lambda_0)) = 0, \qquad j = \overline{1, n-s}. \tag{4.9}$$

We take a fixed non-zero row of the matrix $A_s(\lambda_0)$:

$$[\mathfrak{N}_{\nu,n-s}(\lambda_0), \ldots, \mathfrak{N}_{\nu n}(\lambda_0] \neq [0, \ldots, 0].$$

Since $\operatorname{rank} A_{s-1}(\lambda_0) \leq 1$, it follows that $\mathfrak{N}_{jk}(\lambda_0) = \alpha_j \mathfrak{N}_{\nu k}(\lambda_0)$, $k = \overline{n-s+1, n}$. Then from (4.9) we derive

$$(\mathfrak{N}_{j,n-s}(\lambda_0) - \alpha_j \mathfrak{N}_{\nu,n-s}(\lambda_0)) U_{s+1,T}(\Phi_{n-s,<0>}(x,\lambda_0)) = 0$$

or $\mathfrak{N}_{j,n-s}(\lambda_0) = \alpha_j \mathfrak{N}_{\nu,n-s}(\lambda_0)$. Hence $\operatorname{rank} A_s(\lambda_0) \leq 1$. If $\Delta_{n-s-1,n-s-1}(\lambda_0) \neq 0$ or $\Delta_{n-s,n-s}(\lambda_0) \neq 0$, we get by Lemma 4.1 that

$$\mathfrak{N}_{j,n-s+1}(\lambda_0) = \ldots = \mathfrak{N}_{jn}(\lambda_0) = 0, \qquad j = \overline{1, n-s},$$

i.e. $\operatorname{rank} A_s(\lambda_0) \leq 1$. Lemma 4.2 is proved.

Denote by M the set of meromorphic matrices $\mathfrak{M}(\lambda) = [\mathfrak{M}_{mk}(\lambda)]_{m,k=\overline{1,n}}$, $\mathfrak{M}_{mk}(\lambda) = \delta_{mk}$ $(m \geq k)$, having only simple poles Λ (in general, the set Λ is different for each matrix $\mathfrak{M}(\lambda)$) and such that (4.3) is valid and for each $\lambda_0 \in \Lambda$ the matrix $\mathfrak{M}(\lambda)$ has the properties S_1 and S_2, where the sets $\Lambda_m = \{\lambda_{\ell m}\}_{\ell \geq 1}$, $\lambda_{\ell m} \neq \lambda_{\ell_0, m}$ $(\ell \neq \ell_0)$ are defined as follows: if $\lambda_0 \in \Lambda$, $\mathfrak{N}_{kj}(\lambda_0) \neq 0$, then $\lambda_0 \in \Lambda_k \cap \ldots \cap \Lambda_{j-1}$.

It is clear that if $\mathfrak{M}(\lambda) \in M$, then $\mathfrak{N}(\lambda_0)\mathfrak{N}(\lambda_0) = 0$ for $\lambda_0 \in \Lambda$. If $L \in v_N'$ and $\mathfrak{M}(\lambda)$ is the WM for L, then $\mathfrak{M}(\lambda) \in M$.

Lemma 4.3. *Let a matrix $\mathfrak{M}(\lambda) = [\mathfrak{M}_{mk}(\lambda)]_{m,k=\overline{1,n}}$, $\mathfrak{M}_{mk}(\lambda) = \delta_{mk}$ $(m \geq k)$ have a simple pole at a point λ_0. For the matrix $\mathfrak{M}^*(\lambda) \overset{df}{=} (\mathfrak{M}(\lambda))^{-1}$ to have a simple pole at λ_0 it is necessary and sufficient that $\mathfrak{N}(\lambda_0)\mathfrak{N}(\lambda_0) = 0$.*

The necessity part of the lemma is obvious. We prove the sufficiency. Let $\mathfrak{N}(\lambda_0)\mathfrak{N}(\lambda_0) = 0$. Denote by X_p the set of matrices $A = [A_{\nu j}]_{\nu,j=\overline{1,n}}$ such that $A_{\nu j} = 0$ for $j - \nu < n - p$. It is clear that if $A \in X_p$, $B \in X_q$, then $AB \in X_{p+q-n}$. Since $\mathfrak{M}(\lambda) = \mathfrak{M}^*(\lambda) = E$, it follows that

$$\mathfrak{M}^*(\lambda) = \sum_{k=1-n}^{\infty} (\lambda - \lambda_0)^k \mathfrak{M}^*_{<k>}(\lambda_0),$$

$$\sum_{j=-1}^{n-k-1} \mathfrak{M}^*_{<-j-k>}(\lambda_0) \mathfrak{M}_{<j>}(\lambda_0) = 0, \qquad k = \overline{1, n-1}.$$

From this, in view of the relation $\mathfrak{N}(\lambda_0)\mathfrak{N}(\lambda_0) = 0$, we get

$$\begin{aligned}\mathfrak{M}^*_{<-k>}(\lambda_0) &= -\mathfrak{M}^*_{<1-k>}(\lambda_0)\mathfrak{N}(\lambda_0) \\ &\quad - \left(\sum_{j=1}^{n-k-1} \mathfrak{M}^*_{<-j-k>}(\lambda_0)\mathfrak{M}_{<j>}(\lambda_0)\right)(\mathfrak{M}_{<0>}(\lambda_0))^{-1},\end{aligned}$$

$$\mathfrak{M}^*_{<1-k>}(\lambda_0)\mathfrak{N}(\lambda_0) = -\left(\sum_{j=1}^{n-k} \mathfrak{M}^*_{<-j-k+1>}(\lambda_0)\mathfrak{M}_{<j>}(\lambda_0)\right)(\mathfrak{M}_{<0>}(\lambda_0))^{-1}\mathfrak{N}(\lambda_0),$$

$$k = \overline{2, n-1}.$$

Since $\mathfrak{N}(\lambda_0) \in X_{n-1}$, $\mathfrak{M}^*_{<-k>}(\lambda_0) \in X_{n-k}$, we can find

$$\mathfrak{M}^*_{<1-k>}(\lambda_0)\mathfrak{N}(\lambda_0), \qquad \mathfrak{M}^*_{<-k>}(\lambda_0) \in X_{n-k-2}, \qquad k = \overline{2, n-1}.$$

Repeating this action several times we obtain $\mathfrak{M}^*_{<-k>}(\lambda_0) = 0$, $k = \overline{2, n-1}$. Lemma 4.3 is proved.

Corollary 4.1. *If $\mathfrak{M}(\lambda) \in M$, then the matrix $\mathfrak{M}^*(\lambda) \overset{df}{=} (\mathfrak{M}(\lambda))^{-1}$ has only simple poles.*

4.2. Let $\widetilde{L} \in v'_N$, $\mathfrak{M}(\lambda) \in M$. Denote

$$\widetilde{D}(x, \lambda, \lambda_0) = \left[-\frac{\langle \widetilde{\Phi}(x, \lambda), \widetilde{\Phi}^*(x, \mu)\rangle_{\widetilde{\ell}}}{\lambda - \mu}\right]^{<0>}_{\mu=\lambda_0},$$

$$\widetilde{D}_{<k>}(x, z_0, \lambda_0) = [\widetilde{D}(x, \lambda, \lambda_0)]^{<k>}_{\lambda=z_0}, \qquad k = 0, -1.$$

Lemma 4.4.

$$\left.\begin{aligned}\widetilde{D}(x, \lambda, \lambda_0)\mathfrak{N}(\lambda_0) &= \frac{\langle \widetilde{\Phi}(x, \lambda), \widetilde{\Phi}^*_{<-1>}(x, \lambda_0)\rangle_{\widetilde{\ell}}}{\lambda - \lambda_0}, \\ \widetilde{\mathfrak{N}}(z_0)\widetilde{D}_{<0>}(x, z_0, \lambda_0) &= \left[-\frac{\langle \widetilde{\Phi}_{<-1>}(x, z_0), \widetilde{\Phi}^*(x, \mu)\rangle_{\widetilde{\ell}}}{z_0 - \mu}\right]^{<0>}_{\mu=\lambda_0},\end{aligned}\right\} \tag{4.10}$$

$$\widetilde{D}_{<-1>}(x, z_0, \lambda_0) = \widetilde{\mathfrak{N}}(z_0)\widetilde{D}_{<0>}(x, z_0, \lambda_0) - \delta(z_0, \lambda_0)E, \tag{4.11}$$

$$\left.\begin{aligned} &\widetilde{\mathfrak{N}}(z_0)\widetilde{D}_{<0>}(x,z_0,\lambda_0)\widetilde{\mathfrak{N}}(\lambda_0) \\ &\qquad = \frac{\langle\widetilde{\Phi}_{<-1>}(x,z_0),\widetilde{\Phi}^*_{<-1>}(x,\lambda_0)\rangle_{\widetilde{\ell}}}{z_0-\lambda_0}, \qquad (z_0\neq\lambda_0), \\ &\widetilde{\mathfrak{N}}(z_0)\widetilde{D}_{<0>}(x,z_0,\lambda_0)\widetilde{\mathfrak{N}}(\lambda_0) \\ &\qquad = \widetilde{\mathfrak{N}}(\lambda_0)-\langle\widetilde{\Phi}_{<-1>}(x,\lambda_0),\widetilde{\Phi}^*_{<0>}(x,\lambda_0)\rangle_{\widetilde{\ell}} \quad (z_0=\lambda_0), \end{aligned}\right\} \tag{4.12}$$

where $\delta(z_0,\lambda_0)=0$ $(z_0\neq\lambda_0)$, $\delta(z_0,\lambda_0)=1$ $(z_0=\lambda_0)$.

Proof. By virtue of (1.27) and Lemma 1.1, we compute

$$\langle\widetilde{\Phi}(x,\lambda),\widetilde{\Phi}^*(x,\lambda)\rangle_{\widetilde{\ell}}=\widetilde{\mathfrak{M}}(\lambda)\widetilde{\mathfrak{M}}^*(\lambda)=E,$$

and consequently

$$\sum_{k=-1}^{s}\langle\widetilde{\Phi}_{<k>}(x,\lambda_0),\widetilde{\Phi}^*_{<s-k>}(x,\lambda_0)\rangle_{\widetilde{\ell}}=\delta_{s0}E, \qquad s=0,1. \tag{4.13}$$

Since

$$\widetilde{D}(x,\lambda,\lambda_0)=-\frac{\langle\widetilde{\Phi}(x,\lambda),\widetilde{\Phi}^*_{<0>}(x,\lambda_0)\rangle_{\widetilde{\ell}}}{\lambda-\lambda_0}--\frac{\langle\widetilde{\Phi}(x,\lambda),\widetilde{\Phi}^*_{<-1>}(x,\lambda_0)\rangle_{\widetilde{\ell}}}{(\lambda-\lambda_0)^2}, \tag{4.14}$$

in view of (4.13), we get

$$\left.\begin{aligned} &\widetilde{D}_{<-1>}(x,z_0,\lambda_0) \\ &\quad = -\frac{\langle\widetilde{\Phi}_{<-1>}(x,z_0),\widetilde{\Phi}^*_{<0>}(x,\lambda_0)\rangle_{\widetilde{\ell}}}{z_0-\lambda_0}-\frac{\langle\widetilde{\Phi}_{<-1>}(x,z_0),\widetilde{\Phi}^*_{<-1>}(x,\lambda_0)\rangle_{\widetilde{\ell}}}{(z_0-\lambda_0)^2}, \\ &\widetilde{D}_{<0>}(x,z_0,\lambda_0) \\ &\quad = -\frac{\langle\widetilde{\Phi}_{<0>}(x,z_0),\widetilde{\Phi}^*_{<0>}(x,\lambda_0)\rangle_{\widetilde{\ell}}}{z_0-\lambda_0}+\frac{\langle\widetilde{\Phi}_{<-1>}(x,z_0),\widetilde{\Phi}^*_{<0>}(x,\lambda_0)\rangle_{\widetilde{\ell}}}{(z_0-\lambda_0)^2} \\ &\quad -\frac{\langle\widetilde{\Phi}_{<0>}(x,z_0),\widetilde{\Phi}^*_{<-1>}(x,\lambda_0)\rangle_{\widetilde{\ell}}}{(z_0-\lambda_0)^2}+2\frac{\langle\widetilde{\Phi}_{<-1>}(x,z_0),\widetilde{\Phi}^*_{<-1>}(x,\lambda_0)\rangle_{\widetilde{\ell}}}{(z_0-\lambda_0)^3}, \quad z_0\neq\lambda_0, \\ &\widetilde{D}_{<-1>}(x,\lambda_0,\lambda_0)=-E+\langle\widetilde{\Phi}_{<-1>}(x,\lambda_0),\widetilde{\Phi}^*_{<1>}(x,\lambda_0)\rangle_{\widetilde{\ell}}, \\ &\widetilde{D}_{<0>}(x,\lambda_0,\lambda_0) \\ &\quad = \langle\widetilde{\Phi}_{<0>}(x,\lambda_0),\widetilde{\Phi}^*_{<1>}(x,\lambda_0)\rangle_{\widetilde{\ell}}+\langle\widetilde{\Phi}_{<-1>}(x,\lambda_0),\widetilde{\Phi}^*_{<2>}(x,\lambda_0)\rangle_{\widetilde{\ell}}, \end{aligned}\right\} \tag{4.15}$$

and hence

$$\widetilde{D}_{<0>}(x,z_0,\lambda_0)$$
$$=\left[-\frac{\langle\widetilde{\Phi}_{<0>}(x,z_0),\widetilde{\Phi}^*(x,\mu)\rangle_{\widetilde{\ell}}}{z_0-\mu}+\frac{\langle\widetilde{\Phi}_{<-1>}(x,z_0),\widetilde{\Phi}^*(x,\mu)\rangle_{\widetilde{\ell}}}{(z_0-\mu)^2}\right]^{<0>}_{|\mu=\lambda_0}. \tag{4.16}$$

According to (4.5), we have $\widetilde{\Phi}^*_{<-1>}(x,\lambda_0)\widetilde{\mathfrak{N}}(\lambda_0)=\widetilde{\mathfrak{N}}(\lambda_0)\widetilde{\Phi}_{<-1>}(x,\lambda_0)=0$. From this and (4.14)–(4.16) we obtain the assertion of Lemma 4.4.

From Lemma 4.4, in virtue of the equalities

$$\langle\widetilde{\Phi}(x,\lambda),\widetilde{\Phi}^*(x,\mu)\rangle\big|_{x=0}=\widetilde{\mathfrak{M}}(\lambda)\widetilde{\mathfrak{M}}^*(\mu),$$

$$\langle\widetilde{\Phi}(x,\lambda),\widetilde{\Phi}^*(x,\mu)\rangle\big|_{x=T}=\widetilde{U}_T(\widetilde{\Phi}(x,\lambda))\widetilde{U}_T^*(\widetilde{\Phi}^*(x,\mu)),$$

we get

Corollary 4.2.

$$\widetilde{D}_{k\nu}(0,\lambda,\lambda_0)=-\frac{\delta_{k\nu}}{\lambda-\lambda_0},\qquad k\ge\nu;$$

$$\widetilde{D}_{k\nu}(0,\lambda,\lambda_0)=\mathcal{F}_{k\nu}(\widetilde{\mathfrak{M}}_{j,j+s},\, s=\overline{1,\nu-k},\, j=\overline{1,n-1}),\qquad k<\nu;$$

$$\widetilde{D}_{k\nu}(T,\lambda,\lambda_0)=(\widetilde{D}(T,\lambda,\lambda_0)\widetilde{\mathfrak{N}}(\lambda_0))_{k\nu}=(\widetilde{\mathfrak{N}}(z_0)\widetilde{D}_{<0>}(T,z_0,\lambda_0))_{k\nu}=0,\qquad k<\nu;$$

$$(\widetilde{\mathfrak{N}}(z_0)\widetilde{D}_{<0>}(T,z_0,\lambda_0)\widetilde{\mathfrak{N}}(\lambda_0))_{k\nu}=\delta(z_0,\lambda_0)\widetilde{\mathfrak{N}}_{k\nu}(z_0),\qquad k<\nu;$$

Denote

$$Y=[\delta_{j,k-1}]_{j=\overline{1,n-1};k=\overline{1,n}},\qquad \mathfrak{N}_0(\lambda_0)=\mathfrak{N}(\lambda_0),\qquad \mathfrak{N}_1(\lambda_0)=\widetilde{\mathfrak{N}}(\lambda_0),$$

$$\widetilde{P}_\varepsilon(x,\lambda,\lambda_0)=\widetilde{D}(x,\lambda,\lambda_0)\mathfrak{N}_\varepsilon(\lambda_0)Y^T,\quad \widetilde{G}_\varepsilon(x,z_0,\lambda_0)=Y\widetilde{D}_{<0>}(x,z_0,\lambda_0)\mathfrak{N}_\varepsilon(\lambda_0)Y^T,$$

$$\widetilde{g}^*_\varepsilon(x,\lambda_0)=-\widetilde{\Phi}^*_{<0>}(x,\lambda_0)\mathfrak{N}_\varepsilon(\lambda_0)Y^T,\quad \varepsilon=0,1;\quad P(x,\lambda,\lambda_0)=\widetilde{D}(x,\lambda,\lambda_0)\widehat{\mathfrak{N}}(\lambda_0)Y^T,$$

$$\widetilde{G}(x,z_0,\lambda_0)=Y\widetilde{D}_{<0>}(x,z_0,\lambda_0)\widehat{\mathfrak{N}}(\lambda_0)Y^T,\quad \widetilde{g}^*(x,\lambda_0)=-\widetilde{\Phi}^*_{<0>}(x,\lambda_0)\widehat{\mathfrak{N}}(\lambda_0)Y^T,$$

$$\widetilde{\varphi}(x,\lambda_0) = Y\widetilde{\Phi}_{<0>}(x,\lambda), \quad \widetilde{\Lambda}(\lambda_0) = \lambda_0 E + Y\hat{\widetilde{\mathfrak{N}}}(\lambda_0)Y^T, \quad \Lambda(\lambda_0) = \lambda_0 E + Y\mathfrak{N}(\lambda_0)Y^T.$$

Lemma 4.5.

$$\widetilde{\ell}\widetilde{\varphi}(x,\lambda_0) = \widetilde{\Lambda}(\lambda_0)\widetilde{\varphi}(x,\lambda_0), \tag{4.17}$$

$$\widetilde{P}'_\varepsilon(x,\lambda,\lambda_0) = \widetilde{\Phi}(x,\lambda)\widetilde{g}^*_\varepsilon(x,\lambda_0), \qquad \widetilde{G}'_\varepsilon(x,z_0,\lambda_0) = \widetilde{\varphi}(x,z_0)\widetilde{g}^*_\varepsilon(x,\lambda_0), \tag{4.18}$$

$$\varepsilon = 0,1,$$

$$\widetilde{P}(x,\lambda,\lambda_0)(\lambda E - \Lambda(\lambda_0)) = \langle\widetilde{\Phi}(x,\lambda),\widetilde{g}^*(x,\lambda_0)\rangle_{\widetilde{\ell}}, \tag{4.19}$$

$$\begin{aligned}\widetilde{\Lambda}(z_0)\widetilde{G}(x,z_0,\lambda_0) - \widetilde{G}(x,z_0,\lambda_0)\Lambda(\lambda_0) - \delta(z_0,\lambda_0)Y\hat{\mathfrak{N}}(\lambda_0)Y^T \\ = \langle\widetilde{\varphi}(x,z_0),\widetilde{g}^*(x,\lambda_0)\rangle_{\widetilde{\ell}}.\end{aligned} \tag{4.20}$$

If $\widetilde{\mathfrak{N}}(\lambda_0)\mathfrak{N}(\lambda_0) = 0$, *then*

$$\widetilde{P}_\varepsilon(x,\lambda,\lambda_0)(\lambda E - \Lambda(\lambda_0)) = \langle\Phi(x,\lambda),\widetilde{g}^*_\varepsilon(x,\lambda_0)\rangle_{\widetilde{\ell}}, \qquad \varepsilon = 0,1; \tag{4.21}$$

$$\begin{aligned}\widetilde{\Lambda}(z_0)\widetilde{G}_\varepsilon(x,z_0,\lambda_0) - \widetilde{G}_\varepsilon(x,z_0,\lambda_0)\Lambda(\lambda_0) - \delta(z_0,\lambda_0)Y\mathfrak{N}_\varepsilon(\lambda_0)Y^T \\ = \langle\widetilde{\varphi}(x,z_0),\widetilde{g}^*_\varepsilon(x,\lambda_0)\rangle_{\widetilde{\ell}}, \qquad \varepsilon = 0,1.\end{aligned} \tag{4.22}$$

Proof. In virtue of (4.5) and (4.7) we compute

$$\widetilde{\ell}\widetilde{\Phi}_{<0>}(x,\lambda_0) = (\lambda_0 E + \widetilde{\mathfrak{N}}(\lambda_0))\widetilde{\Phi}_{<0>}(x,\lambda_0).$$

This implies (4.17). Using (1.27) we derive $\widetilde{D}'(x,\lambda,\lambda_0) = -\widetilde{\Phi}(x,\lambda)\widetilde{\Phi}^*_{<0>}(x,\lambda_0)$ and consequently, (4.18) is proved. Further, it follows from (4.10) and (4.14) that

$$(\lambda-\lambda_0)\widetilde{D}(x,\lambda,\lambda_0) + \widetilde{D}(x,\lambda,\lambda_0)\widetilde{\mathfrak{N}}(\lambda_0) = -\langle\widetilde{\Phi}(x,\lambda),\widetilde{\Phi}^*_{<0>}(x,\lambda_0)\rangle_{\widetilde{\ell}}.$$

Multiplying this equality by $\mathfrak{N}_\varepsilon(\lambda_0)Y^T$, we obtain

$$\begin{aligned}\widetilde{P}_\varepsilon(x,\lambda,\lambda_0)(\lambda E - \Lambda(\lambda_0)) + \widetilde{P}_1(x,\lambda,\lambda_0)Y\mathfrak{N}(\lambda_0)Y^T \\ = \langle\widetilde{\Phi}(x,\lambda),\widetilde{g}^*_\varepsilon(x,\lambda_0)\rangle_{\widetilde{\ell}}, \qquad \varepsilon = 0,1.\end{aligned} \tag{4.23}$$

This yields (4.19) and (4.21). It follows from (4.11) that

$$Y[\widetilde{P}_\varepsilon(x,\lambda,\lambda_0)]^{<-1>}_{\lambda=z_0} = Y\widetilde{\mathfrak{N}}(z_0)[\widetilde{P}_\varepsilon(x,\lambda,\lambda_0)]^{<0>}_{\lambda=z_0} - \delta(z_0,\lambda_0)Y\mathfrak{N}_\varepsilon(\lambda_0)Y^T.$$

Using (4.23) we arrive at

$$\widetilde{G}_\varepsilon(x,z_0,\lambda_0)(z_0 E - \Lambda(\lambda_0)) + Y[\widetilde{P}_\varepsilon(x,\lambda,\lambda_0)]^{<-1>}_{\lambda=\lambda_0}$$

$$+\widetilde{G}_1(x,z_0,\lambda_0)Y\mathfrak{N}(\lambda_0)Y^T = \langle \widetilde{\varphi}(x,z_0), \widetilde{g}^*_\varepsilon(x,\lambda_0)\rangle_{\widetilde{\ell}}.$$

From this, in view of the equality

$$Y\widetilde{\mathfrak{N}}(z_0)[P_\varepsilon(x,\lambda,\lambda_0)]^{<0>}_{\lambda=z_0} = Y\widetilde{\mathfrak{N}}(z_0)Y^T \widetilde{G}_\varepsilon(x,z_0,\lambda_0),$$

we have

$$\widetilde{\Lambda}(z_0)\widetilde{G}(x,z_0,\lambda_0) - \widetilde{G}_\varepsilon(x,z_0,\lambda_0)\Lambda(\lambda_0) + \widetilde{G}_1(x,z_0,\lambda_0)Y\mathfrak{N}(\lambda_0)Y^T$$

$$-\delta(z_0,\lambda_0)Y\mathfrak{N}_\varepsilon(\lambda_0)Y^T = \langle \widetilde{\varphi}(x,z_0), \widetilde{g}^*_\varepsilon(x,\lambda_0)\rangle_{\widetilde{\ell}}.$$

This yields (4.20) and (4.22). Lemma 4.5 is proved.

4.3. We consider $L, \widetilde{L} \in v'_N$. Denote

$$\xi_\ell = \sum_{m=1}^{n-1}\left(|\lambda_{\ell m} - \widetilde{\lambda}_{\ell m}| + \sum_{k=m+1}^{n}|\mathfrak{N}_{mk}(\lambda_{\ell m}) - \widetilde{\mathfrak{N}}_{mk}(\widetilde{\lambda}_{\ell m})|\ell\right)\ell^{1-n},$$

and in what follows we will assume that the numbers $\lambda_{\ell m}$ and $\widetilde{\lambda}_{\ell m}$ are numbered such that $\lambda_{\ell m} \neq \lambda_{\ell_0,m_0}$, $\widetilde{\lambda}_{\ell m} \neq \widetilde{\lambda}_{\ell_0,m_0}$, $\lambda_{\ell m} \neq \widetilde{\lambda}_{\ell_0,m_0}$ for $\ell \neq \ell_0$, $|m-m_0|=1$. It is possible and means that "common" poles have the same number ℓ.

Lemma 4.6.

$$\widetilde{\Phi}(x,\lambda) = \Phi(x,\lambda) + \sum_{\lambda_0 \in I} \widetilde{P}(x,\lambda,\lambda_0)\varphi(x,\lambda_0), \qquad (4.24)$$

$$\widetilde{\varphi}(x,z_0) = \varphi(x,z_0) + \sum_{\lambda_0 \in I} \widetilde{G}(x,z_0,\lambda_0)\varphi(x,z_0), \qquad z_0 \in I, \qquad (4.25)$$

$$\widetilde{G}(x,z_0,\kappa_0) - G(x,z_0,\kappa_0) = \sum_{\lambda_0 \in I} \widetilde{G}(x,z_0,\lambda_0)G(x,\lambda_0,\kappa_0), \qquad z_0,\kappa_0 \in I, \qquad (4.26)$$

where $I = \Lambda \cup \widetilde{\Lambda}$,

$$\varphi(x,\lambda_0) = Y\Phi_{<0>}(x,\lambda_0), \qquad G(x,z_0,\lambda_0) = YD_{<0>}(x,z_0,\lambda_0)\widetilde{\mathfrak{N}}(\lambda_0)Y^T,$$

and the series converge "with brackets":

$$\sum_{\lambda_0 \in I} = \lim_{k\to\infty} \sum_{\lambda_0 \in I_k}, \qquad I_k = I \cap \{\lambda : |\lambda| \le R_k\},$$

and the circumferences $|\lambda| = R_k$ are at a positive distance from the set I.

Proof. Using (4.5) and (4.10) we obtain

$$\operatorname*{res}_{\mu=\lambda_0}\left[-\frac{\langle\tilde\Phi(x,\lambda),\tilde\Phi^*(x,\mu)\rangle_{\tilde\ell}}{\lambda-\mu}\Phi(x,\mu)\right] = \tilde D(x,\lambda,\lambda_0)\Phi_{<-1>}(x,\lambda_0)$$

$$-\frac{\langle\tilde\Phi(x,\lambda),\tilde\Phi^*_{<-1>}(x,\lambda_0)\rangle_{\tilde\ell}}{\lambda-\lambda_0}\Phi_{<0>}(x,\lambda_0) = \tilde D(x,\lambda,\lambda_0)\mathfrak{M}(\lambda_0)\Phi_{<0>}(x,\lambda_0)$$

$$= \tilde D(x,\lambda,\lambda_0)\hat{\mathfrak{M}}(\lambda_0)\varphi(x,\lambda_0),$$

$$\operatorname*{res}_{\xi=\lambda_0}\left[\frac{\langle\tilde\Phi(x,\lambda),\tilde\Phi^*(x,\xi)\rangle_{\tilde\ell}}{\lambda-\mu}\,\frac{\langle\Phi(x,\xi),\Phi^*(x,\mu)\rangle_{\ell}}{\xi-\mu}\right] = \tilde D(x,\lambda,\lambda_0)\mathfrak{M}(\lambda_0)$$

$$\times\left[\frac{\langle\Phi_{<0>}(x,\lambda_0),\Phi^*(x,\mu)\rangle_\ell}{\lambda_0-\mu} - \frac{\langle\Phi_{<-1>}(x,\lambda_0),\Phi^*(x,\mu)\rangle_\ell}{(\lambda_0-\mu)^2}\right]$$

$$-\tilde D(x,\lambda,\lambda_0)\frac{\langle\Phi_{<-1>}(x,\lambda_0),\Phi^*(x,\mu)\rangle_\ell}{\lambda_0-\mu}.$$

Hence, using (4.10) and (4.16) we have

$$\operatorname{res}_{\mu=\lambda_0}\left[-\frac{\langle\tilde\Phi(x,\lambda),\tilde\Phi^*(x,\mu)\rangle_{\tilde\ell}}{\lambda-\mu}\Phi(x,\mu)\right] = \tilde P(x,\lambda,\lambda_0)\varphi(x,\lambda_0), \tag{4.27}$$

$$\left[\operatorname{res}_{\xi=\lambda_0}\left(\frac{\langle\tilde\Phi(x,\lambda),\tilde\Phi^*(x,\xi)\rangle_{\tilde\ell}}{\lambda-\xi}\,\frac{\langle\Phi(x,\xi),\Phi^*(x,\mu)\rangle_\ell}{\xi-\mu}\right)\right]^{<0>}_{\mu=\kappa_0} \tag{4.28}$$

$$= \tilde D(x,\lambda,\lambda_0)\hat{\mathfrak{M}}(\lambda_0)D_{<0>}(x,\lambda_0,\kappa_0).$$

In the λ-plane we consider a contour $\gamma = \gamma^+ \cup \gamma^-$, $\gamma^\pm = \{\lambda : \pm\operatorname{Im}\lambda = C_0, -\infty < \mp\operatorname{Re}\lambda < \infty\}$ such that $I \subset \{\lambda : |\operatorname{Im}\lambda| < C_0\}$. Put $J_\gamma = \mathbb{C}\backslash\operatorname{int}\gamma$. Then the relations (2.3) and (2.4) are valid (the proof is the same as for the half-line). Using (4.27), (4.28) snd the residue theorem [94, p. 239] we obtain (4.24) and (4.26). Equality (4.25) follows from (4.24). Lemma 4.6 is proved.

Denote

$$Y_k = \operatorname{diag}[\delta_{\nu k}]_{\nu=\overline{1,n-1}}, \qquad \lambda_{\ell 0k} = \lambda_{\ell k}, \qquad \lambda_{\ell 1k} = \widetilde{\lambda}_{\ell k},$$

$$\widetilde{\Lambda}_{\ell\varepsilon} = \sum_{k=1}^{n-1} Y_k \widetilde{\Lambda}(\lambda_{\ell\varepsilon k}), \qquad \widetilde{\varphi}_{\ell\varepsilon}(x) = \sum_{k=1}^{n-1} Y_k \widetilde{\varphi}(x, \lambda_{\ell\varepsilon k}),$$

$$\widetilde{g}^*_{\ell\varepsilon}(x) = \sum_{k=1}^{n-1} \widetilde{g}^*_\varepsilon(x, \lambda_{\ell\varepsilon k}) Y_k, \qquad \widetilde{P}_{\ell\varepsilon}(x,\lambda) = \sum_{k=1}^{n-1} \widetilde{P}_\varepsilon(x, \lambda, \lambda_{\ell\varepsilon k}) Y_k,$$

$$\widetilde{G}_{(\ell_0,\varepsilon_0),(\ell,\varepsilon)}(x) = \sum_{k,k_0=1}^{n-1} Y_{k0} G_\varepsilon(x, \lambda_{\ell_0,\varepsilon_0,k_0}, \lambda_{\ell\varepsilon k}) Y_k, \qquad \varepsilon, \varepsilon_0 = 0, 1.$$

Analogously we define $\Lambda_{\ell\varepsilon}$, $\varphi_{\ell\varepsilon}(x)$, $G_{(\ell_0,\varepsilon_0),(\ell,\varepsilon)}(x)$. Let V' be a set of indices $v = (\ell, \varepsilon)$, $\ell \geq 1$, $\varepsilon = 0, 1$ (ε changes quicker), and V be a set of indices $j = (v, k) = (\ell, \varepsilon, k)$, $v \in V'$, $k = \overline{1, n-1}$ (k changes quicker). We introduce the matrices

$$\widetilde{\varphi}(x) = [\widetilde{\varphi}_v(x)]_{v \in V'} = [\widetilde{\varphi}_j(x)]_{j \in V}, \qquad \widetilde{g}^*(x) = [\widetilde{g}^*_v(x)]^T_{v \in V'} = [\widetilde{g}^*_j(x)]^T_{j \in V},$$

$$\widetilde{G}(x) = [\widetilde{G}_{v_0,v}(x)]_{v_0,v \in V'} = [\widetilde{G}_{j_0,j}(x)]_{j_0,j \in V},$$

$$v_0 = (\ell_0, \varepsilon_0), \qquad j_0 = (v_0, k_0) = (\ell_0, \varepsilon_0, k_0),$$

$$\widetilde{\Lambda} = \operatorname{diag}[\widetilde{\Lambda}_v]_{v \in V'}, \qquad J = \operatorname{diag}[(-1)^\varepsilon E]_{v \in V'}, \qquad J_1 = [\delta_{\ell_0,\ell} \theta_{v_0,v}]_{v_0,v \in V'},$$

$$\theta_{vv} = E, \qquad \theta_{(\ell,0),(\ell,1)} = -E, \qquad \theta_{(\ell,1),(\ell,0)} = 0, \qquad E = [\delta_{\mu k}]_{\mu,k=\overline{1,n-1}}.$$

Analogously we define the matrices φ, G, Λ. Denote

$$w^*_{\ell k}(x) = \ell^{k-n+1} \exp(-x\ell \cot \frac{k\pi}{n}), \qquad w_{\ell o k}(x) = \xi_\ell w^*_{\ell k}(x),$$

$$w_{\ell 1k}(x) = w^*_{\ell k}(x), \qquad W(x) = \operatorname{diag}[w_j(x)]_{j \in V}.$$

Let us show that $|\widetilde{\varphi}_j^{(\nu)}(x)| < C\ell^\nu w^*_{\ell k}(x)$, $j \in V$. Indeed, by definition, $\widetilde{\varphi}_j(x) = \widetilde{\varphi}_{\ell\varepsilon k}(x) = \widetilde{\Phi}_{k+1,<0>}(x, \lambda_{\ell\varepsilon k})$. It follows from (1.14) that for sufficiently large ℓ ($\ell > \ell^+$) we get $\widetilde{\Delta}_{k+1}(\lambda_{\ell\varepsilon k}) \neq 0$. Hence $\widetilde{\varphi}_j(x) = \widetilde{\Phi}_{k+1}(x, \lambda_{\ell\varepsilon k})$. By virtue of (1.20) we have the estimate

$$|\widetilde{\Phi}^{(\nu)}_{k+1}(x,\lambda)| < C|\rho|^{\nu-n+k+1} \, |\exp(\rho R_{k+1} x)|, \qquad \rho \in S, \qquad \lambda \in G_{\delta,k+1}.$$

Then, using (1.14), we conclude $|\widetilde{\Phi}_{k+1}^{(\nu)}(x,\lambda_{\ell\varepsilon k})| < C\ell^{\nu} w_{\ell k}^{*}(x)$, $\ell > \ell^{+}$. This yields the estimate for $|\widetilde{\varphi}_j^{(\nu)}(x)|$. Using the Schwarz lemma [94, p. 363] we obtain

$$|\widetilde{\varphi}_{\ell 0 k}^{(\nu)}(x) - \widetilde{\varphi}_{\ell 1 k}^{(\nu)}(x)| < C\xi_{\ell}\ell^{\nu} w_{\ell k}^{*}(x).$$

Similarly,

$$|\widetilde{g}_j^{*(\nu)}(x)| < C\ell^{\nu}(w_{\ell k}^{*}(x))^{-1}, \qquad |\widetilde{g}_{\ell 0 k}^{*(\nu)}(x) - \widetilde{g}_{\ell 1 k}^{*(\nu)}(x)| < C\ell^{\nu}\xi_{\ell}(w_{\ell k}^{*}(x))^{-1},$$

$$|\widetilde{G}_{j_0,j}(x)| < \frac{C}{|\ell-\ell_0|+1}\frac{w_{\ell_0,k_0}^{*}(x)}{w_{\ell k}^{*}(x)}, \qquad |\widetilde{G}_{j_0,j}^{(\nu+1)}(x)| < C(\ell+\ell_0)^{\nu}|\frac{w_{\ell_0,k_0}^{*}(x)}{w_{\ell k}^{*}(x)},$$

$$|\widetilde{G}_{j_0,(\ell 0 k)}(x) - \widetilde{G}_{j_0,(\ell 1 k)}(x)| < \frac{C\xi_{\ell}}{|\ell-\ell_0|+1}\frac{w_{\ell_0,k_0}^{*}(x)}{w_{\ell k}^{*}(x)},$$

$$|\widetilde{G}_{(\ell_0,0,k_0),j}(x) - \widetilde{G}_{(\ell_0,1,k_0),j}(x)| < \frac{C\xi_{\ell_0}}{|\ell-\ell_0|+1}\frac{w_{\ell_0,k_0}^{*}(x)}{w_{\ell k}^{*}(x)}.$$

The same estimates is valid for $\varphi(x)$ and $G(x)$.

In view of our notations, (4.24)–(4.26) become

$$\widetilde{\Phi}(x,\lambda) = \Phi(x,\lambda) + \sum_{\ell=1}^{\infty}(\widetilde{P}_{\ell 0}(x,\lambda)\varphi_{\ell 0}(x) - \widetilde{P}_{\ell 1}(x,\lambda)\varphi_{\ell 1}(x)), \tag{4.29}$$

$$\widetilde{\varphi}_{\ell_0,\varepsilon_0}(x) = \varphi_{\ell_0,\varepsilon_0}(x) + \sum_{\ell=1}^{\infty}(\widetilde{G}_{(\ell_0,\varepsilon_0),(\ell,0)}(x)\varphi_{\ell 0}(x) - \widetilde{G}_{(\ell_0,\varepsilon_0),(\ell,1)}(x)\varphi_{\ell 1}(x)),$$

$$\widetilde{G}_{(\ell_0,\varepsilon_0),(\ell_1,\varepsilon_1)}(x) - G_{(\ell_0,\varepsilon_0),(\ell_1,\varepsilon_1)}(x)$$

$$= \sum_{\ell=1}^{\infty}(\widetilde{G}_{(\ell_0,\varepsilon_0),(\ell,0)}(x)G_{(\ell,0),(\ell_1,\varepsilon_1)}(x) - (\widetilde{G}_{(\ell_0,\varepsilon_0),(\ell,1)}(x)G_{(\ell,1),(\ell_1,\varepsilon_1)}(x))$$

or

$$\widetilde{\varphi}(x) = (E + \widetilde{G}(x)J)\varphi(x), \tag{4.30}$$

$$(E + \widetilde{G}(x)J)(E - G(x)J) = E, \tag{4.31}$$

and as above the series in (4.30) and (4.31) converge "with brackets". Furthermore, according to Lemma 4.5, we have

$$\widetilde{\ell}\widetilde{\varphi}(x) = \widetilde{\Lambda}\widetilde{\varphi}(x), \tag{4.32}$$

$$\widetilde{P}'_{\ell\varepsilon}(x,\lambda) = \widetilde{\Phi}(x,\lambda)\widetilde{g}^*_{\ell\varepsilon}(x), \qquad \widetilde{G}'(x) = \widetilde{\varphi}(x)\widetilde{g}^*(x), \tag{4.33}$$

$$\sum_{\varepsilon=0}^{1}(-1)^{\varepsilon}(P_{\ell\varepsilon}(x,\lambda)(\lambda E - \Lambda_{\ell\varepsilon}) - \langle\widetilde{\Phi}(x,\lambda), \widetilde{g}^*_{\ell\varepsilon}(x)\rangle_{\widetilde{\ell}})\varphi_{\ell\varepsilon}(x) = 0, \tag{4.34}$$

$$(\widetilde{\Lambda}(E + \widetilde{G}(x)J) - (E + \widetilde{G}(x)J)\Lambda)\varphi(x) = \langle\widetilde{\varphi}(x), \widetilde{g}^*(x)\rangle_{\widetilde{\ell}} J\varphi(x). \tag{4.35}$$

Let $\sum_{\ell=1}^{\infty} \xi_\ell \ell^{n-1}$. Denote

$$\begin{aligned} \kappa_{\nu s}(x) &= \widetilde{g}^{*(\nu)}(x)J\varphi^{(s)}(x) \\ &= \sum_{\ell=1}^{\infty}(\widetilde{g}^{*(\nu)}_{\ell 0}(x)\varphi^{(s)}_{\ell 0}(x) - \widetilde{g}^{*(\nu)}_{\ell 1}(x)\varphi^{(s)}_{\ell 1}(x)), \qquad \nu + s \le n-1. \end{aligned} \tag{4.36}$$

The functions $t_{j\nu}(x)$, $\xi_\nu(x)$ and $\varepsilon_\nu(x)$ are defined by (2.20)–(2.22).

Lemma 4.7.

$$p_\nu(x) = \widetilde{p}_\nu(x) + \varepsilon_\nu(x), \qquad \widetilde{u}_{\xi\nu a} = \sum_{j=0}^{n-1} u_{\xi j a} t_{j\nu}(a), \qquad a = 0, T. \tag{4.37}$$

Proof. Differentiating (4.29) with respect to x and using (4.33), (4.36) and (2.20), we get

$$\begin{aligned} &\sum_{\nu=0}^{n} t_{j\nu}(x)\widetilde{\Phi}^{(\nu)}(x,\lambda) \\ &\qquad = \Phi^{(j)}(x,\lambda) + \sum_{\ell=1}^{\infty}(\widetilde{P}_{\ell 0}(x,\lambda)\varphi^{(j)}_{\ell 0}(x) - \widetilde{P}_{\ell 1}(x,\lambda)\varphi^{(j)}_{\ell 1}(x)). \end{aligned} \tag{4.38}$$

Furthermore, in view of (4.29), (4.32) and (4.34), we have

$$\begin{aligned} \widetilde{\ell}\widetilde{\Phi}(x,\lambda) &= \ell\Phi(x,\lambda) + \sum_{\ell=1}^{\infty}(\widetilde{P}_{\ell 0}(x,\lambda)\ell\varphi_{\ell 0}(x) - \widetilde{P}_{\ell 1}(x,\lambda)\ell\varphi_{\ell 1}(x) \\ &\quad + \langle\widetilde{\Phi}(x,\lambda), \widetilde{g}^*(x)\rangle_{\widetilde{\ell}} J\varphi(x). \end{aligned} \tag{4.39}$$

From (4.39), by virtue of (4.38) and (1.22) we obtain $p_\nu(x) = \widetilde{p}_\nu(x) + \varepsilon_\nu(x)$, $\nu = \overline{0, n-2}$, as in the proof of Lemma 2.3.

Denote

$$\widetilde{\widetilde{U}}_{\xi a}(y) = \sum_{\nu=0}^{n-1}\left(\sum_{j=0}^{n-1} u_{\xi ja}t_{j\nu}(a)\right) y^{(\nu)}(a).$$

It follows from (4.38) that

$$\widetilde{\widetilde{U}}_{\xi a}(\widetilde{\Phi}(x,\lambda)) = U_{\xi a}(\Phi(x,\lambda)) + \sum_{\ell=1}^{\infty}(\widetilde{P}_{\ell 0}(a,\lambda)U_{\xi a}(\varphi_{\ell 0}(x)) - \widetilde{P}_{\ell 1}(a,\lambda)U_{\xi a}(\varphi_{\ell 1}(x)))$$

or in the coordinates

$$\widetilde{\widetilde{U}}_{\xi a}(\widetilde{\Phi}_k(x,\lambda)) = U_{\xi a}(\Phi_k(x,\lambda))$$

$$+\sum_{\ell=1}^{\infty}\sum_{\nu=2}^{n}\Big((\sum_{j=1}^{\nu-1}\widetilde{D}_{kj}(a,\lambda,\lambda_{\ell,\nu-1})\mathfrak{N}_{j\nu}(\lambda_{\ell,\nu-1}))U_{\xi a}(\Phi_{\nu,<0>}(x,\lambda_{\ell,\nu-1}))$$

$$-\Big(\sum_{j=1}^{\nu-1}\widetilde{D}_{kj}(a,\lambda,\widetilde{\lambda}_{\ell,\nu-1})\widetilde{\mathfrak{N}}_{j\nu}(\widetilde{\lambda}_{\ell,\nu-1})U_{\xi a}(\Phi_{\nu,<0>}(x,\widetilde{\lambda}_{\ell,\nu-1}))\Big)\Big).$$

For $a=0$, using Corollary 4.2, we compute $\widetilde{\widetilde{U}}_{\xi 0}(\widetilde{\Phi}_k) = \delta_{\xi k}$, $\xi \le k$, and consequently $\widetilde{\widetilde{U}}_{\xi 0} = \widetilde{U}_{\xi 0}$. Analogously we find that $\widetilde{\widetilde{U}}_{\xi T} = \widetilde{U}_{\xi T}$. Lemma 4.7 is proved.

Denote

$$\widetilde{\psi}(x) = W^{-1}(x)J_1\widetilde{\varphi}(x), \qquad \widetilde{H}(x) = W^{-1}(x)J_1\widetilde{G}(x)JJ_1^{-1}W(x),$$

$$\psi(x) = W^{-1}(x)J_1\varphi(x), \qquad H(x) = W^{-1}(x)J_1G(x)JJ_1^{-1}W(x).$$

It is obvious that

$$\left.\begin{aligned} &|\widetilde{\psi}_j^{(\nu)}(x)| < C\ell^{\nu}, \qquad |\widetilde{H}_{j_0,j}(x)| < \frac{C\xi_\ell}{|\ell-\ell_0|+1},\\ &|\widetilde{H}_{j_0,j}^{(\nu+1)}(x)| < C(\ell+\ell_0)^{\nu}\xi_\ell, \qquad j,j_0\in V.\end{aligned}\right\} \tag{4.40}$$

Analogous estimates are valid for $\psi(x)$ and $H(x)$.

Then, (4.30) and (4.31) become

$$\widetilde{\psi}(x) = (E+\widetilde{H}(x))\psi(x), \tag{4.41}$$

$$(E+\widetilde{H}(x))(E-H(x)) = E, \tag{4.42}$$

and the series in (4.41) and (4.42) converge absolutely and uniformly for $x\in[0,T]$. Interchanging places for L and $\widetilde{L}$ we obtain analogously

$$\psi(x) = (E - H(x))\widetilde{\psi}(x), \qquad (E - H(x))(E + \widetilde{H}(x)) = E. \tag{4.43}$$

We consider the Banach space m of bounded sequences $\alpha = [\alpha_j]_{j\in V}$ with the norm $\|\alpha\|_m = \sup_j |\alpha_j|$. It follows from (4.40), (4.42) and (4.43) that for each fixed $x \in [0,T]$ the operator $E + \widetilde{H}(x)$ acting from m to m, is a linear bounded operator,

$$\|\widetilde{H}(x)\|_{m\to m} = \sup_{j_0} \sum_j |\widetilde{H}_{j_0,j}(x)| < C \sum_\ell \xi_\ell, \tag{4.44}$$

and $E + \widetilde{H}(x)$ has a bounded inverse operator.

4.4.

Theorem 4.1. *For a matrix $\mathfrak{M}(\lambda) \in M$ to be the WM for $L \in v'_N$ it is necessary and sufficient that the following conditions hold:*

(1) *(asymptotics) there exists $\widetilde{L} \in v'_N$ such that*

$$\sum_{\ell=1}^\infty \xi_\ell \ell^{n-1} < \infty;$$

(2) *(condition P) for each fixed $x \in [0,T]$ the linear bounded operator $E + \widetilde{H}(x)$ acting from m to m, has a bounded inverse operator:*

(3) *$\varepsilon_\nu(x) \in W_{\nu+N}$, $\nu = \overline{0,n-2}$, where the functions $\varepsilon_\nu(x)$ are found by (4.36), (2.20)–(2.22) and $\varphi(x) = J_1^{-1}W(x)(E+\widetilde{H}(x))^{-1}\widetilde{\psi}(x)$.*

Under these conditions the DE and LF $L = (\ell, U)$ are constructed according to (4.37).

The necessity part of Theorem 4.1 was proved above in 4.1–4.3. We now prove the sufficiency. Let $\psi(x) = [\psi_j(x)]_{j\in V}$ be a solution of (4.41). It is easy to see that the functions $\psi_j^{(\nu)}(x)$, $\nu = \overline{0,n-1}$ are absolutely continuous for $x \in [0,T]$, and $|\psi_j^{(\nu)}(x)| < C\ell^\nu$. We define the functions $\varphi(x) = [\varphi_j(x)]_{j\in V} = [\varphi_v(x)]_{v\in V'}$ according to the formula $\varphi(x) = J_1^{-1}W(x)\psi(x)$. Then

$$\widetilde{\varphi}(x) = (E + \widetilde{G}(x)J)\varphi(x), \tag{4.45}$$

and

$$|\varphi_j^{(\nu)}(x)| < C\ell^\nu w^*_{\ell k}(x), \qquad |\varphi_{\ell 0k}^{(\nu)}(x) - \varphi_{\ell 1k}^{(\nu)}(x)| < C\xi_\ell \ell^\nu w^*_{\ell k}(x).$$

We construct the functions $\Phi(x,\lambda) = [\Phi_k(x,\lambda)]_{k=\overline{1,n}}$ by the formula

$$\Phi(x,\lambda) = \widetilde{\Phi}(x,\lambda) - \sum_{\ell=1}^\infty (\widetilde{P}_{\ell 0}(x,\lambda)\varphi_{\ell 0}(x) - \widetilde{P}_{\ell 1}(x,\lambda)\varphi_{\ell 1}(x)). \tag{4.46}$$

It follows from (4.45) and (4.46) that for fixed $x \in [0,T]$ the functions $\Phi^{(\nu)}(x,\lambda)$, $\nu = \overline{0,n}$ are meromorphic in λ, and $\varphi_j(x) - \Phi_{k\,|\,1,<0>}(x,\lambda_{\ell\varepsilon k})$. We note that, by virtue of Lemma 4.5, (4.32)–(4.35) are valid. We construct the DE and LF $L = (\ell, U)$ by (4.37), where the functions $\varepsilon_\nu(x)$ and $t_{j\nu}(x)$ are defined by (4.36) and (2.20)–(2.22). It is clear that $L \in v_N$.

Lemma 4.8.

$$\ell\varphi(x) = \Lambda\varphi(x), \qquad \ell\Phi(x,\lambda) = \lambda\Phi(x,\lambda).$$

Proof. Differentiating (4.45) and (4.46) with respect to x and using (4.33), (4.36) and (2.20), we get

$$\sum_{\nu=0}^{n} t_{j\nu}(x)\widetilde{\varphi}^{(\nu)}(x) = (E + \widetilde{G}(x)J)\varphi^{(j)}(x), \tag{4.47}$$

$$\begin{aligned}\sum_{\nu=0}^{n} t_{j\nu}(x)\widetilde{\Phi}^{(\nu)}(x,\lambda)& \\ = \Phi^{(j)}(x,\lambda) + \sum_{\ell=1}^{\infty}(\widetilde{P}_{\ell 0}(x,\lambda)\varphi_{\ell 0}^{(j)}(x) - \widetilde{P}_{\ell 1}(x,\lambda)\varphi_{\ell 1}^{(j)}(x)).&\end{aligned} \tag{4.48}$$

It follows from (4.47), in view of (1.22), (4.36), (4.37) and (2.20)–(2.22), that

$$\begin{aligned}&(E + \widetilde{G}(x)J)\ell\varphi(x) + \langle\widetilde{\varphi}(x), \widetilde{g}^*(x)\rangle_{\widetilde{\ell}} J\varphi(x) \\ &= \sum_{\nu=0}^{n}\Big(\widetilde{p}_\nu(x) + \sum_{j=\nu+1}^{n} p_j(x)t_{j\nu}(x) + \sum_{j=0}^{n-1}\widetilde{\mathfrak{L}}_{\nu j}(x)\kappa_{j0}(x)\Big)\widetilde{\varphi}^{(\nu)}(x) = \widetilde{\ell}\widetilde{\varphi}(x).\end{aligned} \tag{4.49}$$

Similarly, we obtain

$$\begin{aligned}\widetilde{\ell}\widetilde{\Phi}(x,\lambda) &= \ell\Phi(x,\lambda) + \sum_{\ell=1}^{\infty}(\widetilde{P}_{\ell 0}(x,\lambda)\ell\varphi_{\ell 0}(x) \\ &\quad - \widetilde{P}_{\ell 1}(x,\lambda)\ell\varphi_{\ell 1}(x)) + \langle\widetilde{\Phi}(x,\lambda), \widetilde{g}^*(x)\rangle_{\widetilde{\ell}} J\varphi(x).\end{aligned} \tag{4.50}$$

From (4.49), by virtue of (4.32) and (4.35), we calculate

$$\widetilde{\Lambda}\widetilde{\varphi}(x) = (E + \widetilde{G}(x)J)\ell\varphi(x) + (\widetilde{\Lambda}(E + \widetilde{G}(x)J) - (E + \widetilde{G}(x)J)\Lambda)\varphi(x)$$

or $(E + \widetilde{G}(x)J)(\ell\varphi(x) - \Lambda\varphi(x)) = 0$. According to Condition P of Theorem 4.1 we get $\ell\varphi(x) = \Lambda\varphi(x)$. Furthermore, it follows from (4.50) and (4.34) that

$$\begin{aligned}\lambda\widetilde{\Phi}(x,\lambda) &= \ell\Phi(x,\lambda) + \sum_{\ell=1}^{\infty}(\widetilde{P}_{\ell 0}(x,\lambda)\Lambda_{\ell 0}\varphi_{\ell 0}(x) - \widetilde{P}_{\ell 1}(x,\lambda)\Lambda_{\ell 1}\varphi_{\ell 1}(x)\\ &\quad + \sum_{\ell=1}^{\infty}(\langle\widetilde{\Phi}(x,\lambda),\widetilde{g}^*_{\ell 0}(x)\rangle_{\widetilde{\ell}}\varphi_{\ell 0}(x) - \langle\widetilde{\Phi}(x,\lambda),\widetilde{g}^*_{\ell 1}(x)\rangle_{\widetilde{\ell}}\varphi_{\ell 1}(x)\\ &= \ell\Phi(x,\lambda) + \lambda\sum_{\ell=1}^{\infty}(\widetilde{P}_{\ell 0}(x,\lambda)\varphi_{\ell 0}(x) - \widetilde{P}_{\ell 1}(x,\lambda)\varphi_{\ell 1}(x)),\end{aligned}$$

and hence $\ell\Phi(x,\lambda) = \lambda\Phi(x,\lambda)$. Lemma 4.8 is proved.

The central role in the proof of the sufficiency of Theorem 4.1 plays the following lemma.

Lemma 4.9.

$$U_{\xi 0}(\Phi_k(x,\lambda)) = \delta_{\xi k}, \qquad \xi = \overline{1,k}; \qquad U_{\xi T}(\Phi_k(x,\lambda)) = 0, \qquad \xi = \overline{1,n-k},$$

i.e. $\Phi(x,\lambda)$ *is the WS for L.*

Proof. Using (4.48) and (4.37) we obtain

$$\widetilde{U}_{\xi a}(\widetilde{\Phi}(x,\lambda)) = U_{\xi a}(\Phi(x,\lambda)) + \sum_{\ell=1}^{\infty}(\widetilde{P}_{\ell 0}(a,\lambda)U_{\xi a}(\varphi_{\ell 0}(x)) - \widetilde{P}_{\ell 1}(a,\lambda)U_{\xi a}(\varphi_{\ell 1}(x))),$$

or in coordinates

$$\begin{aligned}\widetilde{U}_{\xi a}(\widetilde{\Phi}_k(x,\lambda)) &= U_{\xi a}(\Phi_k(x,\lambda))\\ &+\sum_{\ell=1}^{\infty}\sum_{\nu=2}^{n}\Big(\sum_{j=1}^{\nu-1}\widetilde{D}_{kj}(a,\lambda,\lambda_{\ell,\nu-1})\mathfrak{N}_{j\nu}(\lambda_{\ell,\nu-1})U_{\xi a}(\Phi_{\nu,<0>}(x,\lambda_{\ell,\nu-1}))\\ &-\sum_{j=1}^{\nu-1}\widetilde{D}_{kj}(a,\lambda,\widetilde{\lambda}_{\ell,\nu-1})\widetilde{\mathfrak{N}}_{j\nu}(\widetilde{\lambda}_{\ell,\nu-1}))U_{\xi a}(\Phi_{\nu,<0>}(x,\widetilde{\lambda}_{\ell,\nu-1}))\Big).\end{aligned} \tag{4.51}$$

Let $a = 0$. Taking $\xi = n, n-1, \dots, 1$ successively in (4.51), and using Corollary 4.2, we find that $U_{\xi 0}(\Phi_k(x,\lambda)) = \delta_{\xi k}$, $\xi = \overline{1,k}$.

Let $a = T$. Here we shall use the properties S_1 and S_2 and Corollary 4.2. We rewrite (4.51) as follows:

$$\widetilde{U}_{\xi T}(\widetilde{\Phi}_k(x,\lambda)) = U_{\xi T}(\Phi_k(x,\lambda)) + \sum_{\ell=1}^{\infty}\Big(\sum_{j=1}^{k}\widetilde{D}_{kj}(T,\lambda,\lambda_{\ell j})\sum_{\nu=j+1}^{n}\mathfrak{M}_{j\nu}(\lambda_{\ell,\nu-1})U_{\xi T}(\Phi_{\nu,<0>}(x,\lambda_{\ell,\nu-1})) - \sum_{\nu=2}^{k}(\widetilde{D}\widetilde{\mathfrak{M}})_{k\nu}(T,\lambda,\widetilde{\lambda}_{\ell,\nu-1})U_{\xi T}(\Phi_{\nu,<0>}(x,\widetilde{\lambda}_{\ell,\nu-1}))\Big). \tag{4.52}$$

From this, taking $\lambda = \lambda_{\ell_0,k-1}$ and $\lambda = \widetilde{\lambda}_{\ell_0,k-1}$, we have

$$\widetilde{U}_{\xi T}(\widetilde{\Phi}_{k,<0>}(x,\lambda_{\ell_0,k-1})) = U_{\xi T}(\Phi_{k,<0>}(x,\lambda_{\ell_0,k-1})) + \sum_{\ell=1}^{\infty}\Big(\sum_{j=1}^{k}\widetilde{D}_{kj,<0>}(T,\lambda_{\ell_0,k-1},\lambda_{\ell j})\sum_{\nu=j+1}^{n}\mathfrak{M}_{j\nu}(\lambda_{\ell,\nu-1})U_{\xi T}(\Phi_{\nu,<0>}(x,\lambda_{\ell,\nu-1})) - \sum_{\nu=2}^{k}(\widetilde{D}_{<0>}\widetilde{\mathfrak{M}})_{k\nu}(T,\lambda_{\ell_0,k-1},\widetilde{\lambda}_{\ell,\nu-1})U_{\xi T}(\Phi_{\nu,<0>}(x,\widetilde{\lambda}_{\ell,\nu-1}))\Big), \quad k=\overline{2,n}, \tag{4.53}$$

$$\widetilde{U}_{\xi T}(\widetilde{\Phi}_{k,<0>}(x,\widetilde{\lambda}_{\ell_0,k-1})) = U_{\xi T}(\Phi_{k,<0>}(x,\widetilde{\lambda}_{\ell_0,k-1})) + \sum_{\ell=1}^{\infty}\Big(\sum_{j=1}^{k}\widetilde{D}_{kj,<0>}(T,\widetilde{\lambda}_{\ell_0,k-1},\lambda_{\ell j}) \times \sum_{\nu=j+1}^{n}\mathfrak{M}_{j\nu}(\lambda_{\ell,\nu-1})U_{\xi T}(\Phi_{\nu,<0>}(x,\lambda_{\ell,\nu-1})) - \sum_{\nu=2}^{k}(\widetilde{D}_{<0>}\widetilde{\mathfrak{M}})_{k\nu}(T,\widetilde{\lambda}_{\ell_0,k-1},\widetilde{\lambda}_{\ell,\nu-1})U_{\xi T}(\Phi_{\nu,<0>}(x,\widetilde{\lambda}_{\ell,\nu-1}))\Big), \quad k=\overline{2,n}. \tag{4.54}$$

We do not change (4.53), but we replace relations (4.54) by their linear combinations. For this we multiply (4.54) by $\widetilde{\mathfrak{M}}_{\mu k}(\widetilde{\lambda}_{\ell_0,k-1})$ for a fixed $\mu=\overline{1,n-1}$ and sum with respect to k from $\mu+1$ to n. We obtain

$$\sum_{\ell=1}^{\infty}\Bigg(\sum_{j=1}^{\mu}(\tilde{\mathfrak{N}}\tilde{D}_{<0>})_{\mu j}(T,\tilde{\lambda}_{\ell_0,\mu},\lambda_{\ell j})\sum_{\nu=j+1}^{n}\mathfrak{N}_{j\nu}(\lambda_{\ell,\nu-1})U_{\xi T}(\Phi_{\nu,<0>}(x,\lambda_{\ell,\nu-1}))$$

$$-\sum_{\nu=2}^{\mu}(\tilde{\mathfrak{N}}\tilde{D}_{<0>}\tilde{\mathfrak{N}})_{\mu\nu}(T,\tilde{\lambda}_{\ell_0,\mu},\tilde{\lambda}_{\ell,\nu-1})U_{\xi T}(\Phi_{\nu,<0>}(x,\tilde{\lambda}_{\ell,\nu-1}))\Bigg)=0,$$

$$\xi=\overline{1,n-\mu}. \tag{4.55}$$

For a fixed s $(1\le s\le n-1)$ we consider the system consisting of (4.55) for $\mu=\overline{1,s}$ and (4.53) for $k=\overline{2,s}$. Using the Condition P of Theorem 4.1 and Property S_2, we get

$$U_{\xi T}(\Phi_{\nu,<0>}(x,\tilde{\lambda}_{\ell,\nu-1}))=U_{\xi T}(\Phi_{\nu,<0>}(x,\lambda_{\ell,\nu-1}))=0,\qquad \nu=\overline{1,s-1},$$

$$\sum_{\nu=s+1}^{n}\mathfrak{N}_{j\nu}(\lambda_{\ell,\nu-1})U_{\xi T}(\Phi_{\nu,<0>}(x,\lambda_{\ell,\nu-1}))=0,\qquad j=\overline{1,s},\qquad \xi=\overline{1,n-s}.$$

Substituting these relations into (4.32) we obtain $U_{\xi T}(\Phi_k(x,\lambda))=0$, $\xi=\overline{1,n-k}$. Lemma 4.9 is proved.

Lemma 4.10. *The matrix $\mathfrak{M}(\lambda)$ is the WM for L.*

Proof. Let $\mathfrak{M}^0_{k\xi}(\lambda)=U_{\xi 0}(\Phi_k)$, and $\mathfrak{M}^0(\lambda)=[\mathfrak{M}^0_{k\xi}(\lambda)]_{k,\xi=\overline{1,n}}$ be the WM for L. Denote $\mathfrak{M}^{0,*}(\lambda)=(\mathfrak{M}^0(\lambda))^{-1}$. Let us show by induction that $\mathfrak{M}_{k,k+\gamma}(\lambda)=\mathfrak{M}^0_{k,k+\gamma}(\lambda)$, $\gamma\ge 1$.

Since $\mathfrak{N}(\lambda_0)=\mathfrak{M}_{<-1>}(\lambda_0)(\mathfrak{M}_{<0>}(\lambda_0))^{-1}$, it follows that

$$\mathfrak{N}_{j\nu}(\lambda_0)=\mathfrak{M}_{j\nu,<-1>}(\lambda_0)+\mathcal{F}_{j\nu}(\mathfrak{M}_{r,r+s},\ s=\overline{1,\nu-j-1},\ r\ge 1). \tag{4.56}$$

From (4.51) for $a=0$, using Lemma 4.9, Corollary 4.2 and (4.56), we get

$$\begin{aligned}\tilde{\mathfrak{M}}_{k,k+\gamma}(\lambda) &= \mathfrak{M}^0_{k,k+\gamma}(\lambda)+\sum_{\ell=1}^{\infty}\left(\frac{\tilde{\beta}_{\ell,k,k+\gamma}}{\lambda-\tilde{\lambda}_{\ell k}}-\frac{\beta_{\ell,k,k+\gamma}}{\lambda-\tilde{\lambda}_{\ell k}}\right)\\ &\quad+\mathcal{F}^0_{k\gamma}(\tilde{\mathfrak{M}}_{r,r+j},\ \mathfrak{M}_{r,r+j},\ j=\overline{1,\gamma-1},\ r\ge 1).\end{aligned} \tag{4.57}$$

In particular, for $\gamma=1$ we have

$$\mathfrak{M}^0_{k,k+1}(\lambda)=\sum_{\ell=1}^{\infty}\frac{\beta_{\ell,k,k+\gamma}}{\lambda-\lambda_{\ell k}}=\mathfrak{M}_{k,k+1}(\lambda).$$

Further, (4.46) can be rewritten to the form

$$\Phi(x,\lambda)=\tilde{\Phi}(x,\lambda)-\sum_{\lambda_0\in I}\tilde{P}(x,\lambda,\lambda_0)\varphi(x,\lambda_0),$$

where $\varphi(x,\lambda_0) = Y\Phi_{<0>}(x,\lambda_0)$, $I = \Lambda \cup \tilde{\Lambda}$. Then

$$\Phi_{<-1>}(x,z_0) = \tilde{\Phi}_{<-1>}(x,z_0) - \sum_{\lambda_0 \in I} \tilde{P}_{<-1>}(x,z_0,\lambda_0)\varphi(x,\lambda_0)$$

By virtue of (4.11) we obtain

$$\tilde{P}_{<-1>}(x,z_0,\lambda_0) = \tilde{\mathfrak{N}}(z_0)\tilde{P}_{<0>}(x,z_0,\lambda_0) - \delta(z_0,\lambda_0)\hat{\mathfrak{N}}(\lambda_0)Y^T,$$

and hence

$$\begin{aligned}\Phi_{<-1>}(x,z_0) &= \tilde{\mathfrak{N}}(z_0)\tilde{\Phi}_{<0>}(x,z_0) - \sum_{\lambda_0 \in I} \tilde{\mathfrak{N}}(z_0)\tilde{P}_{<0>}(x,z_0,\lambda_0) + \hat{\mathfrak{N}}(z_0)Y^T\varphi(x,z_0)\\ &= \tilde{\mathfrak{N}}(z_0)\Phi_{<0>}(x,z_0) + \hat{\mathfrak{N}}(z_0)\Phi_{<0>}(x,z_0) = \mathfrak{N}(z_0)\Phi_{<0>}(x,z_0).\end{aligned}$$

This yields

$$\mathfrak{N}(\lambda_0) = \mathfrak{M}^0_{<-1>}(\lambda_0)(\mathfrak{M}^0_{<0>}(\lambda_0))^{-1} \stackrel{df}{=} \mathfrak{N}^0(\lambda_0).$$

Then, according to Lemma 4.3, the matrix $\mathfrak{M}^{0,*}(\lambda)$ has only simple poles. Furthermore, we obtain $\lambda_{\ell m} = \lambda^0_{\ell m}$, where $\{\lambda^0_{\ell m}\}_{\ell \geq 1}$ are the eigenvalues of the boundary value problems S_m for L.

It follows from the results obtained in 4.3, that

$$\begin{aligned}\tilde{\mathfrak{M}}_{k,k+\gamma}(\lambda) &= \mathfrak{M}^0_{k,k+\gamma}(\lambda) + \sum_{\ell=1}^{\infty} \left(\frac{\tilde{\beta}_{\ell,k,k+\gamma}}{\lambda - \tilde{\lambda}_{\ell k}} - \frac{\beta^0_{\ell,k,k+\gamma}}{\lambda - \lambda_{\ell k}} \right)\\ &\quad + \mathcal{F}^0_{k\gamma}(\tilde{\mathfrak{M}}_{r,r+j},\ \mathfrak{M}^0_{r,r+j},\ j = \overline{1,\gamma-1},\ r \geq 1),\end{aligned} \tag{4.58}$$

where $\beta^0_{\ell k\xi} = \operatorname{res}_{\lambda_{\ell k}} \mathfrak{M}^0_{k\xi}(\lambda)$. If $\mathfrak{M}_{r,r+j}(\lambda) = \mathfrak{M}^0_{r,r+j}(\lambda)$, $j = \overline{1,\gamma-1}$, $r \geq 1$, then, comparing (4.57) and (4.58), we obtain $\mathfrak{M}_{k,k+\gamma}(\lambda) = \mathfrak{M}^0_{k,k+\gamma}(\lambda)$. Lemma 4.10 is proved.

Also, we have proved that $\mathfrak{M}(\lambda)$ and $\mathfrak{M}^*(\lambda)$ have only simple poles, i.e. $L \in v'_N$. Thus, Theorem 4.1 is completely proved.

4.5. The method described above also allows us to study stability of the solution of the IP from the WM. Let $\tilde{L} \in v'_N$ and choose $L \in v'_N$ such that

$$\Lambda^0 = \sum_{\ell=1}^{\infty} \xi_\ell \ell^{n-1} < \infty.$$

The quantity Λ^0 will describe the nearness of the WM $\mathfrak{M}(\lambda)$ and $\tilde{\mathfrak{M}}(\lambda)$.

Theorem 4.2. *There exists $\delta > 0$ (which depends on $\widetilde{L}$) such that if $\Lambda^0 < \delta$, then*

$$\max_{0\le x\le T} |p_\nu^{(j)}(x) - \widetilde{p}_\nu^{(j)}(x)| < C\Lambda^0, \quad 0 \le j \le \nu \le n-2; \quad |u_{\xi\nu a} - \widetilde{u}_{\xi\nu a}| < C\Lambda^0.$$

Here and in the sequel in 4.5, the symbol C will denote various positive constants in estimates which only depend on $\widetilde{L}$.

First we prove an auxiliary statement.

Lemma 4.11. *There exists $\delta > 0$ (which depends on $\widetilde{L}$) such that if*

$$\Lambda^1 \stackrel{df}{=} \sum_{\ell=1}^{\infty} \xi_\ell \ell^{n-2} < \delta,$$

then

$$|\psi_j^{(\nu)}(x)| < C\ell^\nu, \quad |\psi_j^{(\nu)}(x) - \widetilde{\psi}_j^{(\nu)}(x)| < C\ell^{\nu-1}\Lambda^1, \quad j \in V, \quad \nu = \overline{0, n-1}. \tag{4.59}$$

We will prove this lemma by induction and use (4.40). Let us write (4.41) in the coordinates

$$\widetilde{\psi}_{j_0}(x) = \psi_{j_0}(x) + \sum_{j\in V} \widetilde{H}_{j_0,j}(x)\psi_j(x), \qquad j_0 \in V. \tag{4.60}$$

For a fixed $x \in [0,T]$ (4.60) is an equation in the Banach space m. According to (4.44), we can choose $\delta > 0$ such that for $\Lambda^1 < \delta$

$$\|\widetilde{H}(x)\|_{m\to m} = \sup_{j_0\in V} \sum_{j\in V} |\widetilde{H}_{j_0,j}(x)| < \frac{1}{2}.$$

Then from (4.60) we have $|\psi_j(x)| < C$, $j \in V$.

From this and from (4.60) we obtain

$$\begin{aligned} |\widetilde{\psi}_{j_0}(x) - \psi_{j_0}(x)| &< \sum_{j\in V} |\widetilde{H}_{j_0,j}(x)\psi_j(x)| < C\sum_{\ell=1}^{\infty} \frac{\xi_\ell}{|\ell-\ell_0|+1} \\ &< \frac{C}{\ell_0}\sum_{\ell=1}^{\infty} \xi_\ell \ell < \frac{C}{\ell_0}\Lambda^1. \end{aligned}$$

Thus, (4.59) is proved for $\nu = 0$.

Suppose that (4.59) is proved for $\nu = \overline{0, s-1}$. Differentiating (4.60), we get

$$\widetilde{\psi}_{j_0}^{(s)}(x) = \psi_{j_0}^{(s)}(x) + \sum_{j\in V}\sum_{\mu=0}^{s} C_s^\mu \widetilde{H}_{j_0,j}^{(\mu)}(x)\psi^{(s-\mu)}(x), \qquad j_0 \in V, \tag{4.61}$$

or

$$\widetilde{\xi}_{j_0}(x) = \xi_{j_0}(x) + \sum_{j\in V} \widetilde{h}_{j_0,j}(x)\xi_j(x), \qquad j \in V, \tag{4.62}$$

where

$$\widetilde{h}_{j_0,j}(x) = \widetilde{H}_{j_0,j}(x)\frac{\ell^s}{\ell_0^s}, \qquad \xi_j(x) = \frac{1}{\ell^s}\psi_j^{(s)}(x),$$

$$\widetilde{\xi}_{j_0}(x) = \frac{1}{\ell_0^s}\left(\widetilde{\psi}_{j_0}^{(s)}(x) - \sum_{\mu=1}^{s} C_s^\mu \sum_{j\in V} \widetilde{H}_{j_0,j}^{(\mu)}(x)\psi_j^{(s-\mu)}(x)\right).$$

Using (4.59) for $\nu = \overline{0, s-1}$ and (4.40), we obtain

$$|\widetilde{\xi}_{j_0}(x)| < C, \quad \sup_{j_0\in V}\sum_{j\in V}|\widetilde{h}_{j_0,j}(x)| < C\sum_{\ell=1}^{\infty}\frac{\xi_\ell}{|\ell-\ell_0|+1}\frac{\ell^s}{\ell_0^s} < C\Lambda^1.$$

Choose $\delta > 0$ such that for $\Lambda^1 < \delta$

$$||\widetilde{h}(x)||_{m\to m} = \sup_{j_0\in V}\sum_{j\in V}|\widetilde{h}_{j_0,j}(x)| < \frac{1}{2}.$$

Then (4.62) yields $|\xi_j(x)| < C$, $j \in V$ or $|\psi_j^{(s)}(x)| < C\ell^s$, $j \in V$. It follows from (4.61) and (4.40) that

$$|\widetilde{\psi}_{j_0}^{(s)}(x) - \psi_{j_0}^{(s)}(x)| < C\sum_{\mu=0}^{s}\sum_{j\in V}|\widetilde{H}_{j_0,j}^{(\mu)}(x)\psi_j^{(s-\mu)}(x)| < C\ell_0^{s-1}\Lambda^1.$$

Lemma 4.11 is proved.

Proof of Theorem 4.2. Choose $\delta > 0$ as in Lemma 4.11. Let $\Lambda^0 < \delta$. Then, by Lemma 4.11, we get (4.59). Since $J_1\varphi(x) = W(x)\psi(x)$, i.e.

$$\varphi_{\ell 0k}(x) - \varphi_{\ell 1k}(x) = \xi_\ell w_{\ell k}^*(x)\psi_{\ell 0k}(x), \qquad \varphi_{\ell 1k}(x) = w_{\ell k}^*(x)\psi_{\ell 1k}(x),$$

it follows from (4.59) that

$$\left.\begin{aligned} &|\varphi_j^{(\nu)}(x)| < C\ell^\nu w_{\ell k}^*(x), \\ &|\varphi_{\ell 0k}^{(\nu)}(x) - \varphi_{\ell 1k}^{(\nu)}(x)| < C\xi_\ell\ell^\nu w_{\ell k}^*(x), \qquad \nu = \overline{0, n-1}. \end{aligned}\right\} \tag{4.63}$$

Using (4.36), (4.63) and the estimates for the functions $\widetilde{g}_j^{*(\nu)}(x)$, $\widetilde{g}_{\ell 0k}^{*(\nu)}(x) - \widetilde{g}_{\ell 1k}^{*(\nu)}(x)$ obtained in 4.3, we find

$$|\kappa_{\nu s}^{(\mu)}(x)| < C\sum_{\ell=1}^{\infty}\xi_\ell\ell^{\nu+s+\mu}, \qquad \nu+s+\mu \le n-1.$$

Hence, in view of (2.20) and (2.21), we have

$$|t_{j\nu}^{(\mu)}(x)| < C\sum_{\ell=1}^{\infty}\xi_\ell\ell^{j-\nu-1+\mu}, \qquad |\xi_\nu^{(\mu)}(x)| < C\sum_{\ell=1}^{\infty}\xi_\ell\ell^{n-\nu-1+\mu}. \tag{4.64}$$

It follows from (4.37) and (2.22) that

$$\widehat{p}_\nu(x) = \xi_\nu(x) - \sum_{j=\nu+1}^{n-2} \widehat{p}_j(x) t_{j\nu}(x), \qquad \nu = \overline{0, n-2}.$$

From this, using (4.64), we calculate for $\nu = n-2, n-3, \ldots, 0$ successively: $|\widehat{p}_\nu^{(j)}(x)| < C\Lambda^0$, $0 \le j \le \nu$. Similarly, we obtain $|\widehat{u}_{\xi\nu a}| < C\Lambda^0$. Theorem 4.2 is proved.

4.6. Sometimes it is more convenient to work in $\mathcal{L}_2(0,T)$. We will say that $L \in v'_{N2}$ if $L \in v'_N$ and $p_\nu^{(\nu+N)}(x) \in \mathcal{L}_2(0,T)$. We note that for $L \in v'_{N2}$ the asymptotic formulae (4.1) and (4.2) are valid, where $\kappa_{\ell m}$, $\kappa_{\ell mk}$, $\kappa^0_{\ell mk} \in \ell_2$. Similarly to Theorem 4.1, we prove the following theorem.

Theorem 4.3. *For a matrix $\mathfrak{M}(\lambda) \in M$ to be the WM for $L \in v'_{N2}$ it is necessary and sufficient that the following conditions hold:*

(1) *(asymptotics) there exists $\widetilde{L} \in v'_{N2}$ such that*

$$\sum_{\ell=1}^{\infty} (\xi_\ell \ell^{n+N-1})^2 < \infty;$$

(2) *Condition P is fulfilled.*

We note that for "small" perturbations, Condition P is fulfilled automatically, i.e. the following theorem holds.

Theorem 4.4. *Let $\widetilde{L} \in v'_{N2}$ be given. Then there exists $\delta > 0$ (which depends on $\widetilde{L}$) such that if the matrix $\mathfrak{M}(\lambda) \in M$ satisfies the condition*

$$\Lambda^+ \stackrel{df}{=} \left(\sum_{\ell=1}^{\infty} (\xi_\ell \ell^{n+N-1})^2 \right)^{1/2} < \delta,$$

then there exists a unique $L \in v'_{N2}$ for which the matrix $\mathfrak{M}(\lambda)$ is the WM. At that

$$\begin{aligned} &\|p_\nu^{(j)}(x) - \widetilde{p}_j^{(\nu)}(x)\|_{\mathcal{L}_2(0,T)} < C\Lambda^+, \qquad 0 \le j \le \nu + N, \\ &|u_{\xi\nu a} - \widetilde{u}_{\xi\nu a}| < C\Lambda^+, \end{aligned} \tag{4.65}$$

where the constant C depends only on $\widetilde{L}$.

Indeed, by virtue of (4.44), we can choose $\delta > 0$ such that for $\Lambda^+ < \delta$, $\|\widetilde{H}(x)\|_{m \to m} < 1$, $x \in [0,T]$, and consequently Condition P is fulfilled. The proof of (4.65) is similar to the proof of Theorem 4.2.

4.7. Counterexample. For definiteness, let $n = 3$. Let us show that dropping $\mathfrak{M}_{13}(\lambda)$ from the WM $\mathfrak{M}(\lambda)$ leads to non-uniqueness of the solution of the IP. In other words, the WF $\mathfrak{M}_{12}(\lambda)$ and $\mathfrak{M}_{23}(\lambda)$ do not determine the DE and LF L uniquely.

We consider $\widetilde{L} = (\widetilde{\ell}, \widetilde{U})$ of the form

$$\widetilde{\ell}y = y''', \qquad \widetilde{U}_{1a}(y) = y''(a) + \widetilde{\alpha}_a y'(a),$$

$$\widetilde{U}_{2a}(y) = y'(a), \qquad \widetilde{U}_{3a}(y) = y(a), \qquad a = 0, T.$$

Let the functions $\widetilde{X}_k(x,\lambda)$ be solutions of the equation $y''' = \lambda y = \rho^3 y$ under the conditions $\widetilde{X}_k^{(\nu-1)}(0,\lambda) = \delta_{\nu k}$, $\nu, k = 1, 3$. Then

$$\widetilde{X}_k(x,\lambda) = \frac{1}{3}\sum_{j=1}^{3}(\rho R_j)^{1-k}\exp(\rho R_j x). \tag{4.66}$$

In particular, for $\lambda = 0$

$$\widetilde{X}_k(x,0) = \frac{x^{k-1}}{(k-1)!}.$$

It is clear that for $\lambda = 0$

$$\widetilde{\Delta}_{11}(0) = \widetilde{\Delta}_{22}(0) = \widetilde{\Delta}_{12}(0) = 0 \tag{4.67}$$

for any $\widetilde{\alpha}_0$ and $\widetilde{\alpha}_T$. We choose the coefficients $\widetilde{\alpha}_0$ and $\widetilde{\alpha}_T$ such that the functions $\widetilde{\Delta}_{11}(\lambda)$ and $\widetilde{\Delta}_{22}(\lambda)$ have only simple zeroes. Let us show that such a choice is possible. By symmetry, it is sufficient to consider the function $\widetilde{\Delta}_{22}(\lambda) = \widetilde{X}_1''(T,\lambda) + \widetilde{\alpha}_T \widetilde{X}_1'(T,\lambda)$. Using (4.66) we obtain

$$\left.\begin{aligned} \widetilde{\Delta}_{22}(\lambda) &= \lambda\widetilde{X}_2(T,\lambda) + \widetilde{\alpha}_T\lambda\widetilde{X}_3(T,\lambda), \\ 3\dot{\widetilde{\Delta}}_{22}(\lambda) &= (2\widetilde{X}_2(T,\lambda) + T\widetilde{X}_1(T,\lambda)) \\ &\qquad + \widetilde{\alpha}_T(\widetilde{X}_3(T,\lambda) + T\widetilde{X}_2(T,\lambda)). \end{aligned}\right\} \tag{4.68}$$

Denote by B the set of zeros of the function

$$\widetilde{\Delta}(\lambda) \stackrel{df}{=} \lambda\widetilde{X}_2(T,\lambda)(\widetilde{X}_3(T,\lambda) + T\widetilde{X}_2(T,\lambda)) - \lambda\widetilde{X}_3(T,\lambda)(2\widetilde{X}_2(T,\lambda) + T\widetilde{X}_1(T,\lambda)),$$

and by $B(\widetilde{\alpha}_T) = \{\lambda_0 : \widetilde{\Delta}_{22}(\lambda_0) = \dot{\widetilde{\Delta}}_{22}(\lambda_0) = 0\}$ the set of non-simple zeros of $\widetilde{\Delta}_{22}(\lambda)$. It is obvious that $B(\widetilde{\alpha}_T)$ is a finite set. If $\lambda_0 \in B(\widetilde{\alpha}_T)$, then, by virtue of (4.68), $\widetilde{\Delta}(\lambda_0) = 0$, i.e. $B(\widetilde{\alpha}_T) \subset B$. Further, if $\widetilde{\alpha}_T^0 \neq \widetilde{\alpha}_T$, and $\lambda_0 \in B(\widetilde{\alpha}_T) \cap B(\widetilde{\alpha}_T^0)$, then (4.68) implies that

$$\begin{aligned} \lambda_0\widetilde{X}_2(T,\lambda_0) &= \lambda_0\widetilde{X}_3(T,\lambda_0) = 2\widetilde{X}_2(T,\lambda_0) + T\widetilde{X}_1(T,\lambda_0) \\ &= \widetilde{X}_3(T,\lambda_0) + T\widetilde{X}_2(T,\lambda_0) = 0. \end{aligned}$$

Since $2\widetilde{X}_2(T,0)+TX_1(T,0) = 3T \neq 0$, it follows that $\lambda_0 \neq 0$, and hence $\widetilde{X}_1(T,\lambda_0) = \widetilde{X}_2(T,\lambda_0) = \widetilde{X}_3(T,\lambda_0) = 0$. But this is impossible. Thus, if $\widetilde{\alpha}_T^0 \neq \widetilde{\alpha}_T$, then $B(\widetilde{\alpha}_T^0) \cap B(\widetilde{\alpha}_T) = \oslash$. From this and from the relation $B(\widetilde{\alpha}_T) \subset B$ and the continuity of $B(\widetilde{\alpha}_T)$ we conclude that there exists $\widetilde{\alpha}_T$ such that $B(\widetilde{\alpha}_T) = \oslash$.

We define the matrix $\mathfrak{M}(\lambda) = [\mathfrak{M}_{mk}(\lambda)]_{m,k=\overline{1,3}}$, $\mathfrak{M}_{mk}(\lambda) = \delta_{mk}$, $m \geq k$ by

$$\mathfrak{M}_{12}(\lambda) = \widetilde{\mathfrak{M}}_{12}(\lambda), \qquad \mathfrak{M}_{23}(\lambda) = \widetilde{\mathfrak{M}}_{23}(\lambda), \qquad \mathfrak{M}_{13}(\lambda) = \widetilde{\mathfrak{M}}_{13}(\lambda) + \frac{\theta}{\lambda}, \tag{4.69}$$

where θ is a complex number. It follows from (4.67) and (4.69) that for sufficiently small θ $\mathfrak{M}(\lambda) \in M$ and satisfies the conditions of Theorem 4.4. Then, according to Theorem 4.4, there exists $L \in v'_{N2}$ for which $\mathfrak{M}(\lambda)$ is the WM.

5. Inverse problems for the self-adjoint case

We consider the DE and LF $L = (\ell, U)$ of the form (1.1)–(1.2) on the half-line ($T = \infty$). In §1–2 there is a solution of the IP for the general non-self-adjoint case. The central role was played by the main equations of the IP there. One of the conditions under which an arbitrary matrix $\mathfrak{M}(\lambda)$ is the WM for a non-self-adjoint differential operator is the requirement that the main equation must have a unique solution. It is difficult to verify this condition in the general case. In connection with this, an important problem is that of obtaining sufficient conditions for the solvability of the main equation, and the extraction of classes of operators for which unique solvability can be proved. One of such classes is the class of self-adjoint operators. In this item we investigate the IP for the self-adjoint case. We prove unique solvability of the main equation, and obtain necessary and sufficient conditions, along with a procedure for the construction of an operator from its WM. Some difference in the notations is pointed below.

5.1. For definiteness, let $n = 2m$ and $\sigma_\xi = n - \xi$. Denote

$$\langle y(x), z(x)\rangle_\ell = \sum_{\nu+j\leq n-1} \sum_{s=j}^{n-\nu-1} (-1)^s C_s^j p_{s+\nu+1}^{(s-j)}(x) y^{(\nu)}(x)\overline{z^{(j)}(x)}. \tag{5.1}$$

We assume that $L = L^*$, where the adjoint pair $L^* = (\ell^*, U^*)$ is defined by the relations

$$\ell^* z = z^{(n)} + \sum_{\nu=0}^{n-2} (-1)^\nu (\overline{p_\nu(x)} z)^{(\nu)},$$

$$\langle y(x), z(x)\rangle_{\ell|_{x=0}} = \sum_{\xi=1}^{n} (-1)^{\xi-1} U_{\xi 0}(y)\overline{U^*_{n-\xi+1,0}(z)}. \tag{5.2}$$

For any sufficiently smooth functions $y(x)$ and $z(x)$ we have $\ell y\bar{z} - y\overline{\ell z} = \frac{d}{dx}\langle y, z\rangle_\ell$. In particular, if the functions $y(x,\lambda)$ and $z(x,\mu)$ are solutions of the DE $\ell y = \lambda y$ and $\ell z = \mu z$, then

$$\frac{d}{dx}\langle y(x,\lambda), z(x,\bar{\mu})\rangle_\ell = (\lambda-\mu)y(x,\lambda)\overline{z(x,\bar{\mu})}. \tag{5.3}$$

It was proved in §1 that the WF's $\mathfrak{M}_{k\xi}(\lambda)$ are regular in $\Pi_{(-1)^k}$ except for no more than countable bounded sets of poles $\Lambda'_{k\xi}$ and are continuous in $\bar{\Pi}_{(-1)^k}\backslash\{0\}$ except for the bounded sets $\Lambda_{k\xi}$. When $|\lambda|\to\infty$ $\mathfrak{M}_{k\xi}(\lambda)\rho^{\xi-k}=O(1)$. Denote $\Lambda=\bigcup_{k,\xi}\Lambda_{k\xi}$,

$$\mathfrak{M}^{\pm}_{k\xi} = \lim \mathfrak{M}_{k\xi}(\lambda\pm iz), \qquad z\to 0, \qquad \operatorname{Re} z>0, \qquad -\infty<\lambda<\infty;$$

$$Q_{k\xi}(\lambda) = (2\pi i)^{-1}(\mathfrak{M}^{-}_{k\xi}(\lambda)-\mathfrak{M}^{+}_{k\xi}(\lambda)),$$

$$Q_k(\lambda) = Q_{k,k+1}(\lambda).$$

In order to simplify the computations, we confine ourselves to the case in which there is no discrete spectrum. We shall say that $L\in V^0_N$ if $p_\nu(x)\in W_{\nu+N}$, $\Lambda=\oslash$, $\mathfrak{M}_{k\xi}(\lambda)\rho^{\xi-k}=O(1)$, $\lambda\to 0$; $\widehat{Q}_m(\lambda)=O(\rho^{-n-2})$, $|\lambda|\to\infty$, and $L\in V^1_N$ if $L\in V^0_N$, $L=L^*$. We seek the solution of the IP in the classes V^1_N. Along with L, we consider a DE and LF $\widetilde{L}$ of the same form, but with zero coefficients. We note that $\widetilde{L}\in V^1_N$ for any $N\geq 0$. The condition $\widehat{Q}_k(\lambda)=O(\rho^{-\nu-2})$ is not a constraint: it is introduced to simplify the computations.

Theorem 5.1. *Let $L\in V^1_N$. Then the WM $\mathfrak{M}(\lambda)$ has the following properties:*

(1) $\mathfrak{M}_{k\xi}(\lambda)=\delta_{k\xi}$, $k\geq\xi$;

(2) $\mathfrak{M}_{k\xi}(\lambda)$ *is regular in* $\Pi_{(-1)^k}$ *and continuous in* $\bar{\Pi}_{(-1)^k}\backslash\{0\}$;

(3) $\mathfrak{M}_{k\xi}(\lambda)\rho^{\xi-k}$ *is bounded;*

(4) $\mathfrak{M}_{k\xi}(\lambda)-\mathfrak{M}_{k,k+1}(\lambda)\mathfrak{M}_{k+1,\xi}(\lambda)$ *is regular for* $\lambda\in\Gamma_{(-1)^k}$;

(5) $\widehat{Q}_k(\lambda)=O(\rho^{-n-2})$, $|\lambda|\to\infty$;

(6) $\mathfrak{M}_{k,k+1}(\lambda)=\overline{\mathfrak{M}_{n-k,n-k+1}(\bar{\lambda})}$;

(7) $(-1)^m Q_m(\lambda)>0$, $\lambda\in\Gamma_{(-1)^m}$.

We note that $Q_k(\lambda)\equiv 0$ for $\lambda\in\Gamma_{(-1)^{k-1}}$, and the functions $\rho Q_k(\lambda)$ are continuous and bounded for $\lambda\in\Gamma_{(-1)^k}$.

Properties 1–5 are partially obvious, and partially proved in §1–2. They are not associated with the property of being self-adjoint for the operator. We shall prove Properties 6 and 7.

In view of (5.3), we have $\frac{d}{dx}\langle\Phi_k(x,\lambda),\Phi_{n-k}(x,\bar{\lambda})\rangle_\ell=0$. Since $R_{n-k}=-R_{k+1}$, and

$$|\Phi^{(\nu)}_k(x,\lambda)|\leq C|\rho|^{\nu+k-n}\,|\exp(\rho R_k x)|, \tag{5.4}$$

it follows that

$$\lim_{x\to\infty}\langle\Phi_k(x,\lambda),\Phi_{n-k}(x,\bar\lambda)\rangle_\ell=0.$$

Hence

$$\langle\Phi_k(x,\lambda),\Phi_{n-k}(x,\bar\lambda)\rangle_\ell\big|_{x=0}=0.$$

Using (5.2), we obtain

$$\sum_{\xi=1}^{n}(-1)^{\xi-1}U_{\xi 0}(\Phi_k(x,\lambda))\overline{U_{n-\xi+1,0}(\Phi_{n-k}(x,\bar\lambda))}=0,$$

i.e. $\mathfrak{M}_{k,k+1}(\lambda)=\overline{\mathfrak{M}_{n-k,n-k+1}(\bar\lambda)}$. It follows, in particular, that the function $Q_m(\lambda)$ is real.

Let $f(x)\in W_2$ be finite, and $f(0)=0$. Consider the function

$$Y(x,\lambda)=\int_0^\infty G(x,t,\lambda)f(t)\,dt,$$

where

$$G(x,t,\lambda)=\sum_{j=1}^{m}(-1)^{j-1}\begin{cases}\Phi_{n-j+1}(x,\lambda)\overline{\Phi_j(t,\bar\lambda)}, & x<t,\\ \Phi_j(x,\lambda)\overline{\Phi_{n-j+1}(t,\bar\lambda)}, & x>t.\end{cases}$$

We transform $Y(x,\lambda)$ to the form

$$\begin{aligned}Y(x,\lambda) &= \frac{1}{\lambda}\sum_{j=1}^{m}(-1)^{j-1}\Big(\int_0^x \Phi_j(x,\lambda)\overline{\ell\Phi_{n-j+1}(t,\bar\lambda)}f(t)\,dt\\ &\quad+\int_x^\infty \Phi_{n-j+1}(x,\lambda)\overline{\ell\Phi_j(t,\bar\lambda)}f(t)\,dt\Big).\end{aligned}$$

Integrating the terms with highest derivatives twice by parts and using (5.4) and the relation

$$\sum_{j=1}^{n}(-1)^{j-1}\Phi_j^{(\nu)}(x,\lambda)\overline{\Phi_{n-j+1}(x,\bar\lambda)}=\delta_{\nu,n-1},$$

we compute $Y(x,\lambda)=\lambda^{-1}(-f(x)+Z(x,\lambda))$, where $Z(x,\lambda)=O(\rho^{-1})$, $|\rho|\to\infty$ uniformly with respect to $x\geq 0$. Consequently,

$$\lim_{R\to\infty}\left|\frac{1}{2\pi i}\oint_{|\lambda|=R} Y(x,\lambda)\,d\lambda + f(x)\right| = 0. \tag{5.5}$$

Furthermore, we have

$$\Phi_k(x,\lambda) = C_k(x,\lambda) + \sum_{\xi=k+1}^{n} \mathfrak{M}_{k\xi}(\lambda)C_\xi(x,\lambda).$$

Since the functions $\mathfrak{M}_{k\xi}(\lambda)\rho^{\xi-k}$ are bounded, we get $G(x,t,\lambda) = O(\rho^{1-n})$ uniformly on compacts. Hence

$$\lim_{\varepsilon\to 0}\frac{1}{2\pi i}\int_{|\lambda|=\varepsilon} Y(x,\lambda)\,d\lambda = 0. \tag{5.6}$$

The functions $\Phi_k(x,\lambda) - \mathfrak{M}_{k,k+1}(\lambda)\Phi_{k+1}(x,\lambda)$ are regular for $\lambda \in \Gamma_{(-1)^k}$. Then from (5.5) and (5.6), and applying the contour integral method we obtain

$$\begin{gathered} f(x) = \int_{\Gamma_{(-1)^m}} F(\lambda)\Phi_{m+1}(x,\lambda)(-1)^m Q_m(\lambda)\,d\lambda, \\ F(\lambda) = \int_0^\infty f(x)\overline{\Phi_{m+1}(x,\lambda)}\,dx, \end{gathered} \tag{5.7}$$

uniformly on compacts. Hence

$$\int_0^\infty |f(x)|^2\,dx = \int_{\Gamma_{(-1)^m}} |F(\lambda)|^2(-1)^m Q_m(\lambda)\,d\lambda. \tag{5.8}$$

Known methods of the spectral theory of differential operators can be used to show that (5.7) and (5.8) are valid for all functions

$$f(x) \in \mathcal{L}_2(0,\infty) \qquad \text{and} \qquad (-1)^m Q_m(\lambda) > 0, \qquad \lambda \in \Gamma_{(-1)^m}.$$

Note that the relation $(-1)^m Q_m(\lambda) > 0$ can also be derived directly from the properties of the WM. Theorem 5.1 is proved.

Theorem 5.2. *Let $L \in V_N^1$. The WM $\mathfrak{M}(\lambda)$ is determined uniquely by the functions $Q_1(\lambda), \dots, Q_m(\lambda)$ according the formulas*

$$Q_{n-k}(\lambda) = \overline{Q_k(\lambda)}, \qquad k = \overline{1, m-1}, \tag{5.9}$$

$$\mathfrak{M}_{k\xi}(\lambda) = \int\limits_{\Gamma_{(-1)^k}} \frac{Q_k(\mu)\mathfrak{M}_{k+1,\xi}(\mu)}{\lambda - \mu}\, d\mu, \qquad k < \xi, \qquad \lambda \in \Pi_{(-1)^k}. \tag{5.10}$$

Indeed, the functions $\mathfrak{M}_{k\xi}(\lambda)$ are regular in $\Pi_{(-1)^k}$ and continuous in $\bar{\Pi}_{(-1)^k}\setminus\{0\}$, and $\mathfrak{M}_{k\xi}(\lambda)\rho^{\xi-k}$ is bounded. Hence

$$\mathfrak{M}_{k\xi}(\lambda) = \int\limits_{\Gamma_{(-1)^k}} \frac{Q_{k\xi}(\mu)}{\lambda - \mu}\, d\mu. \tag{5.11}$$

Since the functions $\mathfrak{M}_{k\xi}(\lambda) - \mathfrak{M}_{k,k+1}(\lambda)\mathfrak{M}_{k+1,\xi}(\lambda)$ are regular for $\lambda \in \Gamma_{(-1)^k}$, we have $Q_{k\xi}(\lambda) = Q_k(\lambda)\mathfrak{M}_{k+1,\xi}(\lambda)$, and (5.10) is proved. Relation (5.9) follows from Property 6 of Theorem 5.1.

Remark. Formula (5.10) is also true for the case $L \in V_N^0$ as well as for any matrix $\mathfrak{M}(\lambda)$ having the Properties 1–4 of Theorem 5.1.

Notation: $\varphi(x,\lambda) = [\chi((-1)^{k-1}\lambda)\Phi_k(x,\lambda)]_{k=\overline{2,n}}$ is a column vector,

$$\overline{q(x,\lambda)} = [(-1)^{k-1}\chi((-1)^{k-1}\lambda)\overline{\widetilde{\Phi}_{n-k+2}(x,\lambda)\widehat{Q}_{k-1}(\lambda)}]^T_{k=\overline{2,n}}$$

is a row vector, $\chi(\lambda)$ is the Heaviside function, and

$$r(x,\lambda,\mu) = \frac{\langle\widetilde{\varphi}(x,\lambda), q(x,\mu)\rangle_{\widetilde{\ell}}}{\lambda - \mu}, \tag{5.12}$$

$$\gamma_\nu(x) = \int\limits_{-\infty}^{\infty} |\overline{q^{(\nu)}(x,\lambda)}\varphi(x,\lambda)|\, d\lambda, \qquad \gamma_{\nu s}(x) = \int\limits_{-\infty}^{\infty} \overline{q^{(\nu)}(x,\lambda)}\varphi^{(s)}(x,\lambda)\, d\lambda, \tag{5.13}$$

$$\left.\begin{aligned} t_{j\nu}(x) &= -\sum_{\beta=\nu+1}^{j} C_j^\beta C_{\beta-1}^\nu \gamma_{\beta-\nu-1,j-\beta}(x), \qquad j > \nu \\ t_{j\nu}(x) &= \delta_{j\nu}, \qquad j \le \nu; \end{aligned}\right\} \tag{5.14}$$

$$\xi_\nu(x) = (-1)^\nu \gamma_{n-\nu-1,0}(x) + \sum_{s=0}^{n-\nu-1} C_n^s C_{n-s-1}^\nu \gamma_{n-\nu-s-1,s}(x), \tag{5.15}$$

$$\psi_\nu(x) = \xi_\nu(x) - \sum_{j=\nu+1}^{n-2} \psi_j(x) t_{j\nu}(x), \qquad \nu = \overline{0, n-2}. \tag{5.16}$$

Theorem 5.3. *For a fixed $x \geq 0$, the vector function $\varphi(x,\lambda)$ is the solution of the linear integral equation*

$$\widetilde{\varphi}(x,\lambda) = \varphi(x,\lambda) + \int_{-\infty}^{\infty} r(x,\lambda,\mu)\varphi(x,\mu)\,d\mu. \tag{5.17}$$

Equation (5.17) is called the main equation of the IP.

Theorem 5.4.

$$p_\nu(x) = \psi_\nu(x), \qquad u_{\xi\nu 0} + \sum_{j=\nu+1}^{n-1} u_{\xi j 0} t_{j\nu}(0) = 0. \tag{5.18}$$

Theorems 5.3 and 5.4 are obvious corollaries of Theorem 2.1 and Lemma 2.3. We note that (5.17) is the Fredholm equation of the second kind, and

$$\sup_{-\infty<\lambda<\infty} \int_{-\infty}^{\infty} |(\Omega(x,\lambda)r(x,\lambda,\mu)\Omega^{-1}(x,\mu))_{jk}|\,d\mu < \infty,$$

where $\Omega(x,\lambda) = \operatorname{diag}[\rho^{n-k}\exp(-\rho R_k x)]_{k=\overline{2,n}}$. Hence, the operator

$$Af(\lambda) = f(\lambda) + \int_{-\infty}^{\infty} \Omega(x,\lambda)r(x,\lambda,\mu)\Omega^{-1}(x,\mu)f(\mu)\,d\mu$$

is a linear bounded operator that maps B into B, where $B = \mathcal{L}_\infty^{n-1}(-\infty,\infty)$ is the Banach space of vector functions $z(\lambda) = [z_j(\lambda)]_{j=\overline{1,n-1}}$, $\lambda \in (-\infty,\infty)$, $z_j(\lambda) \in \mathcal{L}_\infty(-\infty,\infty)$ with the norm

$$||z||_B = \sum_{j=1}^{n-1} ||z_j||_{\mathcal{L}_\infty(-\infty,\infty)}.$$

5.2. Solution of the IP. In this section we present a theorem on the solvability of the main equation in the self-adjoint case, and give the solution of the IP. Denote the set of matrices $\mathfrak{M}(\lambda) = [\mathfrak{M}_{k\xi}(\lambda)]_{k,\xi=\overline{1,n}}$, having the Properties 1–7 of Theorem 5.1 by M.

Theorem 5.5. *Let $\mathfrak{M}(\lambda) \in M$. Then, for each fixed $x \geq 0$ Eq. (5.17) has a unique solution in the class $\Omega(x,\lambda)\varphi(x,\lambda) \in B$.*

Proof. It is sufficient to show that the homogeneous equation

$$h(x,\lambda) + \int_{-\infty}^{\infty} r(x,\lambda,\mu)h(x,\mu)\,d\mu = 0, \tag{5.19}$$

where $\Omega(x,\lambda)h(x,\lambda) \in B$, $h(x,\lambda) = [h_k(x,\lambda)]_{k=\overline{2,n}}$ has only the zero solution.

Consider the function

$$B(x,\lambda) = \sum_{j=1}^{m}(-1)^{j-1}H_j(x,\lambda)\overline{H_{n-j+1}(x,\bar{\lambda})}, \tag{5.20}$$

where the vector function $H(x,\lambda) = [H_j(x,\lambda)]_{j=\overline{1,n}}$ is defined by the relation

$$H(x,\lambda) = -\int_{-\infty}^{\infty} \frac{\langle\widetilde{\Phi}(x,\lambda), q(x,\mu)\rangle_{\widetilde{\ell}}}{\lambda-\mu} h(x,\mu)\,d\mu. \tag{5.21}$$

We now rewrite (5.19) and (5.21) in the coordinate form:

$$h_k(x,\lambda) + \sum_{j=2}^{n}(-1)^{j-1}\int_{\Gamma_{(-1)^{j-1}}} \frac{\langle\widetilde{\varphi}_k(x,\lambda), \widetilde{\Phi}_{n-j+2}(x,\mu)\rangle_{\widetilde{\ell}}}{\lambda-\mu}\widehat{Q}_{j-1}(\mu)h_j(x,\mu)d\mu = 0, \tag{5.22}$$

$$H_k(x,\lambda) + \sum_{j=2}^{n}(-1)^{j-1}\int_{\Gamma_{(-1)^{j-1}}} \frac{\langle\widetilde{\Phi}_k(x,\lambda), \widetilde{\Phi}_{n-j+2}(x,\mu)\rangle_{\widetilde{\ell}}}{\lambda-\mu}\widehat{Q}_{j-1}(\mu)h_j(x,\mu)d\mu = 0. \tag{5.23}$$

It is clear that for a fixed $x \geq 0$ the functions $H_k(x,\lambda)$ are regular in $\Pi_{(-1)^k}$ and $H_k(x,\lambda) = h_k(x,\lambda)$ for $\lambda \in \Gamma_{(-1)^{k-1}}$, $k = \overline{2,n}$. It also follows from (5.23) that

$$H_k(x,\lambda) - \mathfrak{M}_{k,k+1}(\lambda)H_{k+1}(x,\lambda) = -\widehat{\mathfrak{M}}_{k,k+1}(\lambda)H_{k+1}(x,\lambda)$$

$$+\sum_{j=2}^{n}(-1)^j\int_{\Gamma_{(-1)^{j-1}}} \frac{\langle\widetilde{\Phi}_k(x,\lambda) - \widetilde{\mathfrak{M}}_{k,k+1}(\lambda)\widetilde{\Phi}_{k+1}(x,\lambda), \widetilde{\Phi}_{n-j+2}(x,\mu)\rangle_{\widetilde{\ell}}}{\lambda-\mu}\widehat{Q}_{j-1}(\mu)h_j(x,\mu)d\mu.$$

Since the functions $\widetilde{\Phi}_k(x,\lambda) - \widetilde{\mathfrak{M}}_{k,k+1}(\lambda)\widetilde{\Phi}_{k+1}(x,\lambda)$ are regular for $\lambda \in \Gamma_{(-1)^k}$, and since, by (5.3), for $|\lambda-\mu| \leq \delta_0$

$$\frac{\langle\widetilde{\Phi}_k(x,\lambda), \widetilde{\Phi}_{n-j+2}(x,\mu)\rangle_{\widetilde{\ell}}}{\lambda-\mu} = \frac{(-1)^{k+1}\delta_{j,k+1}}{\lambda-\mu}\int_a^x \widetilde{\Phi}_k(t,\lambda)\overline{\widetilde{\Phi}_{n-j+2}(t,\mu)}\,dt \tag{5.24}$$

($a = 0$ for $j \leq k+1$, and $a = \infty$ for $j > k+1$), we get that all of the terms in the sum with $j \neq k+1$ are regular for $\lambda \in \Gamma_{(-1)^k}$. Therefore

$$H_k(x,\lambda) - \mathfrak{M}_{k,k+1}(\lambda)H_{k+1}(x,\lambda) = -\widehat{\mathfrak{M}}_{k,k+1}(\lambda)H_{k+1}(x,\lambda)$$

$$+(-1)^{k+1}\int\limits_{\Gamma_{(-1)^k}} \frac{\langle \widetilde{\Phi}_k(x,\lambda) - \widetilde{\mathfrak{M}}_{k,k+1}(\lambda)\widetilde{\Phi}_{k+1}(x,\lambda), \widetilde{\Phi}_{n-k+1}(x,\mu)\rangle_{\widetilde{\ell}}}{\lambda-\mu}\widehat{Q}_k(\mu)H_{k+1}(x,\mu)d\mu.$$

$$+\Omega_{k1}(x,\lambda) = \left(-\widehat{\mathfrak{M}}_{k,k+1}(\lambda) + \int\limits_{\Gamma_{(-1)^k}} \frac{\widehat{Q}_k(\mu)}{\lambda-\mu}\,d\mu\right)H_{k+1}(x,\lambda) + \Omega_{k2}(x,\lambda),$$

where the functions $\Omega_{ks}(x,\lambda)$ are regular for $\lambda \in \Gamma_{(-1)^k}$. Hence, the functions $H_k(x,\lambda) - \mathfrak{M}_{k,k+1}(\lambda)H_{k+1}(x,\lambda)$ are regular for $\lambda \in \Gamma_{(-1)^k}$.

It follows from the above-noted properties of the functions $H_k(x,\lambda)$ that for a fixed $x \geq 0$ the function $B(x,\lambda)$ is regular in $\Pi_{(-1)^m}$ and continuous in $\bar{\Pi}_{(-1)^m}\setminus\{0\}$, and the function

$$B(x,\lambda) + (-1)^m\mathfrak{M}_{m,m+1}(\lambda)H_{m+1}(x,\lambda)\overline{H_{m+1}(x,\bar{\lambda})}$$

is regular for $\lambda \in \Gamma_{(-1)^m}$. In addition, since $\Omega(x,\lambda)h(x,\lambda) \in B$, it follows from (5.20), (5.22) and (5.23) that

$$\lim_{R\to\infty}\frac{1}{2\pi i}\int\limits_{|\lambda|=R} B(x,\lambda)\,d\lambda = 0, \qquad \lim_{\varepsilon\to 0}\frac{1}{2\pi i}\int\limits_{|\lambda|=\varepsilon} B(x,\lambda)\,d\lambda = 0.$$

From this we get

$$\int\limits_{\Gamma_{(-1)^m}} (-1)^m Q_m(\lambda)|h_{m+1}(x,\lambda)|^2\,d\lambda = 0.$$

Since $(-1)^m Q_m(\lambda) > 0$, we obtain $h_{m+1}(x,\lambda) \equiv 0$.

Let us show by induction that $h_j(x,\lambda) \equiv 0$ for $j < m+1$. We assume that for some $k < m$, the relations $h_{m+1}(x,\lambda) = \ldots = h_{k+1}(x,\lambda) \equiv 0$ have been proved already. It now follows from (5.23) and the induction hypothesis that the function $\rho^{n-k}H_k(x,\lambda)$ is entire in ρ and $\lim \rho^{n-k}H_k(x,\lambda) = 0$ as $|\rho| \to \infty$. It follows from this that $H_k(x,\lambda) \equiv 0$, and, consequently, $h_k(x,\lambda) \equiv 0$.

We now show by induction that $h_j(x,\lambda) \equiv 0$ for $j > m+1$. Assume that for some $k \geq m+1$ the relations $h_2(x,\lambda) = \ldots = h_k(x,\lambda) \equiv 0$ are already proved. But then $H_j(x,\lambda) \equiv 0$ for $j = 2,\ldots,k$. Taking into account that the function $H_k(x,\lambda) - \mathfrak{M}_{k,k+1}(\lambda)H_{k+1}(x,\lambda)$ is regular for $\lambda \in \Gamma_{(-1)^k}$, we get $Q_k(\lambda)H_{k+1}(x,\lambda) \equiv 0$, $\lambda \in \Gamma_{(-1)^k}$. Since $Q_k(\lambda) = Q_k^0\rho^{-1} + O(\rho^{-2})$, $Q_k^0 \neq 0$ as $|\rho| \to \infty$, it follows from analycity that $H_{k+1}(x,\lambda) \equiv 0$, and hence $h_{k+1}(x,\lambda) \equiv 0$. Theorem 5.5 is proved.

Theorem 5.6. *A matrix $\mathfrak{M}(\lambda) \in M$ is the WM for $L \in V_N^1$ if and only if*

$$\sup_{x \geq 0} \gamma_\nu(x) < \infty, \qquad \psi_\nu(x) \in W_{\nu+N}, \qquad \nu = \overline{0, n-2}, \tag{5.25}$$

where the functions $\gamma_\nu(x)$ and $\psi_\nu(x)$ are constructed by (5.13)–(5.16), *and the vector function $\varphi(x, \lambda)$ is the solution of* (5.17). *At that, the DE and LF $L = (\ell, U)$ are constructed via* (5.18).

Proof. The necessity part of the theorem is obvious, so we shall prove the sufficiency. Let $\varphi(x, \lambda)$ be the solution of (5.17) and $\Omega(x, \lambda)\varphi(x, \lambda) \in B$. We construct the functions $\Phi(x, \lambda) = [\Phi_k(x, \lambda)]_{k=\overline{1,n}}$ by the formula

$$\Phi(x, \lambda) = \widetilde{\Phi}(x, \lambda) - \int_{-\infty}^{\infty} \frac{\langle \widetilde{\Phi}(x, \lambda), q(x, \mu)\rangle_{\widetilde{\ell}}}{\lambda - \mu} \varphi(x, \mu)\, d\mu \tag{5.26}$$

and we construct the DE and LF $L = (\ell, U)$ by (5.18). Let us show that $\Phi(x, \lambda)$ is the WS, and $\mathfrak{M}(\lambda)$ is the WM for L.

The functions $\Phi_k(x, \lambda)$ are regular in $\Pi_{(-1)^k}$, and $\Phi_k(x, \lambda) = \varphi_k(x, \lambda)$, $\lambda \in \Gamma_{(-1)^{k-1}}$. It is easy to see that the functions $\varphi^{(\nu)}(x, \lambda)$, $\nu = \overline{0, n-1}$ are absolutely continuous with respect to x, and for a fixed $x \geq 0$ we have $\rho^{-\nu}\Omega(x, \lambda)\varphi(x, \lambda) \in B$. Differentiating (5.17) and (5.26) with respect to x and taking (5.3) into account, we obtain

$$\sum_{\nu=0}^{n} t_{j\nu}(x)\widetilde{\varphi}^{(\nu)}(x, \lambda) = \varphi^{(j)}(x, \lambda) + \int_{-\infty}^{\infty} r(x, \lambda, \mu)\varphi^{(j)}(x, \mu)\, d\mu,$$

$$\sum_{\nu=0}^{n} t_{j\nu}(x)\widetilde{\Phi}^{(\nu)}(x, \lambda) = \Phi^{(j)}(x, \lambda) + \int_{-\infty}^{\infty} \frac{\langle \widetilde{\Phi}(x, \lambda), q(x, \mu)\rangle_{\widetilde{\ell}}}{\lambda - \mu} \varphi^{(j)}(x, \mu)\, d\mu.$$

Hence, in view of (5.12)–(5.17), we have

$$\widetilde{U}_{\xi 0}(\widetilde{\Phi}_k(x, \lambda)) = U_{\xi 0}(\Phi_k(x, \lambda)) + \int_{-\infty}^{\infty} \frac{\langle \widetilde{\Phi}_k(x, \lambda), q(x, \mu)\rangle_{\widetilde{\ell}}}{\lambda - \mu}\Bigg|_{x=0} U_{\xi 0}(\varphi(x, \mu))\, d\mu \tag{5.27}$$

$$\begin{aligned} \widetilde{\ell}\widetilde{\varphi}(x, \lambda) &= \ell\varphi(x, \lambda) + \int_{-\infty}^{\infty} r(x, \lambda, \mu)\ell\varphi(x, \mu)\, d\mu \\ &\quad + \int_{-\infty}^{\infty} \langle \widetilde{\varphi}(x, \lambda), q(x, \mu)\rangle_{\widetilde{\ell}}\, \varphi(x, \mu)\, d\mu, \end{aligned} \tag{5.28}$$

$$\begin{aligned}\widetilde{\ell}(\widetilde{\Phi}(x,\lambda)) &= \ell(\Phi(x,\lambda)) + \int\limits_{-\infty}^{\infty} \frac{\langle \widetilde{\Phi}(x,\lambda), q(x,\mu)\rangle_{\widetilde{\ell}}}{\lambda-\mu} \ell\varphi(x,\mu)\, d\mu \\ &+ \int\limits_{-\infty}^{\infty} \langle \widetilde{\Phi}(x,\lambda), q(x,\mu)\rangle_{\widetilde{\ell}} \varphi(x,\mu)\, d\mu.\end{aligned} \tag{5.29}$$

Denote $h(x,\lambda) = \ell\varphi(x,\lambda) - \lambda\varphi(x,\lambda)$. It follows from (5.28), (5.17) and (5.12) that

$$h(x,\lambda) + \int\limits_{-\infty}^{\infty} r(x,\lambda,\mu) h(x,\mu)\, d\mu = 0,$$

and $\Omega(x,\lambda)h(x,\lambda) \in B$. By Theorem 5.5, we have $h(x,\lambda) \equiv 0$, i.e. $\ell\varphi(x,\lambda) = \lambda\varphi(x,\lambda)$. It now follows from (5.29) and (5.26) that

$$\ell\Phi_k(x,\lambda) = \lambda\Phi_k(x,\lambda), \qquad k = \overline{1,n}. \tag{5.30}$$

We now rewrite (5.27) in the form

$$\begin{aligned}&\widetilde{U}_{\xi 0}(\widetilde{\Phi}_k(x,\lambda)) = U_{\xi 0}(\Phi_k(x,\lambda)) \\ &+ \sum_{j=2}^{n} (-1)^{j-1} \int\limits_{\Gamma_{(-1)^{j-1}}} \frac{\langle \widetilde{\Phi}_k(x,\lambda), \widetilde{\Phi}_{n-j+2}(x,\mu)\rangle_{\widetilde{\ell}}}{\lambda-\mu}\Big|_{x=0} \widehat{Q}_{j-1}(\mu) U_{\xi 0}(\Phi_j(x,\mu))\, d\mu.\end{aligned} \tag{5.31}$$

Using (5.24) for $j \le k+1$ and (5.31), we compute

$$U_{\xi 0}(\Phi_k) = \delta_{\xi k}, \qquad \xi \le k, \tag{5.32}$$

$$U_{k+1,0}(\Phi_k) = \widetilde{\mathfrak{M}}_{k,k+1}(\lambda) + \int\limits_{\Gamma_{(-1)^k}} \frac{\widehat{Q}_k(\mu)}{\lambda-\mu}\, d\mu = \mathfrak{M}_{k,k+1}(\lambda). \tag{5.33}$$

In view of (5.1), we write (5.26) in the form

$$\begin{aligned}\Phi_k(x,\lambda) &= \widetilde{\Phi}_k(x,\lambda) - \sum_{\nu+j\le n-1} \left(\sum_{s=j}^{n-\nu-1} (-1)^s C_s^j p_{s+\nu+1}^{(s-j)}(x) \right) \\ &\times \widetilde{\Phi}_k^{(\nu)}(x,\lambda) \int\limits_{-\infty}^{\infty} \frac{\overline{q^{(j)}(x,\mu)}\varphi(x,\mu)}{\lambda-\mu}\, d\mu.\end{aligned} \tag{5.34}$$

Denote $G_\varepsilon = \{\lambda : |\lambda| \ge \varepsilon, \arg(\pm\lambda) \notin (-\varepsilon, \varepsilon)\}$, $\varepsilon > 0$. In the domain G_ε we have $|\lambda - \mu| \ge C_\varepsilon |\lambda|$. We separate the term with $j = n-1$ in (5.34), and obtain

$$\Phi_k(x,\lambda) = \widetilde{\Phi}_k(x,\lambda)(1 + A(x,\lambda)) + B_k(x,\lambda) \tag{5.35}$$

where

$$A(x,\lambda) = \int_{-\infty}^{\infty} \frac{\overline{q^{(n-1)}(x,\mu)}\varphi(x,\mu)}{\lambda - \mu}\, d\mu.$$

In view of (5.25), for the functions $B_k(x,\lambda)$ we have the estimate

$$|B_k(x,\lambda)| \le C|\rho|^{n-k+1}\, |\exp(\rho R_k x)|, \qquad \lambda \in G_\varepsilon. \tag{5.36}$$

By virtue of (5.30) and (5.32), the function $\Phi_n(x,\lambda)$ is the WS for L. Therefore, $\Phi_n(x,\lambda) = a_n^0 \exp(\rho R_n x)(1 + O(\rho^{-1}))$, $a_n^0 \ne 0$ as $|\rho| \to \infty$. The same asymptotics holds for $\widetilde{\Phi}_n(x,\lambda)$. It follows from (5.35) for $k = n$ that $A(x,\lambda) = (\widetilde{\Phi}_n(x,\lambda))^{-1}(\Phi_n(x,\lambda) - B_n(x,\lambda)) - 1$. Consequently, $A(x,\lambda) = O(\rho^{-1})$, $|\rho| \to \infty$. It now follows from (5.35) and (5.36) that

$$|\Phi_k(x,\lambda)| \le C\, |\rho|^{k-n}\, |\exp(\rho R_k x)|, \qquad x \ge x_0 > 0, \qquad \lambda \in G_\varepsilon.$$

Thus, the functions $\Phi_k(x,\lambda)$, $k = \overline{1,n}$ are the WS's for L.

Denote by $\mathfrak{M}^0(\lambda)$ the WM for L. It follows from (5.33) that $\mathfrak{M}_{k,k+1}(\lambda) = \mathfrak{M}^0_{k,k+1}(\lambda)$, and hence $Q_k(\lambda) = Q_k^0(\lambda)$. As in the proof of Theorem 5.5, one can show that the functions $\Phi_k(x,\lambda) - \mathfrak{M}_{k,k+1}(\lambda)\Phi_{k+1}(x,\lambda)$ are regular for $\lambda \in \Gamma_{(-1)^k}$. The functions $\Phi_k(x,\lambda)$ and $\mathfrak{M}^0_{k\xi}(\lambda)$ are continuous in $\overline{\Pi}_{(-1)^k}\setminus\{0\}$, and $\mathfrak{M}^0_{k\xi}(\lambda)\rho^{\xi-k} = O(1)$. Thus, $L \in V_N^0$. By virtue of the remark on Theorem 5.2, we have

$$\mathfrak{M}^0_{k\xi}(\lambda) = \int_{\Gamma_{(-1)^k}} \frac{Q_k(\mu)\mathfrak{M}^0_{k+1,\xi}(\mu)}{\lambda - \mu}\, d\mu. \tag{5.37}$$

Since $\mathfrak{M}(\lambda) \in M$, formula (5.10) holds. Comparing (5.10) with (5.37), we obtain $\mathfrak{M}^0(\lambda) = \mathfrak{M}(\lambda)$, i.e. $\mathfrak{M}(\lambda)$ is the WM for L.

It remains to be shown that $L = L^*$. It is clear that $L^* \in V_N^0$. Let $\mathfrak{N}(\lambda)$ be the WM for L^*. Then $\mathfrak{N}_{k,k+1}(\lambda) = \overline{\mathfrak{M}_{n-k,n-k+1}(\bar{\lambda})}$, which is analogous to Property 6 of Theorem 5.1, but for the non-self-adjoint case. On the other hand, by hypothesis, we have $\mathfrak{M}_{k,k+1}(\lambda) = \overline{\mathfrak{M}_{n-k,n-k+1}(\bar{\lambda})}$. Hence, $\mathfrak{M}_{k,k+1}(\lambda) = \mathfrak{N}_{k,k+1}(\lambda)$, $k = \overline{1,n-1}$, and, by virtue of (5.10), $\mathfrak{M}(\lambda) = \mathfrak{N}(\lambda)$. From this, by virtue of the uniqueness theorem for the IP from the WM, we have $L = L^*$, i.e. $L \in V_N^1$. Theorem 5.6 is proved.

6. Differential operators with singularities

6.1. Let us consider the DE

$$\ell y \equiv y^{(n)} + \sum_{j=0}^{n-2} \left(\frac{\nu_j}{x^{n-j}} + q_j(x)\right) y^{(j)} = \lambda y, \qquad x > 0 \tag{6.1}$$

on the half-line. Let $\mu_1, \dots, \mu_n$ be the roots of the characteristic polynomial

$$\Delta(\mu) = \sum_{j=0}^{n} \nu_j \prod_{k=0}^{j-1} (\mu - k), \qquad \nu_n = 1, \qquad \nu_{n-1} = 0.$$

It is clear that $\mu_1 + \ldots + \mu_n = n(n-1)/2$. For definiteness, we assume that $\mu_k - \mu_j \neq sn$ $(s = 0, \pm 1, \pm 2, \ldots)$; $\mathrm{Re}\,\mu_1 < \ldots < \mathrm{Re}\,\mu_n$, $\mu_k \neq 0, 1, \ldots, n-3$ (the other cases require minor modifications). Let $\theta_n = n - 1 - \mathrm{Re}\,(\mu_n - \mu_1)$. Denote $q_{0j}(x) = q_j(x)$ for $x \geq 1$, and $q_{0j}(x) = q_j(x) x^{\min(\theta_n - j, 0)}$ for $x \leq 1$ and assume that $q_{0j}(x) \in \mathcal{L}(0, \infty)$, $j = \overline{0, n-2}$.

In this section we construct special FSS's for the DE (6.1) and use them to investigate the IP. The presence of a singularity in the DE introduces essential qualitative modifications in the investigation of the operator. Fundamental difficulties arise when $n > 2$. In the construction of the special FSS's for (6.1) the elementary solutions of the simplest equation are no longer exponentials, but functions that are a generalization of the Hankel solution of the Bessel equation. An important and technically difficult problem is the determination of the asymptotic behaviuor of the Stokes multipliers for the constructed FSS's. Using properties of the FSS's and the Stokes multipliers, we introduce and study the WS's and the WM for (6.1), and investigate the IP: to construct the operator ℓ from its WM.

We mention that DE's with singularities arise in various areas of mathematics as well as in applications. In addition, various DE's with a turning point, for example, the equation

$$z^{(n)}(t) = \lambda r(t) z(t), \qquad t > 0; \qquad r(t) \sim \alpha t^{\gamma}, \qquad t \to +0, \qquad \gamma > 0,$$

and other more general equations, can be reduced to (6.1). We also note that for $n = 2$ inverse problems for operators with a singularity have been studied by several authors (see, for example, [97]–[99]).

6.2. First of all, we consider the DE

$$\ell_0 y \equiv y^{(n)} + \sum_{j=0}^{n-2} \frac{\nu_j}{x^{n-j}} y^{(j)} = y. \tag{6.2}$$

Let $x = r * \exp(i\varphi)$, $r > 0$, $\varphi \in (-\pi, \pi]$, $x^{\mu} = \exp(\mu(\ln r + i\varphi))$ and Π_- be the x-plane with a cut along the semi-axis $x \leq 0$. Take numbers C_{j0}, $j = \overline{1, n}$ from the condition

$$\prod_{j=1}^{n} C_{j0} = (\det [\mu_j^{\nu - 1}]_{j,\nu = \overline{1,n}})^{-1}.$$

Then the functions

$$C_j(x) = x^{\mu_j} \sum_{k=0}^{\infty} C_{jk} x^{nk}, \qquad C_{jk} = C_{j0}(\prod_{s=1}^{k} \Delta(\mu_j + sn))^{-1}, \tag{6.3}$$

are solutions of (6.2), and $\det [C_j^{(\nu-1)}(x)]_{j,\nu=\overline{1,n}} \equiv 1$. Furthermore, the functions $C_j(x)$ are regular in Π_-.

Denote

$$\varepsilon_k = \exp\left(\frac{2\pi i(k-1)}{n}\right), \qquad S_\nu = \left\{x : \arg x \in \left(\frac{\nu\pi}{n}, \frac{(\nu+1)\pi}{n}\right)\right\},$$

$$S_1^* = \overline{S}_{n-1}, \qquad S_k^* = \overline{S}_{n-2k+1} \cup \overline{S}_{n-2k+2}, \qquad k = \overline{2,n};$$

$$Q_k = \left\{x : \arg x \in [\max\left(-\pi, (-2k+2)\frac{\pi}{n}\right), \min\left(\pi, (2n-2k+2)\frac{\pi}{n}\right)]\right\}, \quad k = \overline{1,n}.$$

For $x \in S_k^*$ there are solutions of (6.2) $e_k(x)$, $k = \overline{1,n}$ of the form $e_k^{(\nu-1)}(x) = \varepsilon_k^\nu \exp(\varepsilon_k x)\, z_{k\nu}(x)$, $\nu = \overline{0, n-1}$, where $z_{k\nu}(x)$ are solutions of the integral equations

$$z_{k\nu}(x) = 1 + \frac{1}{n}\int_x^\infty \left(\sum_{j=1}^{n} \varepsilon_j^{\nu+1}\varepsilon_k^{-\nu}\exp((\varepsilon_k - \varepsilon_j)(t-x))\right)\left(\sum_{m=0}^{n-2} \nu_m \varepsilon_k^m t^{m-n} z_{km}(t)\right) dt,$$

(where $\arg t = \arg x$, $|t| > |x|$). Using the FSS $\{C_j(x)\}_{j=\overline{1,n}}$ we can write

$$e_k(x) = \sum_{j=1}^{n} \beta_{kj}^0 C_j(x). \tag{6.4}$$

In particular, this gives the analytic continuation for $e_k(x)$ on Π_-.

Lemma 6.1. *The system $\{e_k(x)\}_{k=\overline{1,n}}$, $x \in \Pi_-$ is a FSS of (6.2), and*

$$\det [e_k^{(\nu-1)}(x)]_{k,\nu=\overline{1,n}} = \det [\varepsilon_k^{\nu-1}]_{k,\nu=\overline{1,n}}.$$

The asymptotics

$$e_k^{(\nu-1)}(x) = \varepsilon_k^{\nu-1}\exp(\varepsilon_k x)(1 + O(x^{-1})), \qquad |x| \to \infty, \qquad x \in Q_k, \tag{6.5}$$

are valid.

We observe that the asymptotics (6.5) holds in the sectors Q_k which are wider than the sectors S_k^*. Next we obtain connections between the Stokes multipliers β_{kj}^0.

Lemma 6.2.

$$\beta_{kj}^0 = \beta_{1j}^0 \varepsilon_k^{\mu_j}, \qquad j,k = \overline{1,n} \tag{6.6}$$

$$\prod_{j=1}^{n}\beta_{1j}^{0}=(\det[\varepsilon_k^{\mu_j}]_{k,j=\overline{1,n}})^{-1}\det[\varepsilon_k^{j-1}]_{k,j=\overline{1,n}}\neq 0. \tag{6.7}$$

Indeed, for $\arg x\in(-\pi,\pi-\frac{2\pi s}{n})$ we have, by virtue of (6.3) and (6.4),

$$e_k(\varepsilon^s x)=\sum_{j=1}^{n}\beta_{kj}^{0}(\varepsilon^s)^{\mu_j}C_j(x). \tag{6.8}$$

It is easily seen from construction of the functions $e_k(x)$ that $e_1(\varepsilon^s x)=e_{s+1}(x)$. Substituting (6.4) and (6.8) in this equality and comparing the corresponding coefficients, we obtain (6.6). After that (6.7) becomes obvious.

Now we consider the DE

$$\ell_0 y=\lambda y=\rho^n y,\qquad x>0. \tag{6.9}$$

It is evident that if $y(x)$ is a solution of (6.2), then $y(\rho x)$ satisfies (6.9). Define $C_j(x,\lambda)$ by

$$C_j(x,\lambda)=\rho^{-\mu_j}C_j(\rho x)=x^{\mu_j}\sum_{k=0}^{\infty}C_{jk}(\rho x)^{nk}.$$

The functions $C_j(x,\lambda)$ are entire in λ and $\det[C_j^{(\nu-1)}(x,\lambda)]_{j,\nu=\overline{1,n}}\equiv 1$. From Lemmas 6.1 and 6.2 we get the following theorem.

Theorem 6.1. *In each sector $S_{k_0}=\{\rho:\arg\rho\in(\frac{k_0\pi}{n},\frac{(k_0+1)\pi}{n})\}$ equation (6.9) has a FSS $B_0=\{y_k(x,\rho)\}_{k=\overline{1,n}}$ such that $y_k(x,\rho)=y_k(\rho x)$,*

$$|y_k^{(\nu)}(x,\rho)(\rho R_k)^{-\nu}\exp(-\rho R_k x)-1|\leq\frac{M_0}{|\rho|x}, \tag{6.10}$$

$$\rho\in\bar{S}_{k_0},\qquad |\rho|x\geq 1,\qquad \nu=\overline{0,n-1},$$

$$\det[y_k^{(\nu-1)}(x,\rho)]_{k,\nu=\overline{1,n}}\equiv\rho^{n(n-1)/2}\Omega,\qquad \Omega=\det[R_k^{\nu-1}]_{k,\nu=\overline{1,n}}\neq 0, \tag{6.11}$$

$$y_k(x,\rho)=\sum_{j=1}^{n}b_{kj}^{0}\rho^{\mu_j}C_j(x,\lambda),\qquad b_{kj}^{0}=\beta_j^{0}R_k^{\mu_j},\qquad \beta_j^{0}\neq 0, \tag{6.12}$$

where the constant M_0 depends only on ν_j.

The functions $y_k(x,\rho)$ are analogues of the Hankel functions for the Bessel equation. Denote

$$C_j^*(x,\lambda)=\det[C_k^{(\nu)}(x,\lambda)]_{\nu=\overline{0,n-2};k=\overline{1,n}\setminus n-j+1},$$

$$y_j^*(x,\rho) = (-1)^{n-j}\left(\rho^{(n-1)(n-2)/2}\Omega\right)^{-1}\det[y_k^{(\nu)}(x,\rho)]_{\nu=\overline{0,n-2};k=\overline{1,n}\backslash j},$$

$$F_{k\nu}(\rho x) = \begin{cases} R_k^\nu \exp(\rho R_k x), & |\rho|x > 1, \\ (\rho x)^{\mu_1-\nu}, & |\rho|x \le 1, \end{cases} \qquad F_k^*(\rho x) = \begin{cases} \exp(-\rho R_k x), & |\rho|x > 1, \\ (\rho x)^{n-1-\mu_n}, & |\rho|x \le 1, \end{cases}$$

$$U_{k\nu}^0(x,\rho) = y_k^{(\nu)}(x,\rho)(\rho^\nu F_{k\nu}(\rho x))^{-1}, \qquad U_k^{0,*}(x,\rho) = y_k^*(x,\rho)(F_k^*(\rho x))^{-1},$$

$$g(x,t,\lambda) = \sum_{j=1}^n (-1)^{n-j} C_j(x,\lambda)C_{n-j+1}^*(t,\lambda) = \frac{1}{\rho^{n-1}}\sum_{j=1}^n y_j(x,\rho)y_j^*(t,\rho).$$

The function $g(x,t,\lambda)$ is the Green's function of the Cauchy problem $\ell_0 y - \lambda y = f(x)$, $y^{(\nu)}(0) = 0$, $\nu = \overline{0, n-1}$. Using (6.10)–(6.12), we obtain

$$|U_{k\nu}^0(x,\rho)| \le M_1, \qquad |U_k^{0,*}(x,\rho)| \le M_1, \qquad x \le 0, \qquad \rho \in \bar{S}_{k_0}, \tag{6.13}$$

$$|C_j^{(\nu)}(x,\lambda)| \le M_2|x^{\mu_j-\nu}|, \tag{6.14}$$

$$\left|\frac{\partial^\nu}{\partial x^\nu} g(x,t,\lambda)\right| \le M_2 \sum_{j=1}^n |x^{\mu_j-\nu} t^{n-1-\mu_j}|, \qquad |\rho x| \le C_0, \qquad t \le x$$

where M_1 depends on ν_j, and M_2 on ν_j and C_0.

6.3. In this item we construct FSS's of Eq. (6.1). Denote

$$J(\rho) = \sum_{m=0}^{n-2} J_m(\rho),$$

$$J_m(\rho) = |\rho|^{\mathrm{Re}(\mu_1-\mu_n)} \int_0^{|\rho|^{-1}} t^{\theta_n - m}|q_m(t)|\,dt + |\rho|^{m-n+1}\int_{|\rho|^{-1}}^\infty |q_m(t)|\,dt.$$

Lemma 6.3.

$$J(\rho) \le \frac{Q}{|\rho|}, \qquad |\rho| \ge 1, \qquad Q \overset{df}{=} \sum_{m=0}^{n-2}\int_0^\infty |q_{0m}(t)|\,dt.$$

Indeed, if $\theta_m - m \le 0$, then $\operatorname{Re}(\mu_n - \mu_1) \ge n - m - 1$, and so

$$J_m(\rho) \le |\rho|^{m-n+1}\left(\int_0^{|\rho|^{-1}} t^{\theta_n - m}|q_m(t)|\,dt + \int_{|\rho|^{-1}}^{\infty} |q_m(t)|\,dt\right) \le |\rho|^{m-n+1}\int_0^\infty |q_{0m}(t)|\,dt.$$

But if $\theta_n - m > 0$, then

$$J_m(\rho) \le |\rho|^{m-n+1}\int_0^\infty |q_m(t)|\,dt \le |\rho|^{m-n+1}\int_0^\infty |q_{0m}(t)|\,dt.$$

Hence $J(\rho) \le Q|\rho|^{-1}$, $|\rho| \ge 1$, and the proof is complete.

We now construct the functions $S_j(x,\lambda)$, $j = \overline{1,n}$ from the system of integral equations

$$S_j^{(\nu)}(x,\lambda) = C_j^{(\nu)}(x,\lambda) - \int_0^x \frac{\partial^\nu}{\partial x^\nu} g(x,t,\lambda)\left(\sum_{m=0}^{n-2} q_m(t) S_j^{(m)}(t,\lambda)\right) dt, \qquad (6.15)$$

$$\nu = \overline{0, n-1}.$$

By (6.14), system (6.15) has a unique solution; moreover the functions $S_j^{(\nu)}(x,\lambda)$ are entire in λ for each $x > 0$, the functions $\{S_j(x,\lambda)\}_{j=\overline{1,n}}$ form an FSS for (6.1), $\det[S_j^{(\nu-1)}(x,\lambda)]_{j,\nu=\overline{1,n}} \equiv 1$, and

$$S_j^{(\nu)}(x,\lambda) = O(x^{\mu_j - \nu}), \quad (S_j(x,\lambda) - C_j(x,\lambda))x^{-\mu_j} = o(x^{\mu_n - \mu_1}), \quad x \to 0. \quad (6.16)$$

Let

$$S_{k_0,\alpha} = \{\rho : \rho \in S_{k_0}, |\rho| > \alpha\}, \qquad \rho_0 = 2M_1 Q + 1.$$

For $k = \overline{1,n}$, $\rho \in \overline{S}_{k_0,\rho_0}$ consider the system of integral equations

$$U_{k\nu}(x,\rho) = U_{k\nu}^0(x,\rho) + \sum_{m=0}^{n-2}\int_0^\infty A_{k\nu m}(x,t,\rho) U_{km}(t,\rho)\,dt, \qquad (6.17)$$

$$x \ge 0, \qquad \nu = \overline{0, n-1},$$

where

$$A_{k\nu m}(x,t,\rho)=\frac{q_m(t)F_{km}(\rho t)}{\rho^{n-1-m}F_{k\nu}(\rho x)}\begin{cases}-\sum_{j=1}^{k}F_{j\nu}(\rho x)U^0_{j\nu}(x,\rho)F_j^*(\rho t)U_j^{0,*}(t,\rho), & t\le x,\\ \sum_{j=k+1}^{n}F_{j\nu}(\rho x)U^0_{j\nu}(x,\rho)F_j^*(\rho t)U_j^{0,*}(t,\rho), & t>x.\end{cases}$$

Using (6.13) and Lemma 6.3, we obtain

$$\sum_{m=0}^{n-2}\int_0^\infty |A_{k\nu m}(x,t,\rho)|\,dt\le M_1J(\rho)\le\frac{M_1Q}{|\rho|}.$$

Consequently, system (6.17) with $\rho\in\bar S_{k_0,\rho_0}$ has a unique solution and, uniformly in $x\ge 0$

$$U_{k\nu}(x,\rho)-U^0_{k\nu}(x,\rho)=O(\rho^{-1}),\qquad \rho\in\bar S_{k_0,\rho_0}. \tag{6.18}$$

Theorem 6.2. *For $x>0$, $\rho\in S_{k_0,\rho_0}$ there exists an FSS of (6.1), $B=\{\mathcal{Y}_k(x,\rho)\}_{k=\overline{1,n}}$ of the form*

$$\mathcal{Y}_k^{(\nu)}(x,\rho)=\rho^\nu F_{k\nu}(\rho x)U_{k\nu}(x,\rho),$$

where the functions $U_{k\nu}(x,\rho)$ are solutions of (6.17), and (6.18) is true.

The functions $\mathcal{Y}_k^{(\nu)}(x,\rho)$ considered for each $x>0$, are regular in $\rho\in S_{k_0,\rho_0}$, continuous in $\rho\in\bar S_{k_0,\rho_0}$, and $\det[\mathcal{Y}_k^{(\nu-1)}(x,\rho)]_{k,\nu=\overline{1,n}}=\rho^{n(n-1)/2}\Omega(1+O(\rho^{-1}))$ as $|\rho|\to\infty$. The functions $\mathcal{Y}_k(x,\rho)$ satisfy the equality

$$\begin{aligned}\mathcal{Y}_k(x,\rho)&=y_k(x,\rho)-\frac{1}{\rho^{n-1}}\int_0^x\left(\sum_{j=1}^{k}y_j(x,\rho)y_j^*(t,\rho)\right)\left(\sum_{m=0}^{n-2}q_m(t)\mathcal{Y}_k^{(m)}(t,\rho)\right)dt\\&+\frac{1}{\rho^{n-1}}\int_x^\infty\left(\sum_{j=k+1}^{n}y_j(x,\rho)y_j^*(t,\rho)\right)\left(\sum_{m=0}^{n-2}q_m(t)\mathcal{Y}_k^{(m)}(t,\rho)\right)dt.\end{aligned}$$

Moreover, one has a representation

$$\mathcal{Y}_k(x,\rho)=\sum_{j=1}^{n}b_{kj}(\rho)S_j(x,\lambda), \tag{6.19}$$

where

$$b_{kj}(\rho)=b^0_{kj}\rho^{\mu_j}(1+O(\rho^{-1})),\qquad |\rho|\to\infty,\qquad \rho\in\bar S_{k_0}. \tag{6.20}$$

The only part of the theorem that needs proof is the asymptotic formula (6.20). Let ρ be fixed, $x \leq |\rho|^{-1}$. Then (6.12) and (6.19) become

$$\left.\begin{aligned} U^0_{k0}(x,\rho) &= \sum_{j=1}^{n} b^0_{kj}(\rho x)^{\mu_j-\mu_1}\widehat{C}_j(x,\lambda), \\ U_{k0}(x,\rho) &= \sum_{j=1}^{n} b_{kj}(\rho)\rho^{-\mu_1}x^{\mu_j-\mu_1}\widehat{S}_j(x,\lambda), \end{aligned}\right\} \tag{6.21}$$

where

$$\widehat{C}_j(x,\lambda) = x^{-\mu_j}C_j(x,\lambda), \qquad \widehat{S}_j(x,\lambda) = x^{-\mu_j}S_j(x,\lambda),$$

$$\widehat{S}_j(0,\lambda) = \widehat{C}_j(0,\lambda) = C_{j0} \neq 0.$$

It follows from (6.21) that

$$\begin{aligned} U_{k0}(x,\rho) - U^0_{k0}(x,\rho) &= \sum_{j=1}^{n}(b_{kj}(\rho)\rho^{-\mu_1} - b^0_{kj}\rho^{\mu_j-\mu_1})x^{\mu_j-\mu_1}\widehat{S}_j(x,\lambda) \\ &\quad + \sum_{j=1}^{n} b^0_{kj}(\rho x)^{\mu_j-\mu_1}(\widehat{S}_j(x,\lambda) - \widehat{C}_j(x,\lambda)). \end{aligned} \tag{6.22}$$

Denote

$$\left.\begin{aligned} &\mathcal{F}_{k1}(x,\rho) = U_{k0}(x,\rho) - U^0_{k0}(x,\rho), \\ &\mathcal{F}_{k,s+1}(x,\rho) = (\mathcal{F}_{ks}(x,\rho) - \mathcal{F}_{ks}(0,\rho)\widehat{S}_s(x,\lambda)C^{-1}_{s0})x^{\mu_s-\mu_{s+1}}, \quad s = \overline{1,n-1}. \end{aligned}\right\} \tag{6.23}$$

Lemma 6.4.

$$(b_{ks}(\rho)\rho^{-\mu_1} - b^0_{ks}\rho^{\mu_s-\mu_1})C_{s0} = \mathcal{F}_{ks}(0,\rho), \qquad s = \overline{1,n}, \tag{6.24}$$

$$\begin{aligned} \mathcal{F}_{ks}(x,\rho) &= ((U_{k0}(x,\rho) - U^0_{k0}(x,\rho)) - \sum_{j=1}^{s-1}(b_{kj}(\rho)\rho^{-\mu_1} \\ &\quad - b^0_{kj}\rho^{\mu_j-\mu_1})x^{\mu_j-\mu_1}\widehat{S}_j(x,\lambda))x^{\mu_1-\mu_s}, \qquad s = \overline{1,n}. \end{aligned} \tag{6.25}$$

Proof. When $s = 1$ equality (6.24) follows from (6.22) for $x = 0$, while (6.25) is obviously true. Suppose now that (6.24) and (6.25) have been proved for

$s = 1, \ldots, N-1$. Then

$$\left((U_{k0}(x,\rho) - U^0_{k0}(x,\rho)) - \sum_{j=1}^{N-1} (b_{kj}(\rho)\rho^{-\mu_1} - b^0_{kj}\rho^{\mu_j-\mu_1})x^{\mu_j-\mu_1}\widehat{S}_j(x,\lambda)\right)x^{\mu_1-\mu_N}$$

$$= \left((U_{k0}(x,\rho) - U^0_{k0}(x,\rho)) - \sum_{j=1}^{N-2} (b_{kj}(\rho)\rho^{-\mu_1} - b^0_{kj}\rho^{\mu_j-\mu_1})x^{\mu_j-\mu_1}\widehat{S}_j(x,\lambda)\right)x^{\mu_1-\mu_{N-1}}x^{\mu_{N-1}-\mu_N}$$

$$-(b_{k,N-1}(\rho)\rho^{-\mu_1} - b^0_{k,N-1}\rho^{\mu_{N-1}-\mu_1})\widehat{S}_{N-1}(x,\lambda)x^{\mu_{N-1}-\mu_N} = \mathcal{F}_{kN}(x,\rho),$$

which gives (6.25) for $s = N$. We now write (6.22) as

$$\mathcal{F}_{kN}(x,\rho) = \sum_{j=N}^{n} (b_{kj}(\rho)\rho^{-\mu_1} - b^0_{kj}\rho^{\mu_j-\mu_1})x^{\mu_j-\mu_N}\widehat{S}_j(x,\lambda)$$

$$+\sum_{j=1}^{n} b^0_{kj}(\rho x)^{\mu_j-\mu_1}(\widehat{S}_j(x,\lambda) - \widehat{C}_j(x,\lambda))x^{\mu_1-\mu_N}.$$

Hence, using (6.16), we obtain $\mathcal{F}_{kN}(0,\rho) = (b_{kN}(\rho)\rho^{-\mu_1} - b^0_{kN}\rho^{\mu_N-\mu_1})C_{N0}$, which gives (6.24) for $s = N$ and completes the proof of Lemma 6.4.

Now write (6.17) for $\nu = 0$ as

$$\begin{aligned}\mathcal{F}_{k1}(x,\rho) = \frac{1}{\rho^{n-1}}\Bigg(&-\int_0^x \Big(\sum_{j=1}^{n} U^0_{j0}(x,\rho)U^{0,*}_j(t,\rho)\Big)(\rho t)^{n-1-\mu_n}V_k(t,\rho)\,dt \\ &+\int_0^\infty \left(\sum_{j=k+1}^{n} U^0_{j0}(x,\rho)U^{0,*}_j(t,\rho)F^*_j(\rho t)\right)V_k(t,\rho)\,dt\Bigg),\end{aligned} \tag{6.26}$$

where

$$V_k(t,\rho) = \sum_{m=0}^{n-2} q_m(t)\rho^m F_{km}(\rho t)U_{km}(t,\rho).$$

Since for $t \le x \le |\rho|^{-1}$ we have

$$\sum_{j=1}^{n} U^0_{j0}(x,\rho)U^{0,*}_j(t,\rho) = \rho^{\mu_n-\mu_1}x^{-\mu_1}t^{1-n+\mu_n}g(x,t,\lambda),$$

it follows by way of (6.14) that

$$\left|\sum_{j=1}^{n} U^0_{j0}(x,\rho)U^{0,*}_j(t,\rho)\right| \le M_3|(\rho x)^{\mu_n-\mu_1}|, \qquad 0 \le t \le x \le |\rho|^{-1}. \tag{6.27}$$

Lemma 6.5.

$$\mathcal{F}_{ks}(0,\rho) = \rho^{\mu_s-\mu_1-n+1}C_{s0}\int_0^\infty\left(\sum_{j=k+1}^n b_{js}^0 F_j^*(\rho t)U_j^{0,*}(t,\rho)\right)V_k(t,\rho)\,dt, \qquad (6.28)$$

$$\mathcal{F}_{ks}(x,\rho) = \frac{1}{\rho^{n-1}}\Bigg(-x^{\mu_1-\mu_s}\int_0^x\Big(\sum_{j=1}^n U_{j0}^0(x,\rho)U_j^{0,*}(t,\rho)\Big)(\rho t)^{n-1-\mu_n}V_k(t,\rho)\,dt$$
$$-\sum_{\ell=1}^{s-1}x^{\mu_\ell-\mu_s}\int_0^\infty\Big(\sum_{j=k+1}^n b_{j\ell}^0\rho^{\mu_\ell-\mu_1}(\widehat{S}_\ell(x,\lambda)-\widehat{C}_\ell(x,\lambda))F_j^*(\rho t)U_j^{0,*}(t,\rho)\Big)V_k(t,\rho)\,dt$$
$$+\int_0^\infty\Big(\sum_{j=k+1}^n\Big(\sum_{\xi=s}^n b_{j\xi}^0\rho^{\mu_\xi-\mu_1}x^{\mu_\xi-\mu_s}\widehat{C}_\xi(x,\lambda)\Big)F_j^*(\rho t)U_j^{0,*}(t,\rho)\Big)V_k(t,\rho)\,dt\Bigg),$$
$$x \le |\rho|^{-1}. \qquad (6.29)$$

Proof. For $s = 1$, (6.28) and (6.29) follow from (6.26), in view of (6.21). Suppose now that (6.28) and (6.29) have been proved for $s = 1,\ldots,N$. Then, using (6.23), we obtain

$$\mathcal{F}_{k,N+1}(x,\rho) = (\mathcal{F}_{kN}(x,\rho) - \mathcal{F}_{kN}(0,\rho)\widehat{S}_N(x,\lambda)C_{N0}^{-1})x^{\mu_N-\mu_{N+1}}$$

$$= \frac{1}{\rho^{n-1}}\Bigg(-x^{\mu_1-\mu_{N+1}}\int_0^x\Big(\sum_{j=1}^n U_{j0}^0(x,\rho)U_j^{0,*}(t,\rho)\Big)(\rho t)^{n-1-\mu_n}V_k(t,\rho)\,dt$$

$$-\sum_{\ell=1}^{N-1}x^{\mu_\ell-\mu_{N+1}}\int_0^\infty\Big(\sum_{j=k+1}^n b_{j\ell}^0\rho^{\mu_\ell-\mu_1}(\widehat{S}_\ell(x,\lambda)-\widehat{C}_\ell(x,\lambda))F_j^*(\rho t)U_j^{0,*}(t,\rho)\Big)$$

$$\times V_k(t,\rho)dt + x^{\mu_N-\mu_{N+1}}\int_0^\infty\Big(\sum_{j=k+1}^n\Big(\sum_{\xi=N}^n b_{j\xi}^0\rho^{\mu_\xi-\mu_1}x^{\mu_\xi-\mu_N}\widehat{C}_\xi(x,\lambda)$$

$$-b_{jN}^0\rho^{\mu_N-\mu_1}\widehat{C}_N(x,\lambda) - b_{jN}^0\rho^{\mu_N-\mu_1}(\widehat{S}_N(x,\lambda)-\widehat{C}_N(x,\lambda))\Big)F_j^*(\rho t)U_j^{0,*}(t,\rho))V_k(t,\rho)\,dt\Bigg)$$

$$= \frac{1}{\rho^{n-1}}\Bigg(-x^{\mu_1-\mu_{N+1}}\int_0^x\Big(\sum_{j=1}^n U_{j0}^0(x,\rho)U_j^{0,*}(t,\rho)\Big)(\rho t)^{n-1-\mu_n}V_k(t,\rho)\,dt$$

$$-\sum_{\ell=1}^{N}x^{\mu_\ell-\mu_{N+1}}\int_0^\infty\Big(\sum_{j=k+1}^n b_{j\ell}^0\rho^{\mu_\ell-\mu_1}(\widehat{S}_\ell(x,\lambda)-\widehat{C}_\ell(x,\lambda))F_j^*(\rho t)U_j^{0,*}(t,\rho)\Big)V_k(t,\rho)\,dt$$

$$+\int_0^\infty\Big(\sum_{j=k+1}^n\Big(\sum_{\xi=N+1}^n b_{j\xi}^0\rho^{\mu_\xi-\mu_1}x^{\mu_s-\mu_{N+1}}\widehat{C}_\xi(x,\lambda)\Big)F_j^*(\rho t)U_j^{0,*}(t,\rho)\Big)V_k(t,\rho)\,dt\Bigg),$$

giving (6.29) for $s = N+1$. We now let $x \to 0$ in (6.29) for $s = N+1$, using (6.27), to obtain (6.28) for $s = N+1$. This proves Lemma 6.3.

It follows from (6.24) and (6.28) that

$$b_{ks}(\rho)\rho^{-\mu_s} - b_{ks}^0 = \frac{1}{\rho^{n-1}} \int_0^\infty \left(\sum_{j=k+1}^n b_{js}^0 F_j^*(\rho t) U_j^{0,*}(t,\rho) \right) V_k(t,\rho)\, dt. \qquad (6.30)$$

Using (6.30), (6.13), (6.18) and Lemma 6.3, we obtain

$$b_{ks}(\rho)\rho^{-\mu_s} - b_{ks}^0 = O(J(\rho)) = O(\rho^{-1}), \qquad |\rho| \to \infty, \qquad \rho \in S_{k_0},$$

i.e. (6.20) is valid. Theorem 6.2 is proved.

Note that, as a consequence of (6.10) and (6.18),

$$|\mathcal{Y}_k^{(\nu)}(x,\rho)(\rho R_k)^{-\nu} \exp(-\rho R_k x) - 1| \le \frac{M_4}{|\rho|}, \qquad x \ge 1, \qquad \rho \in \bar{S}_{k_0,\rho_0}. \qquad (6.31)$$

6.4. Let the functions $\Phi_m(x,\lambda)$, $m = \overline{1,n}$ be solutions of (6.1) satisfying the conditions $\Phi_m(x,\lambda) \sim c_{m0}x^{\mu_m}$, $x \to 0$; $\Phi_m(x,\lambda) = O(\exp(\rho R_m x))$, $x \to \infty$, $\rho \in S_{k_0}$. We call $\Phi_m(x,\lambda)$ the WS's for (6.1). Let $\{\mathcal{Y}_k(x,\rho)\}_{k=\overline{1,n}}$ be the FSS B of (6.1) in S_{k_0,ρ_0}. We will look for the WS's in the form

$$\Phi_m(x,\lambda) = \sum_{k=1}^n a_{mk}(\rho)\mathcal{Y}_k(x,\rho) = \sum_{j=1}^n S_j(x,\lambda) \sum_{k=1}^n b_{kj}(\rho) a_{mk}(\rho).$$

The conditions imposed on the WS's, combined with (6.16) and (6.31), imply that for $|\rho| \ge 2M_4$

$$a_{mk}(\rho) = 0, \qquad k > m; \qquad \sum_{k=1}^m b_{kj}(\rho) a_{mk}(\rho) = \delta_{jm}, \qquad j = \overline{1,m}.$$

Hence we obtain

$$\Phi_m(x,\lambda) = \sum_{k=1}^m a_{mk}(\rho)\mathcal{Y}_k(x,\rho) = S_m(x,\lambda) + \sum_{j=m+1}^n \mathfrak{M}_{mj}(\lambda) S_j(x,\lambda), \qquad (6.32)$$

$$a_{mk}(\rho) = (-1)^{m-k}(\Delta_{mm}(\rho))^{-1} \det [b_{\xi\nu}(\rho)]_{\xi=\overline{1,m}\setminus k;\, \nu=\overline{1,m-1}}, \qquad (6.33)$$

$$\mathfrak{M}_{mj}(\lambda) = \sum_{k=1}^m b_{kj}(\rho) a_{mk}(\rho) = (\Delta_{mm}(\rho))^{-1}\Delta_{mj}(\rho), \qquad j > m, \qquad (6.34)$$

where $\Delta_{mj}(\rho) = \det[b_{k\nu}(\rho)]_{k=\overline{1,m};\nu=1,m-1,j}$. Denote

$$\Delta^0_{00} = 1, \qquad \Delta^0_{mj} = \det[b^0_{k\nu}]_{k=\overline{1,m};\nu=\overline{1,m-1},j}, \qquad j \geq m \geq 1,$$

$$a^0_{mk} = \det[b^0_{\xi\nu}]_{\xi=\overline{1,m}\setminus k;\nu=\overline{1,m-1}}(-1)^{m-k}(\Delta^0_{mm})^{-1};$$

and $\Pi_{\pm 1}$ is the λ- plane with the cut $\pm\lambda \geq 0$. Since $b^0_{k\nu} = \beta^0_\nu R_k^{\mu_\nu}$, we have $\Delta^0_{mj} \neq 0$. Clearly, $a^0_{mm} = (\Delta^0_{mm})^{-1}\Delta^0_{m-1,m-1} \neq 0$. Using (6.20), (6.32)–(6.34), we see that for $|\rho| \to \infty$, $\rho \in S_{k_0}$, $\arg\rho = \varphi$

$$a_{mk}(\rho) = \rho^{-\mu_m}(a^0_{mk} + O(\rho^{-1})), \tag{6.35}$$

$$\mathfrak{M}_{mj}(\lambda) = \rho^{\mu_j - \mu_m}\mathfrak{M}^0_{mj}(1 + O(\rho^{-1})), \quad \mathfrak{M}^0_{mj} = (\Delta^0_{mm})^{-1}\Delta^0_{mj} \neq 0, \tag{6.36}$$

$$\Phi^{(\nu)}_m(x,\lambda) = \rho^{-\mu_m} a^0_{mm}(\rho R_m)^\nu \exp(\rho R_m x)(1 + O(\rho^{-1})) \tag{6.37}$$

for every fixed $x > 0$.

Repeating the preceding arguments for the FSS $B_{\alpha m} = \{\mathcal{Y}_{mk}(x,\rho)\}_{k=\overline{1,n}}$, we obtain

$$\mathcal{Y}_{mk}(x,\rho) = \sum_{j=1}^n B_{mkj}(\rho)S_j(x,\lambda),$$

$$\mathfrak{M}_{mj}(\lambda) = (\Delta^1_{mm}(\rho))^{-1}\Delta^1_{mj}(\rho), \quad \Delta^1_{mj}(\rho) = \det[B_{mk\nu}(\rho)]_{k=\overline{1,m};\nu=\overline{1,m-1},j}.$$

Denote $G = \{\rho : \arg\rho \in (((-1)^{n-m} - 1)\frac{\pi}{2n}, ((-1)^{n-m} + 3)\frac{\pi}{2n})\}$. The domain G is the union of two sectors with the same $\{R_\xi\}_{\xi=\overline{1,m}}$. Consequently, the functions $\Delta^1_{mj}(\rho)$ are regular for $\rho \in G$, $|\rho| > \rho_\alpha$, and continuous for $\rho \in \bar{G}$, $|\rho| \geq \rho_\alpha$. We have thus proved

Theorem 6.3. *The WS's $\Phi_m(x,\lambda)$ can be written as*

$$\Phi_m(x,\lambda) = S_m(x,\lambda) + \sum_{j=m+1}^n \mathfrak{M}_{mj}(\lambda)S_j(x,\lambda), \tag{6.38}$$

where the functions $\mathfrak{M}_{mj}(\lambda)$ are regular are in $\Pi_{(-1)^{n-m}}$ with the exception of an at most countable bounded set of poles Λ'_{mj} and continuous in $\bar{\Pi}_{(-1)^{n-m}}$ with the exception of bounded sets Λ_{mj}. The WS's $\{\Phi_m(x,\lambda)\}_{m=\overline{1,n}}$ form an FSS for (6.1), such that $\det[\Phi^{(\nu-1)}_m(x,\lambda)]_{m,\nu=\overline{1,n}} \equiv 1$. For $|\rho| \to \infty$, $\rho \in S_{k_0}$, $\arg\rho = \varphi$ and fixed $x > 0$, one has the symptotic formulas (6.36) and (6.37).

The functions $\mathfrak{M}_{mj}(\lambda)$ are called the WF's, and the matrix

$$\mathfrak{M}(\lambda) = [\mathfrak{M}_{mj}(\lambda)]_{m,j=\overline{1,n}}, \qquad \mathfrak{M}_{mj}(\lambda) = \delta_{mj} \qquad (m \geq j)$$

is called the WM for ℓ.

The IP is formulated as follows: given the WM $\mathfrak{M}(\lambda)$ construct the differential operator ℓ.

Let us prove the uniqueness theorem for the solutuion of the IP.

Theorem 6.4. *If $\mathfrak{M}(\lambda) = \widetilde{\mathfrak{M}}(\lambda)$, then $\ell = \tilde{\ell}$.*

Proof. Denote $S(x,\lambda) = [S_j^{(\nu)}(x,\lambda)]$, $\Phi(x,\lambda) = [\Phi_j^{(\nu)}(x,\lambda)]$.

The (6.38) becomes

$$\Phi(x,\lambda) = S(x,\lambda)\mathfrak{M}^T(\lambda). \tag{6.39}$$

Moreover, $\det\Phi(x,\lambda) = \det S(x,\lambda) \equiv 1$. Define a matrix $P(x,\lambda) = [P_{jk}(x,\lambda)]_{j,k=\overline{1,n}}$ by the formula $P(x,\lambda) = \Phi(x,\lambda)(\tilde{\Phi}(x,\lambda))^{-1}$ or

$$\begin{aligned} P_{jk}(x,\lambda) = \det[\tilde{\Phi}_m(x,\lambda),\dots,\tilde{\Phi}_m^{(k-2)}(x,\lambda),& \\ \Phi_m^{(j-1)}(x,\lambda),\tilde{\Phi}_m^{(k)}(x,\lambda),\dots,\tilde{\Phi}_m^{(n-1)}(x,\lambda)]_{m=\overline{1,n}}&. \end{aligned} \tag{6.40}$$

From (6.40) and the asymptotic properties of the WS's $\Phi_m(x,\lambda)$ and $\tilde{\Phi}_m(x,\lambda)$ we see that for a fixed $x > 0$ and $|\lambda| \to \infty$

$$P_{jk}(x,\lambda) = O(\rho^{j-k}), \qquad P_{1k}(x,\lambda) - \delta_{1k} = O(\rho^{-1}). \tag{6.41}$$

Using (6.39), we transform the matrix $P(x,\lambda)$ as follows:

$$P(x,\lambda) = \Phi(x,\lambda)(\tilde{\Phi}(x,\lambda))^{-1} = S(x,\lambda)(\tilde{S}(x,\lambda))^{-1}.$$

Hence we conclude that for each fixed $x > 0$ the functions $P_{jk}(x,\lambda)$ are entire in λ. Using (6.41) and Liouville's theorem, we obtain $P_{11}(x,\lambda) \equiv 1$, $P_{1k}(x,\lambda) \equiv 0$, $k = \overline{2,n}$. But then $\Phi_m(x,\lambda) \equiv \tilde{\Phi}_m(x,\lambda)$ for all x, λ, m, and hence $\ell = \tilde{\ell}$. Theorem 6.4 is proved.

Using the results, obtained above, and the contour integral method, one can obtain an algorithm for the solution of the IP from the WM, along with necessary and sufficient conditions of its solvability, in analogous manner as in §2.

PART II

RECOVERY OF DIFFERENTIAL OPERATORS FROM THE WEYL FUNCTIONS

7. Differential operators with a "separate" spectrum

We consider the IP of recovering differential operators of the form (1.1) under the condition of "separation" of the spectrum. In this case we do not need all the WM but only its part to construct the diffirential operator. More exactly, the differential operator is uniquely determined from given $n-1$ Weyl functions. We provide a rule how to choose sets of the WF's which guarantee uniqueness of the solution of the IP. We give the solution of the IP from chosen WF's. It is shown that theorems obtained contain results of Leibenzon [53]–[54]. Further, we give a solution of the IP of recovering the differential operator from a system of $2n-2$ spectra. It is shown that this problem can be reduced to the IP from the WF's.

7.1. We consider the DE and LF (1.1)–(1.2). For definiteness, we confine ourselves to the case of $T < \infty$. Let Λ_{mk} $(1 \le m \le k \le n)$ be the set of zeroes (with multiplicity) of the entire function

$$\Delta_{mk}(\lambda) = (-1)^{m+k} \det [U_{\eta,T}(C_\nu)]_{\nu=\overline{1,n-m};\nu=\overline{m,n}\setminus k}.$$

The set Λ_{mk} coincides with the set of eigenvalues of the boundary value problem S_{mk} for the DE (1.1) under the conditions $U_{\xi 0}(y) = U_{\eta T}(y) = 0$, $\eta = \overline{1,n-m}$, $\xi = \overline{1,m-1},k$. In particular, $\Lambda_{mm} = \Lambda_m$.

Let r $(2 \le r \le n)$ be a fixed natural, and $\theta_m = \max(r, m+1)$. Suppose that

$$\Lambda_{mm} \cap \Lambda_{m+1,\theta_m} = \oslash, \qquad m = \overline{1,n-2}. \tag{7.1}$$

We suppose everywhere in §7 that the condition (7.1) of "separation" of the spectrum is fulfilled. In this case we need $n-1$ WF's for recovering the DE and LF. The IP is formulated as follows

Problem 7.1. Given the WF's $\{\mathfrak{M}_{m,\theta_m}(\lambda)\}_{m=\overline{1,n-1}}$, construct the DE and LF $L = (\ell, U)$.

First of all we study the uniqueness of the solution of the IP.

Theorem 7.1. *If* $\mathfrak{M}_{m,\theta_m}(\lambda) = \widetilde{\mathfrak{M}}_{m,\theta_m}(\lambda)$, $m = \overline{1,n-1}$, *then* $L = \widetilde{L}$.

For definiteness, we prove Theorem 7.1 for $r = 2$. For $r = 2$ the "separation" condition (7.1) means that

$$\Lambda_{mm} \cap \Lambda_{m+1,m+1} = \oslash, \qquad m = \overline{1,n-2}, \tag{7.2}$$

and the condition of Theorem 7.1 become

$$\mathfrak{M}_{m,m+1}(\lambda) = \widetilde{\mathfrak{M}}_{m,m+1}(\lambda), \qquad m = \overline{1, n-1}. \tag{7.3}$$

First we prove an auxiliary statement.

Lemma 7.1. *Suppose that for a certain m $(1 \le m \le n-1)$ a number λ_0 is a zero of $\Delta_{mm}(\lambda)$ of multiplicity $\kappa_m \ge 1$ and $\Delta_{m+1,m+1}(\lambda_0) \ne 0$. Then in a neighbourhood of the point $\lambda = \lambda_0$ we have the representation*

$$\Phi_m(x, \lambda) = \xi_m(x, \lambda) + \sum_{\nu=1}^{\kappa_m} \frac{C_{\nu m}}{(\lambda - \lambda_0)^\nu} \Phi_{m+1}(x, \lambda), \tag{7.4}$$

where the function $\xi_m(x, \lambda)$ is regular at $\lambda = \lambda_0$.

Proof. Since $\Delta_{m+1,m+1}(\lambda_0) \ne 0$, it follows from (1.13) that the function $\Phi_{m+1}(x, \lambda)$ is regular at $\lambda = \lambda_0$, and for $\lambda = \lambda_0$

$$U_{\xi 0}(\Phi_{m+1}(x, \lambda_0)) = U_{\xi T}(\Phi_{m+1}(x, \lambda_0)) = 0, \qquad \xi = \overline{1, m}, \qquad \eta = \overline{1, n-m}.$$

Moreover, $U_{m+1,0}(\Phi_{m+1}(x, \lambda_0)) = 1$. Thus, the function $\Phi_{m+1}(x, \lambda_0) \not\equiv 0$ is an eigenfunction of S_{mm} for the eigenvalue λ_0.

The number λ_0 is a one-fold [92, p. 24] eigenvalue of S_{mm}. Indeed, if there were two linear independent eigenfunctions $f_1(x)$ and $f_2(x)$ of S_{mm} for the eigenvalue λ_0, then they could be chosen such that $U_{m+1,0}(f_2) = 0$. But this would mean that $f_2(x)$ is an eigenfunction of $S_{m+1,m+1}$, i.e. $\Delta_{m+1,m+1}(\lambda_0) = 0$. This contradicts the condition of the lemma.

Let us show by induction that there exist sequences of functions $\xi_{\nu m}(x, \lambda)$ and numbers $C_{\nu m}$ $(\nu = \overline{0, \kappa_m})$ such that the functions $\xi_{\nu m}(x, \lambda)$ are regular at $\lambda = \lambda_0$, and

$$\left.\begin{aligned} &\xi_{\nu m}(x, \lambda) = (\lambda - \lambda_0)^{-1}(\xi_{\nu+1,m}(x, \lambda) - C_{\nu+1,m}\Phi_{m+1}(x, \lambda)), \\ &\xi_{\kappa_m,m}(x, \lambda) = (\lambda - \lambda_0)^{\kappa_m}\Phi_m(x, \lambda), \end{aligned}\right\} \tag{7.5}$$

$$\left.\begin{aligned} &U_{i0}(\xi_{\nu m}(x, \lambda)) = U_{jT}(\xi_{\nu m}(x, \lambda)) = 0, \quad i = \overline{1, m-1},\ j = \overline{1, n-m-1}, \\ &U_{m0}(\xi_{\nu m}(x, \lambda)) = (\lambda - \lambda_0)^\nu,\ \ U_{n-m,T}(\xi_{\nu m}(x, \lambda)) = (\lambda - \lambda_0)^\nu \eta_{\nu m}(\lambda), \end{aligned}\right\} \tag{7.6}$$

where the functions $\eta_{\nu m}(\lambda)$ are regular at $\lambda = \lambda_0$. It is clear that this yields (7.4) where $\xi_m(x, \lambda) = \xi_{0m}(x, \lambda)$.

(1) In a neighbourhood of the point $\lambda = \lambda_0$ we have $\Phi_m(x, \lambda) = (\lambda - \lambda_0)^{-\kappa_m}\xi_{\kappa_m,m}(x, \lambda)$ by virtue of (1.13), where the function $\xi_{\kappa_m,m}(x, \lambda)$ is regular at $\lambda = \lambda_0$ and satisfies (7.6) for $\nu = \kappa_m$. This means that the function $\xi_{\kappa_m,m}(x, \lambda_0)$ is an eigenfunction of S_{mm} for the eigenvalue λ_0. Since λ_0 is a one-fold eigenvalue it follows that $\xi_{\kappa_m,m}(x, \lambda_0) = C_{\kappa_m,m}\Phi_{m+1}(x, \lambda_0)$, and hence the function $\xi_{\kappa_m-1,m}(x, \lambda)$, defined

by (7.5) for $\nu = \kappa_m - 1$ is regular at $\lambda = \lambda_0$. The relations (7.6) for $\nu = \kappa_m - 1$ are obvious.

(2) Suppose that our assertion is proved for $\nu = \kappa_m - 1, \ldots, s$ $(s \geq 1)$. By the assumption of induction the function $\xi_{sm}(x, \lambda)$ is regular at $\lambda = \lambda_0$ and satisfies (7.6) for $\nu = s$. It means that $\xi_{sm}(x, \lambda_0) = C_{sm}\Phi_{m+1}(x, \lambda_0)$, and hence the function $\xi_{s-1,m}(x, \lambda)$ defined by (7.5) for $\nu = s - 1$ is regular at $\lambda = \lambda_0$. Relations (7.6) for $\nu = s - 1$ follows from (7.5) and (7.6) for $\nu = s$. Lemma 7.1 is proved.

Proof of Theorem 7.1. Suppose that for a certain m $(1 \leq m \leq n-1)$ a number λ_0 is a zero of $\Delta_{mm}(\lambda)$ of multiplicity κ_m, i.e. $\lambda_0 \in \Lambda_{mm}$. Then it follows from (7.2) that $\lambda_0 \notin \Lambda_{m+1,m+1}$, i.e. $\Delta_{m+1,m+1}(\lambda_0) \neq 0$, and, by Lemma 7.1, we have the representation (7.4) in a neighbourhood of $\lambda = \lambda_0$. Applying the LF $U_{m+1,0}$ to the both sides of (7.4) and taking into account the relations $U_{m+1,0}(\Phi_m(x, \lambda)) = \mathfrak{M}_{m,m+1}(\lambda)$, $U_{m+1,0}(\Phi_{m+1}(x, \lambda)) = 1$, we obtain

$$\mathfrak{M}_{m,m+1}(\lambda) = U_{m+1,0}(\xi_m(x, \lambda)) + \sum_{\nu=1}^{\kappa_m} \frac{C_{\nu m}}{(\lambda - \lambda_0)^\nu} .$$

Hence $C_{\nu m} = [\mathfrak{M}_{m,m+1}(\lambda)]^{<-\nu>}_{\lambda=\lambda_0}$. By virtue of (7.3) we get $C_{\nu m} = \widetilde{C}_{\nu m}$. It follows from Lemma 7.1 and (1.35) that for each fixed $\lambda_0 \in \Lambda$ we have the following reperesentation in a neighbourhood of the point $\lambda = \lambda_0$

$$[\Phi_m^{(\nu-1)}(x, \lambda)]_{\nu,m=\overline{1,n}} = [\xi_m^{(\nu)}(x, \lambda)]_{\nu,m=\overline{1,n}}[\theta_{\nu m}(\lambda)]_{\nu,m=\overline{1,n}},$$

where the functions $\xi_m(x, \lambda)$ are regular at $\lambda = \lambda_0$,

$$\det [\xi_m^{(\nu-1)}(x, \lambda)]_{\nu,m=\overline{1,n}} = (-1)^{n(n-1)/2},$$

and $\theta_{\nu m}(\lambda) = \widetilde{\theta}_{\nu m}(\lambda)$. Hence, for each fixed $x \in [0, T]$ the matrix $\mathcal{P}(x, \lambda)$, defined in §1, is entire in λ. Further, as in the proof of Theorem 1.2, we obtain that $L = \widetilde{L}$. Theorem 7.1 is proved.

The counterexample from 4.7 shows that waiving the requirement of "separation" of the spectrum leads to a violation of uniqueness for solution of the IP.

7.2. Here we provide necessary and sufficient conditions and an algorithm for the solution of the IP. For simplicity, we confine ourselves to the case of $L \in v'_{N2}$.

Lemma 7.2. *If* $\lambda_0 \in \Lambda_m \cap \ldots \cap \Lambda_{\mu-1}$, $\lambda_0 \notin \Lambda_\mu$, *then* $\mathfrak{N}_{\xi\mu}(\lambda_0) \neq 0$, $\mathfrak{N}_{\xi j}(\lambda_0) = 0$, $j = \overline{\xi+1, n}\backslash\mu$ *for* $\xi = \overline{m, \mu-1}$.

Indeed, by the condition of the lemma $\Delta_{\xi\xi}(\lambda_0) = 0$ for $\xi = \overline{m, \mu-1}$. Denote $h_\xi(x, \lambda) = \Delta_{\xi\xi}(\lambda)\Phi_\xi(x, \lambda)$. It follows from (1.13) that the functions $h_\xi(x, \lambda)$ are entire in λ. If $h_\xi(x, \lambda_0) \equiv 0$, then from (1.13) follows that $\Delta_{\xi-1,\xi-1}(\lambda_0) = \Delta_{\xi s}(\lambda_0) = 0$, $s = \overline{\xi, n}$. But this is impossible by virtue of (7.1). Thus $h_\xi(x, \lambda_0) \equiv 0$, and hence $\Phi_{\xi,<-1>}(x, \lambda_0) \not\equiv 0$. According to (4.5) we get

$$\Phi_{\xi,<-1>}(x, \lambda_0) = \sum_{j=\xi+1}^{n} \mathfrak{N}_{\xi j}(\lambda_0)\Phi_{j,<0>}(x, \lambda_0).$$

Therefore

$$\sum_{j=\xi+1}^{n} |\mathfrak{N}_{\xi j}(\lambda_0)| \neq 0.$$

Further, since $\Delta_{\mu-1,\mu-1}(\lambda_0) = 0$, $\Delta_{\mu\mu}(\lambda_0) \neq 0$, it follows from Lemma 4.1 that $\mathfrak{N}_{\mu-1,j}(\lambda_0) = 0$, $j = \overline{\mu+1,n}$; $\mathfrak{N}_{\mu-1,\mu}(\lambda_0) \neq 0$. From this and from Lemma 4.2 we obtain the assertion of Lemma 7.2.

By (7.1) and Lemma 7.2, there exist natural $\mu_{\ell m}$ for each $\ell \geq 1$, $m = \overline{1,n-1}$ $(m+1 \leq \mu_{\ell m} \leq \theta_m)$ such that

$$\Delta_{kk}(\lambda_{\ell m}) = 0, \qquad k = \overline{m+1, \mu_{\ell m} - 1}; \qquad \Delta_{\mu_{\ell m}, \mu_{\ell m}}(\lambda_{\ell m}) \neq 0.$$

Moreover $\Delta_{\mu_{\ell m}, \theta_m}(\lambda_{\ell m}) \neq 0$, and consequently $\mathfrak{M}_{\mu_{\ell m}, \theta_m, <0>}(\lambda_{\ell m}) \neq 0$. It follows from Lemma 7.2 that

$$\mathfrak{N}_{m,\mu_{\ell m}}(\lambda_{\ell m}) \neq 0, \qquad \mathfrak{N}_{mj}(\lambda_{\ell m}) = 0, \qquad j = \overline{m+1,n} \backslash \mu_{\ell m}. \tag{7.7}$$

Furthermore, from the equality $\mathfrak{N}\mathfrak{M}_{<0>} = \mathfrak{M}_{<-1>}$ we obtain

$$\mathfrak{N}_{mk}(\lambda_{\ell m}) = \mathfrak{M}_{mk,<-1>}(\lambda_{\ell m}) - \sum_{j=m+1}^{k-1} \mathfrak{N}_{mj}(\lambda_{\ell m}) \mathfrak{M}_{jk,<0>}(\lambda_{\ell m}). \tag{7.8}$$

Hence $\mathfrak{M}_{m,\theta_m,<-1>}(\lambda_{\ell m}) - \mathfrak{N}_{m,\mu_{\ell m}}(\lambda_{\ell m}) \mathfrak{M}_{\mu_{\ell m},\theta_m,<0>}(\lambda_{\ell m}) = 0$ or

$$\mathfrak{N}_{m,\mu_{\ell m}}(\lambda_{\ell m}) = (\mathfrak{M}_{\mu_{\ell m},\theta_m,<0>}(\lambda_{\ell m}))^{-1} \beta_{\ell,m,\theta_m}. \tag{7.9}$$

Relations (7.7) and (7.9) give us the connections allowing us to find the WM $\mathfrak{M}(\lambda)$ from the given WF's $\{\mathfrak{M}_{m,\theta_m}(\lambda)\}_{m=\overline{1,n-1}}$ (or, which is the same, from their poles and residues $\{\lambda_{\ell m}, \beta_{\ell,m,\theta_m}\}$). Thus, our IP can be reduced to the IP of recovering L from the given WM $\mathfrak{M}(\lambda)$.

First we formulate necessary and sufficient conditions for $r = 2$. In this case we have

$$\theta_m = m+1, \qquad \mathfrak{N}_{m,m+1}(\lambda_{\ell m}) = \mathfrak{M}_{m,m+1,<-1>}(\lambda_{\ell m}) \neq 0,$$

$$\mathfrak{N}_{mj}(\lambda_{\ell m}) = 0, \qquad j = \overline{m+2,n}.$$

So for $r = 2$ numbers ξ_ℓ, defined in 4.3, have the form

$$\xi_\ell = \sum_{m=1}^{n-1} (|\lambda_{\ell m} - \tilde{\lambda}_{\ell m}| + |\beta_{\ell,m,m+1} - \tilde{\beta}_{\ell,m,m+1}|\ell) \ell^{1-n},$$

where

$$\beta_{\ell mk} = \beta_{\ell,m,m+1} \mathfrak{M}_{m+1,k,<0>}(\lambda_{\ell m}). \tag{7.10}$$

From the given $\{\mathfrak{M}_{m,m+1}(\lambda)\}_{m=\overline{1,n-1}}$ (or, which is the same, from $\{\lambda_{\ell m}, \beta_{\ell,m,m+1}\}_{\ell\geq 1})$, using (4.4) and (7.10) one can construct the WF's $\mathfrak{M}_{mk}(\lambda)$ for $k > m+1$ reccurently. Thus, the WM $\mathfrak{M}(\lambda)$ is constructed. At that, the properties S_1 and S_2 for $\mathfrak{M}(\lambda)$ are clearly fulfilled. Thus, the following Theorems 7.2 and 7.3 follows from Theorem 4.3 and 4.4.

Theorem 7.2. *For meromorphic functions $\{\mathfrak{M}_{m,m+1}(\lambda)\}_{m=\overline{1,n-1}}$ with simple poles $\{\lambda_{\ell m}\}_{\ell\geq 1}$, $\lambda_{\ell m} \neq \lambda_{\ell_0,m+1}$ $(\ell, \ell_0 \geq 1)$ and residues $\beta_{\ell,m,m+1} \neq 0$ to be the WF's for $L \in v'_{N2}$ it is necessary and sufficient that the following conditions hold:*

(1) $|\mathfrak{M}_{m,m+1}(\lambda)| < C|\rho|^{-1}, \quad \lambda \in G_\delta$;

(2) *there exists* $\widetilde{L} \in v'_{N2}$ *such that* $\sum_{\ell}(\xi_\ell \ell^{n+N-1})^2 < \infty$;

(3) *condition P of Theorem 4.1 is fulfilled.*

Theorem 7.3. *Let $\widetilde{L} \in v'_{N2}$ be given. Then there exists $\delta > 0$ (which depends on $\widetilde{L}$) such that if numbers $\{\lambda_{\ell m}, \beta_{\ell,m,m+1}\}_{\ell\geq 1, m=\overline{1,n-1}}$ $\lambda_{\ell m} \neq \lambda_{\ell_0,m}$ $(\ell \neq \ell_0)$, $\lambda_{\ell m} \neq \lambda_{\ell_0,m+1}$ $(\ell, \ell_0 \geq 1)$, $\beta_{\ell,m,m+1} \neq 0$ satisfy the condition*

$$\Lambda^+ := \left(\sum_{\ell=1}^{\infty}(\xi_\ell \ell^{n+N-1})^2\right)^{1/2} < \delta,$$

then there exists a unique $L \in v'_{N2}$ for which $\{\lambda_{\ell m}, \beta_{\ell,m,m+1}\}_{\ell\geq 1}$ are poles and residues of the WF's $\mathfrak{M}_{m,m+1}(\lambda)$. Moreover

$$\|p_j^{(\nu)}(x) - \widetilde{p}_j^{(\nu)}(x)\|_{\mathcal{L}_2(0,T)} < C\Lambda^+, \qquad 0 \leq \nu \leq j+N; \qquad |u_{\xi\nu a} - \widetilde{u}_{\xi\nu a}| < C\Lambda^+,$$

where the constant C depends only on $\widetilde{L}$.

We note that to solve the IP it is not necessary to find the WM $\mathfrak{M}(\lambda)$, since the main equation of the IP can be constructed directly from $\lambda_{\ell m}$ and $\beta_{\ell,m,m+1}$ for $r = 2$.

Remark. From Theorems 7.2 and 7.3 the results of Leibenzon [53]–[54] follow. Indeed, in [53]–[54] the IP of recovering the DE and LF is studied from the given $\{\lambda_{\ell m}, \alpha_{\ell m}\}_{\ell\geq 1}^{m=\overline{1,n-1}}$ under the "separation" condition (7.2), where $\alpha_{\ell m}$ are "weight" numbers connected with the residues $\beta_{\ell,m,m+1}$ of the WF's $\mathfrak{M}_{m,m+1}(\lambda)$ by the formula

$$\beta_{\ell,m,m+1} = (\dot{\Delta}_{mm}(\lambda_{\ell m}))^{-1}\Delta_{m,m+1}(\lambda_{\ell m}) = (-1)^{n-m}(\alpha_{\ell m})^{-1}.$$

Thus, the specification of the numbers $\{\lambda_{\ell m}, \alpha_{\ell m}\}$ is equivalent to the specification of the WF's $\{\mathfrak{M}_{m,m+1}(\lambda)\}_{m=\overline{1,n-1}}$, and the problem of Leibenzon is a particular case of Problem 7.1.

Now we consider the IP for $r > 2$. In this case it is necessary to find the asymptotics for $\beta_{\ell,m,m+1}$ or for $\mathfrak{N}_{m,\mu_{\ell m}}(\lambda_{\ell m})$ using the asymptotics for $\{\lambda_{\ell m}, \beta_{\ell,m,\theta_m}\}$. To obtain the asymptotics we use the IP for $r = 2$ which was solved above.

Lemma 7.3. *Let meromorphic functions $\mathfrak{M}_{m,\theta_m}$, $m = \overline{1, n-1}$ with simple poles $\{\lambda_{\ell m}\}_{\ell \geq 1}$ and residues $\beta_{\ell,m,\theta_m} \neq 0$ be given, with $|\mathfrak{M}_{m,\theta_m}(\lambda)| < C|\rho|^{-1}$, $\lambda \in G_\delta$. Then there exists $\tilde{\tilde{L}} \in v'_{N2}$ such that*

$$\sum_{\ell=1}^{\infty} (\xi_\ell^1 \ell^{n+N-1})^2 < \infty,$$

where

$$\xi_\ell^1 = \sum_{m=1}^{n-1} (|\lambda_{\ell m} - \tilde{\tilde{\lambda}}_{\ell m}| |\beta_{\ell,m,\theta_m} - \tilde{\tilde{\beta}}_{\ell,m,\theta_m}| \ell^{\theta_m - m}) \ell^{1-n}.$$

Define $\mathfrak{N}_{m,\mu_{\ell m}}(\lambda_{\ell m})$ by (7.9). Then

$$\mathfrak{N}_{m,\mu_{\ell m}}(\lambda_{\ell m}) = \ell^{n-2} \left(\sum_{\nu=0}^{n+N-1} \frac{\chi^1_{m\nu}}{\ell^\nu} + \frac{\kappa^1_{\ell,m}}{\ell^{n+N-1}} \right), \qquad \chi^1_{m0} \neq 0, \qquad \kappa^1_{\ell m} \in \ell_2. \quad (7.11)$$

Proof. According to (4.2) we have

$$\tilde{\tilde{\beta}}_{\ell,m,\theta_m} = \ell^{n+m-\theta_m-1} \left(\sum_{\nu=0}^{n+N-1} \frac{\tilde{\tilde{\chi}}_{m,\theta_m,\nu}}{\ell^\nu} + \frac{\tilde{\tilde{\kappa}}_{\ell,m,\theta_m}}{\ell^{n+N-1}} \right), \quad (7.12)$$

$$\tilde{\tilde{\kappa}}_{\ell,m,\theta_m} \in \ell_2, \qquad \tilde{\tilde{\chi}}_{m,\theta_m,0} \neq 0.$$

For $m \geq r-1$ we obtain

$$\theta_m = m+1, \qquad \mu_{\ell m} = m+1, \qquad \mathfrak{N}_{m,\mu_{\ell m}}(\lambda_{\ell m}) = \beta_{\ell,m,m+1} = \beta_{\ell,m,\theta_m},$$

and (7.11) follows clearly from (7.12).

Let $m < r-1$. Then $\theta_m = r$. According to (1.14) we have $\mu_{\ell m} = m+1$, for sufficiently large ℓ ($\ell > \ell^+$) and, in view of (7.9) we get

$$\mathfrak{N}_{m,\mu_{\ell m}}(\lambda_{\ell m}) = \beta_{\ell,m,m+1} = \beta_{\ell m r}(\mathfrak{M}_{m+1,r,<0>}(\lambda_{\ell m}))^{-1}, \qquad m = \overline{1, r-2}. \quad (7.13)$$

It follows from the conditions of the lemma and from (7.12) that

$$\beta_{\ell m r} = \ell^{n+m-r-1} \left(\sum_{\nu=0}^{n+N-1} \frac{\chi_{mr\nu}}{\ell^\nu} + \frac{\kappa_{\ell m r}}{\ell^{n+N-1}} \right), \qquad \chi_{mr0} \neq 0, \qquad \kappa_{\ell m r} \in \ell_2. \quad (7.14)$$

Further, from the conditions of the lemma follows that

$$\mathfrak{M}_{m+1,r}(\lambda) = \sum_{\ell_0=1}^{\infty} \frac{\beta_{\ell_0,m+1,r}}{\lambda - \lambda_{\ell_0,m+1}},$$

and hence

$$\mathfrak{M}_{m+1,r,<0>}(\lambda_{\ell m}) = \sum_{\ell_0=1}^{\infty} \frac{\beta_{\ell_0,m+1,r}}{\lambda_{\ell m} - \lambda_{\ell_0,m+1}}, \qquad \ell > \ell^+. \tag{7.15}$$

To obtain the asymptotics for $\mathfrak{M}_{m+1,r,<0>}(\lambda_{\ell m})$ we use Theorem 7.3. Fix m. According to Theorem 7.3, there exists $\widetilde{L} \in v'_{N2}$ and a number q such that for $\ell > q$ $\lambda_{\ell,m+1} = \widetilde{\lambda}_{\ell,m+1}$, $\beta_{\ell,m+1,r} = \widetilde{\beta}_{\ell,m+1,r}$. Hence from (7.15) we get

$$\mathfrak{M}_{m+1,r,<0>}(\lambda_{\ell m}) = \widetilde{\mathfrak{M}}_{m+1,r,<0>}(\widetilde{\lambda}_{\ell m})$$
$$+ \sum_{\ell_0=1}^{q} \left(\frac{\beta_{\ell_0,m+1,r}}{\lambda_{\ell m} - \lambda_{\ell_0,m+1}} - \frac{\beta_{\ell_0,m+1,r}}{\lambda_{\ell m} - \widetilde{\lambda}_{\ell_0,m+1}} \right), \qquad \ell > \max(\ell^+, q). \tag{7.16}$$

Since, by (4.2),

$$\widetilde{\mathfrak{M}}_{m+1,r,<0>}(\widetilde{\lambda}_{\ell m}) = \ell^{m+1-r} \left(\sum_{\nu=0}^{n+N-1} \frac{\widetilde{\eta}_{mk\nu}}{\ell^{\nu}} + \frac{\widetilde{\kappa}^0_{\ell mk}}{\ell^{n+N-1}} \right), \quad \widetilde{\eta}_{mk0} \neq 0, \quad \widetilde{\kappa}^0_{\ell mk} \in \ell_2,$$

it follows from (7.16) that

$$\mathfrak{M}_{m+1,r,<0>}(\lambda_{\ell m}) = \ell^{m+1-r} \left(\sum_{\nu=0}^{n+N-1} \frac{\eta_{mk\nu}}{\ell^{\nu}} + \frac{\kappa^0_{\ell mk}}{\ell^{n+N-1}} \right), \quad \eta_{mk0} \neq 0, \quad \kappa^0_{\ell mk} \in \ell_2.$$

This and (7.14) yield Lemma 7.3 is proved.

From Theorem 4.3 and Lemma 7.3 we get Theorem 7.4.

Theorem 7.4. *For meromorphic functions $\{\mathfrak{M}_{m,\theta_m}(\lambda)\}_{m=\overline{1,n-1}}$ with simple poles $\{\lambda_{\ell m}\}_{\ell \geq 1}$ and residues $\beta_{\ell,m,\theta_m} \neq 0$ to be the WF's for $L \in v'_{N2}$ it is necessary and sufficient that the following conditions hold:*

(1) $|\mathfrak{M}_{m,\theta_m}(\lambda)| < C|\rho|^{-1}, \quad \lambda \in G_\delta$;

(2) *(asymptotics) there exists $\widetilde{\widetilde{L}} \in v'_{N2}$ such that*

$$\sum_{\ell=1}^{\infty} (\xi^1_\ell \ell^{n+N-1})^2 < \infty;$$

3) *for* $\widetilde{L} \in v'_{N2}$ *such that*

$$\sum_{\ell=1}^{\infty}\left(\sum_{m=1}^{n-1}(|\lambda_{\ell m}-\widetilde{\lambda}_{\ell m}|+|\mathfrak{N}_{m,\mu_{\ell m}}(\lambda_{\ell m})-\widetilde{\mathfrak{N}}_{m,\mu_{\ell m}}(\widetilde{\lambda}_{\ell m})|\ell)\ell^N\right)^2<\infty,$$

condition P *is fulfilled.*

7.3. We consider the IP of recovering the DE and LF (1.1)–(1.2) from a system of $2n-2$ spectra. Denote by $\{\lambda^1_{\ell m}\}_{\ell\ge 1}$ the eigenvalues of S_{m,θ_m}. The IP is formulated as follows.

Problem 7.2. Given the spectra $\{\lambda_{\ell m},\lambda^1_{\ell m}\}_{\ell\ge 1, m=\overline{1,n-1}}$, construct the DE and LF $L=(\ell,U)$.

Let us show that this IP can be reduced to Problem 7.1 of recovering L from the WF's $\{\mathfrak{M}_{m,\theta_m}(\lambda)\}_{m=\overline{1,n-1}}$.

Let $\Lambda_{mk}=\{\lambda_{\ell mk}\}_{\ell\ge 1}$, i.e. the numbers $\{\lambda_{\ell mk}\}_{\ell\ge 1}$ are eigenvalues of S_{mk}. It follows from (1.21) that the function $\Delta_{mk}(\lambda)$ is entire in λ of the order $\frac{1}{n}$. Since Λ_{mk} is the set of zeroes of $\Delta_{mk}(\lambda)$, we have by the Borel theorem [96, p. 31]

$$\Delta_{mk}(\lambda)=B_{mk}\prod_{\ell=1}^{\infty}\left(1-\frac{\lambda}{\lambda_{\ell mk}}\right),\qquad B_{mk}\ \text{— const.}$$

(the case when $\lambda=0$ is the eigenvalue of S_{mk} requires minor modifications). Then

$$\frac{\Delta_{mk}(\lambda)}{\widetilde{\Delta}_{mk}(\lambda)}=\frac{B_{mk}}{\widetilde{B}_{mk}}\prod_{\ell=1}^{\infty}\frac{\widetilde{\lambda}_{\ell mk}}{\lambda_{\ell mk}}\prod_{\ell=1}^{\infty}\left(1-\frac{\widetilde{\lambda}_{\ell mk}-\lambda_{\ell mk}}{\widetilde{\lambda}_{\ell mk}-\lambda}\right).$$

Since for $(-1)^{n-m+1}\ \lambda\to\infty$

$$\lim\frac{\Delta_{mk}(\lambda)}{\widetilde{\Lambda}_{mk}(\lambda)}=1,\qquad \lim\prod_{\ell=1}^{\infty}\left(1-\frac{\widetilde{\lambda}_{\ell mk}-\lambda_{\ell mk}}{\widetilde{\lambda}_{\ell mk}-\lambda}\right)=1,$$

it follows that

$$\frac{B_{mk}}{\widetilde{B}_{mk}}\prod_{\ell=1}^{\infty}\frac{\widetilde{\lambda}_{\ell mk}}{\lambda_{\ell mk}}=1,$$

and hence

$$\Delta_{mk}(\lambda)=\widetilde{\Delta}_{mk}(\lambda)\prod_{\ell=1}^{\infty}\left(1-\frac{\widetilde{\lambda}_{\ell mk}-\lambda_{\ell mk}}{\widetilde{\lambda}_{\ell mk}-\lambda}\right). \tag{7.17}$$

In particular, from this we get that the characteristic function $\Delta_{mk}(\lambda)$ of the boundary value problem S_{mk} is uniquely determined by its zeros. Moreover, the function $\Delta_{mk}(\lambda)$ can be constructed by (7.17), where $\widetilde{L} = (\widetilde{\ell}, \widetilde{U})$ are known DE and LF (for example, with zero coefficients). Then, in view of (1.7) we get the following statement.

Lemma 7.4. *If* $\Lambda_m = \widetilde{\Lambda}_m$, $\Lambda_{mk} = \widetilde{\Lambda}_{mk}$, *then* $\mathfrak{M}_{mk}(\lambda) = \widetilde{\mathfrak{M}}_{mk}(\lambda)$.

Thus, the specification of two spectra of S_{mm} and S_{mk} uniquely determines the WF $\mathfrak{M}_{mk}(\lambda)$.

From Theorem 7.1 and Lemma 7.4 we obtain the following uniqueness theorem of the solution of the IP from a system of $2n-2$ spectra.

Theorem 7.5. *If* $\lambda_{\ell m} = \widetilde{\lambda}_{\ell m}$, $\lambda^1_{\ell m} = \widetilde{\lambda}^1_{\ell m}$, $\ell \geq 1$, $m = \overline{1, n-1}$, *then* $L = \widetilde{L}$.

Thus, the DE and LF are uniquely determined from the given $2n-2$ spectra of the boundary value problems S_{mm}, S_{m,θ_m}, $m = \overline{1, n-1}$.

To solve the IP of recovering from $2n-2$ spectra $\{\lambda_{\ell m}, \lambda^1_{\ell m}\}_{\ell \geq 1}^{m=\overline{1,n-1}}$ we can construct the characteristic functions $\Delta_{mm}(\lambda)$, $\Delta_{m,\theta_m}(\lambda)$ and then the WF's $\mathfrak{M}_{m,\theta_m}(\lambda)$ by the formula

$$\mathfrak{M}_{m,\theta_m}(\lambda) = (\Delta_{mm}(\lambda))^{-1} \Delta_{m,\theta_m}(\lambda)$$

and the residues β_{ℓ,m,θ_m} of the WF's $\mathfrak{M}_{m,\theta_m}(\lambda)$ by the formula

$$\beta_{\ell,m,\theta_m} = (\dot{\Delta}_{mm}(\lambda_{\ell m}))^{-1} \Delta_{m,\theta_m}(\lambda_{\ell m})$$

and use the results of 7.2. Thus, the IP from $2n-2$ spectra can be reduced to the IP of recovering the DE and LF L from the WF's $\{\mathfrak{M}_{m,\theta_m}(\lambda)\}_{m=\overline{1,n-1}}$ or, which is the same, from the poles and residues of the WF's $\{\lambda_{\ell,m}, \beta_{\ell,m,\theta_m}\}_{\ell \geq 1, m=\overline{1,n-1}}$.

8. Stability of the solution of the inverse problem

Stability of the solution of the IP from the WM was studied in Part I. Things are more complicated for the IP from a system of spectra. Here we study stability of the solution of the IP in the uniform norm from spectra. It is shown that small perturbations of the spectra lead to small perturbations of the operator. In 8.1 we consider second-order differential operators. Higher-order differential operators are studied in 8.2.

8.1. Let $\{\lambda_{ni}\}_{n \geq 0}$, $i = 1, 2$ be eigenvalues of the boundary value problems L_i of the form

$$-y'' + q(x)y = \lambda y = \rho^2 y, \qquad 0 \leq x \leq \pi, \tag{8.1}$$

$$y'(0) - hy(0) = y^{(i-1)}(\pi) = 0, \tag{8.2}$$

where $q(x)$ is a real function which is continuous on $[0,\pi]$ and h is a real number. Let $\varphi(x,\lambda)$ be a solution of (8.1) under the conditions $\varphi(0,\lambda)=1$, $\varphi'(0,\lambda)=h$. It is known [8] that the following representation is valid

$$\varphi(x,\lambda)=\cos\rho x+\int_0^x K(x,t)\cos\rho t\,dt. \tag{8.3}$$

Denote $\Delta_i(\lambda)=\varphi^{(i-1)}(\pi,\lambda)$, $i=1,2$. The eigenvalues $\{\lambda_{ni}\}_{n\geq 0}$ of L_i are zeroes of the entire functions $\Delta_i(\lambda)$ and have the asymptotics

$$\rho_{n1}=\sqrt{\lambda_{n1}}=n+\frac{1}{2}+\frac{a}{n+\frac{1}{2}}+\frac{\kappa_{n1}}{n},\qquad \rho_{n2}=\sqrt{\lambda_{n2}}=n+\frac{a}{n}+\frac{\kappa_{n2}}{n}, \tag{8.4}$$

where

$$a=\frac{1}{\pi}\left(h+\frac{1}{2}\int_0^\pi q(t)\,dt\right),\qquad \{\kappa_{ni}\}\in\ell_2.$$

Without loss of generality we assume in the following that $a=0$. In this case we have

$$\sum_{n=0}^{\infty}(|\rho_{n1}-(n+\frac{1}{2})|+|\rho_{n2}-n|)<\infty. \tag{8.5}$$

Stability of recovering Sturm–Liouville operators was studied in [2], [100]–[106] and other works. In particular, it was shown by H. Hochstadt [104]–[105] that if $\lambda_{n1}=\tilde\lambda_{n1}$ $(n\geq 0)$, $\lambda_{n2}=\tilde\lambda_{n2}$ $(n>N)$, $|\lambda_{n2}-\tilde\lambda_{n2}|<\varepsilon$ $(n=\overline{0,N})$, $h=\tilde h$, then $|q(x)-\tilde q(x)|<C\varepsilon$, where the constant C depends only on L_i and N.

Here we study the case when there are perturbations for all eigenvalues of two spectra. Let the boundary value problems $\tilde L_i$ be chosen such that

$$\Lambda=\sum_{n=0}^{\infty}(|\lambda_{n1}-\tilde\lambda_{n1}|+|\lambda_{n2}-\tilde\lambda_{n2}|)<\infty. \tag{8.6}$$

The quantity Λ will describe nearness of the spectra.

Theorem 8.1. *There exists $\delta>0$ (which depends on L_i) such that if $\Lambda<\delta$ then*

$$\max_{0\leq x\leq\pi}|q(x)-\tilde q(x)|<C\Lambda,\qquad |h-\tilde h|<C\Lambda.$$

Here and in the following, one and the same symbol C denotes various positive constants in the estimates which only depend on L_i.

We first prove some auxiliary propositions. Denote

$$\alpha_n=\int_0^\pi\varphi^2(x,\lambda_{n2})\,dx.$$

Lemma 8.1. *There exists* $\delta_1 > 0$ *such that if* $\Lambda < \delta_1$, *then*

$$\sum_{n=0}^{\infty} |\alpha_n - \widetilde{\alpha}_n| < C\Lambda. \tag{8.7}$$

Proof. Since $\varphi(x,\lambda)$ is a solution of (8.1) it follows that

$$(\lambda - \lambda_{n2}) \int_0^{\pi} \varphi(x,\lambda)\varphi(x,\lambda_{n2})\,dx$$

$$= \Big|_0^{\pi} (\varphi(x,\lambda)\varphi'(x,\lambda_{n2}) - \varphi(x,\lambda_{n2})\varphi'(x,\lambda)) = -\Delta_2(\lambda)\Delta_1(\lambda_{n2}).$$

Hence

$$\alpha_n = -\dot{\Delta}_2(\lambda_{n2})\Delta_1(\lambda_{n2}). \tag{8.8}$$

The functions $\Delta_i(\lambda)$ are entire in λ of the order $1/2$, and consequently by the Borel theorem, we have

$$\Delta_i(\lambda) = B_i \prod_{k=0}^{\infty} \left(1 - \frac{\lambda}{\lambda_{ki}}\right) \tag{8.9}$$

(the case when $\lambda = 0$ is the eigenvalue of L_i requires minor modifications). Then

$$\frac{\widetilde{\Delta}_i(\lambda)}{\Delta_i(\lambda)} = \frac{\widetilde{B}_i}{B_i} \prod_{k=0}^{\infty} \frac{\lambda_{ki}}{\widetilde{\lambda}_{ki}} \prod_{k=0}^{\infty} \left(1 + \frac{\widetilde{\lambda}_{ki} - \lambda_{ki}}{\lambda_{ki} - \lambda}\right).$$

Since for $\lambda \to -\infty$

$$\lim \frac{\widetilde{\Delta}_i(\lambda)}{\Delta_i(\lambda)} = 1, \quad \lim \prod_{k=0}^{\infty} \left(1 + \frac{\widetilde{\lambda}_{ki} - \lambda_{ki}}{\lambda_{ki} - \lambda}\right) = 1,$$

it follows that

$$\frac{\widetilde{B}_i}{B_i} \prod_{k=0}^{\infty} \frac{\lambda_{ki}}{\widetilde{\lambda}_{ki}} = 1. \tag{8.10}$$

From (8.8) we get, in view of (8.9) and (8.10)

$$\frac{\widetilde{\alpha}_n}{\alpha_n} = \prod_{k=0}^{\infty} \left(\frac{\widetilde{\lambda}_{k1} - \widetilde{\lambda}_{n2}}{\lambda_{k1} - \lambda_{n2}}\right) \prod_{\substack{k=0 \\ k \neq n}}^{\infty} \left(\frac{\widetilde{\lambda}_{k2} - \widetilde{\lambda}_{n2}}{\lambda_{k2} - \lambda_{n2}}\right)$$

or

$$\frac{\widetilde{\alpha}_n}{\alpha_n} = \prod_{k=0}^{\infty}(1-\theta_{kn}^{(1)}) \prod_{\substack{k=0\\k\neq n}}^{\infty}(1-\theta_{kn}^{(2)}), \tag{8.11}$$

where

$$\theta_{kn}^{(i)} = \frac{\lambda_{ki}-\widetilde{\lambda}_{ki}}{\lambda_{ki}-\lambda_{n2}} - \frac{\widetilde{\lambda}_{n2}-\lambda_{n2}}{\lambda_{ki}-\lambda_{n2}}.$$

Denote

$$\theta_n = \sum_{k=0}^{\infty} |\theta_{kn}^{(1)}| + \sum_{\substack{k=0\\k\neq n}}^{\infty} |\theta_{kn}^{(2)}|.$$

Then

$$\begin{aligned}\sum_{n=0}^{\infty} Q_n \leq & \sum_{n=0}^{\infty} |\lambda_{n2}-\widetilde{\lambda}_{n2}|\Big\{2\sum_{\substack{k=0\\k\neq n}}^{\infty}\frac{1}{|\lambda_{k2}-\lambda_{n2}|} + \sum_{k=0}^{\infty}\frac{1}{|\lambda_{k1}-\lambda_{n2}|}\Big\} \\ & + \sum_{n=0}^{\infty} |\lambda_{n1}-\widetilde{\lambda}_{n1}| \sum_{k=0}^{\infty}\frac{1}{|\lambda_{n1}-\lambda_{k2}|}.\end{aligned} \tag{8.12}$$

Since

$$\begin{aligned}\sum_{\substack{k=1\\k\neq n}}^{\infty}\frac{1}{|k^2-n^2|} \leq & \frac{1}{2n-1}+\frac{1}{2n+1}-\frac{1}{n^2}+\int_0^{n-1}\frac{dx}{n^2-x^2}+\int_{n+1}^{\infty}\frac{dx}{x^2-n^2} \\ \leq & \frac{1}{2n-1}+\frac{1}{2n+1}-\frac{1}{n^2}+\frac{1}{2n}\ln(2n-1)+\frac{1}{2n}\ln(2n+1),\end{aligned}$$

it follows that

$$\sum_{\substack{k=1\\k\neq n}}^{\infty}\frac{1}{|k^2-n^2|} < 1, \qquad n \geq 1. \tag{8.13}$$

Using (8.4) we obtain

$$\frac{1}{\lambda_{k2}-\lambda_{n2}} = \frac{1}{k^2-n^2} + \frac{\eta_{kn}}{(k^2-n^2)(\lambda_{k2}-\lambda_{n2})}, \qquad |\eta_{kn}| < C,$$

and hence

$$\frac{1}{\lambda_{k2}-\lambda_{n2}} < \frac{C}{|k^2-n^2|}, \qquad k \neq n.$$

From this and from (8.13) we have

$$\sum_{\substack{k=0\\k\neq n}}^{\infty} \frac{1}{|\lambda_{k2}-\lambda_{n2}|} < C. \tag{8.14}$$

Similarly,

$$\sum_{k=0}^{\infty} \left(\frac{1}{|\lambda_{k1}-\lambda_{n2}|} + \frac{1}{|\lambda_{n1}-\lambda_{k2}|} \right) < C. \tag{8.15}$$

It follows from (8.12), (8.14) and (8.15) that

$$\sum_{n=0}^{\infty} \theta_n < C\Lambda. \tag{8.16}$$

Choose $\delta_1 > 0$ such that if $\Lambda < \delta_1$ then $\theta_n < \frac{1}{4}$. Since for $|\xi| < \frac{1}{2}$

$$|\ln(1-\xi)| \leq \sum_{k=1}^{\infty} \frac{|\xi|^k}{k} \leq \sum_{k=1}^{\infty} |\xi|^k = \frac{|\xi|}{1-|\xi|} < 2|\xi|,$$

it follows that

$$\left| \ln \frac{\widetilde{\alpha}_n}{\alpha_n} \right| \leq \sum_{k=0}^{\infty} |\ln(1-\theta_{kn}^{1)})| + \sum_{\substack{k=0\\k\neq n}}^{\infty} |\ln(1-\theta_{kn}^{(2)})| < 2\theta_n.$$

Hence $|(\alpha_n)^{-1}\widetilde{\alpha}_n - 1| < 4\theta_n$ or $|\widetilde{\alpha}_n - \alpha_n| < C\theta_n$. From this and from (8.16) we obtain (8.7). Lemma 8.1 is proved.

Lemma 8.2. *There exists $\delta_2 > 0$ such that if $\Lambda < \delta_2$ then*

$$|K(x,t) - \widetilde{K}(x,t)| < C\Lambda, \tag{8.17}$$

$$|\varphi'(\pi,\lambda_{n1})\widetilde{\varphi}(\pi,\lambda_{n1})| < C|\lambda_{n1} - \widetilde{\lambda}_{n1}|, \tag{8.18}$$

$$|\varphi'(\pi,\widetilde{\lambda}_{n2})\widetilde{\varphi}(\pi,\widetilde{\lambda}_{n2})| < C|\lambda_{n2} - \widetilde{\lambda}_{n2}|. \tag{8.19}$$

Proof. The function $K(x,t)$ is the solution of the Gel'fand–Levitan equation [10]

$$F(x,t) + K(x,t) + \int_0^x K(x,s)F(s,t)\,ds = 0, \qquad 0 < t < x, \tag{8.20}$$

where

$$F(x,t)=\sum_{n=0}^{\infty}\left(\frac{1}{\alpha_n}\cos\,\rho_{n2}x\cos\,\rho_{n2}t-\frac{1}{\alpha_n^0}\cos\,nx\,\cos\,nt\right),$$

$$\alpha_n^0=\frac{\pi}{2}\quad(n\geq 1),\qquad \alpha_0^0=\pi.$$

It is clear that the series converges absolutely and uniformly for $0\leq x,\,t\leq\pi$, since

$$\sum_{n=0}^{\infty}(|\rho_{n2}-n|+|\alpha_n-\frac{\pi}{2}|)<\infty.$$

We compute

$$\widehat{F}(x,t)=\sum_{n=0}^{\infty}\left(\frac{1}{\alpha_n}\cos\,\rho_{n2}x\cos\,\rho_{n2}t-\frac{1}{\widetilde{\alpha}_n}\cos\,\widetilde{\rho}_{n2}x\,\cos\,\widetilde{\rho}_{n2}t\right),$$

or

$$\widehat{F}(x,t)=\frac{1}{2}\sum_{n=0}^{\infty}(b_n(x+t)-b_n(x-t)),$$

where

$$b_n(\xi)=\frac{1}{\alpha_n}\cos\,\rho_{n2}\xi-\frac{1}{\widetilde{\alpha}_n}\cos\widetilde{\rho}_{n2}\xi.$$

For $\Lambda<\delta_1$ we have

$$\begin{aligned}|b_n(\xi)| &\leq \left|\left(\frac{1}{\alpha_n}-\frac{1}{\widetilde{\alpha}_n}\right)\cos\widetilde{\rho}_{n2}\xi\right|+\frac{1}{|\alpha_n|}\,|\cos\widetilde{\rho}_{n2}\xi-\cos\,\rho_{n2}\xi|\\ &\leq C(|\alpha_n-\widetilde{\alpha}_n|+|\rho_{n2}-\widetilde{\rho}_{n2}|).\end{aligned}$$

Hence, by Lemma 8.1, $|\widehat{F}(x,t)|<C\Lambda$. From this and from the unique solvability of the Gel'fand–Levitan equation (8.20) we obtain (8.17). Furthermore, it follows from (8.3) that

$$\varphi'(\pi,\lambda_{n1})=-\rho_{n1}\sin\,\rho_{n1}\,\pi+K(\pi,\pi)\cos\,\rho_{n1}\,\pi+\int_0^{\pi}K_x(\pi,t)\cos\,\rho_{n1}\,t\,dt,$$

$$\begin{aligned}\widetilde{\varphi}(\pi,\lambda_{n1}) &= \widetilde{\varphi}(\pi,\lambda_{n1})-\widetilde{\varphi}(\pi,\widetilde{\lambda}_{n1})=(\cos\,\rho_{n1}\pi-\cos\,\widetilde{\rho}_{n1}\pi)\\ &\quad+\int_0^{\pi}\widetilde{K}(\pi,t)(\cos\,\rho_{n1}t-\cos\widetilde{\rho}_{n1}t)\,dt.\end{aligned}$$

Hence, for $\Lambda<\delta_2$ we have in view of (8.17) that

$$|\varphi'(\pi, \lambda_{n1})| < C(n+1), \qquad |\tilde{\varphi}(\pi, \lambda_{n1})| < C|\rho_{n1} - \tilde{\rho}_{n1}|.$$

This yields (8.18). Analogously we obtain (8.19).

Lemma 8.3. *Let $g(x)$ be a continuous function on $[0, \pi]$, and let ρ_n, $n \geq 0$ be numbers such that*

$$\sum_{n=0}^{\infty} |\rho_n - n| < \infty.$$

Denote

$$\varepsilon_n = \int_0^{\pi} g(x) \cos \rho_n x \, dx,$$

and suppose that

$$\Lambda_1 \stackrel{df}{=} \sum_{n=0}^{\infty} |\varepsilon_n| < \infty.$$

Then $g(x)| < M\Lambda_1$, where the constant M only depends on the set $\{\rho_n\}_{n\geq 0}$.

Proof. Since the system of the functions $\{\cos \rho_n x\}_{n\geq 0}$ is complete in $\mathcal{L}_2(0, \pi)$, it follows that ε_n uniquely determine the function $g(x)$. From the equality

$$\int_0^{\pi} g(x) \cos nx \, dx = \varepsilon_n + \int_0^{\pi} g(x)(\cos nx - \cos \rho_n x) \, dx,$$

we obtain

$$g(x) = \sum_{n=0}^{\infty} \frac{\varepsilon_n}{\alpha_n^0} \cos nx + \sum_{n=0}^{\infty} \frac{1}{\alpha_n^0} \cos nx \int_0^{\pi} g(t)(\cos nt - \cos \rho_n t) \, dt.$$

Hence, the function $g(x)$ is the solution of the integral equation

$$g(x) = \varepsilon(x) + \int_0^{\pi} H(x, t) g(t) \, dt, \tag{8.21}$$

where

$$\varepsilon(x) = \sum_{n=0}^{\infty} \frac{\varepsilon_n}{\alpha_n^0} \cos nx, \qquad H(x, t) = \sum_{n=0}^{\infty} \frac{1}{\alpha_n^0} \cos nx \, (\cos nt - \cos \rho_n t),$$

and the series converge absolutely and uniformly for $0 \le x, t \le \pi$, and

$$|\varepsilon(x)| < \frac{2}{\pi}\Lambda_1, \qquad |H(x,t)| < C\sum_{n=0}^{\infty} |\rho_n - n|.$$

We show that the homogeneous equation

$$y(x) = \int_0^\pi H(x,t)y(t)\,dt, \qquad y(x) \in C[0,\pi] \tag{8.22}$$

has only the trivial solution. Indeed, it follows from (8.22) that

$$y(x) = \sum_{n=0}^{\infty} \frac{1}{\alpha_n^0} \cos nx \int_0^\pi (\cos nt - \cos \rho_n t)y(t)\,dt.$$

Hence

$$\int_0^\pi y(x)\cos nx\,dx = \int_0^\pi (\cos nt - \cos \rho_n t)y(t)\,dt$$

or

$$\int_0^\pi y(x)\cos \rho_n x\,dx = 0.$$

From this, by the completeness of the system $\{\cos \rho_n x\}_{n\ge 0}$ it follows that $y(x) = 0$.

Thus, the integral equation (8.21) is uniquely solvable, and hence $|g(x)| < M\Lambda_1$. Lemma 8.3 is proved.

Remark. It is easy to see from the proof that if we consider numbers $\{\tilde\rho_n\}$ instead of $\{\rho_n\}$ such that

$$\sum_{n=0}^{\infty} |\tilde\rho_n - \rho_n| < \delta,$$

for sufficiently small δ, then the constant M will not depend on $\tilde\rho_n$.

Proof of Theorem 8.1. Since the functions $\varphi(x,\lambda)$ and $\tilde\varphi(x,\lambda)$ satisfy the equations

$$-\varphi'' + q(x)\varphi = \lambda\varphi, \qquad -\tilde\varphi'' + \tilde q(x)\tilde\varphi = \lambda\tilde\varphi,$$

it follows that

$$\int_0^\pi \widehat{q}(x)\varphi(x,\lambda)\widetilde{\varphi}(x,\lambda)\,dx = \varphi'(\pi,\lambda)\widetilde{\varphi}(\pi,\lambda) - \varphi(\pi,\lambda)\widetilde{\varphi}'(\pi,\lambda) + \widetilde{h} - h. \tag{8.23}$$

Since

$$\widehat{h} + \frac{1}{2}\int_0^\pi \widehat{q}(x)\,dx = 0,$$

(8.23) become

$$\int_0^\pi \widehat{q}(x)\left(\varphi(x,\lambda)\widetilde{\varphi}(x,\lambda) - \frac{1}{2}\right)dx = \varphi'(\pi,\lambda)\widetilde{\varphi}(\pi,\lambda) - \varphi(\pi,\lambda)\widetilde{\varphi}'(\pi,\lambda). \tag{8.24}$$

We continue the functions $K(x,t)$ and $\widetilde{K}(x,t)$ by the formulas $K(x,-t) = K(x,t)$, $\widetilde{K}(x,-t) = \widetilde{K}(x,t)$. Then

$$\varphi(x,\lambda) = \cos\rho x + \frac{1}{2}\int_{-x}^{x} K(x,t)\cos\,\rho t\,dt,$$

$$\widetilde{\varphi}(x,\lambda) = \cos\rho x + \frac{1}{2}\int_{-x}^{x} \widetilde{K}(x,t)\cos\,\rho t\,dt,$$

and hence

$$\begin{aligned}\varphi(x,\lambda)\widetilde{\varphi}(x,\lambda) &= \cos^2\rho x + \frac{1}{2}\int_{-x}^{x}(K(x,t)+\widetilde{K}(x,t))\cos\,\rho(x-t)\,dt \\ &+ \frac{1}{4}\int_{-x}^{x}\int_{-x}^{x} K(x,s)\widetilde{K}(x,t)\cos\,\rho(t-s)\,dt\,ds.\end{aligned}$$

The change of variables: $\tau = \frac{1}{2}(x-t)$ and $\tau = \frac{1}{2}(s-t)$ in the integrals yields

$$\varphi(x,\lambda)\widetilde{\varphi}(x,\lambda) - \frac{1}{2} = \frac{1}{2}\left(\cos\,2\rho x + \int_0^x V(x,\tau)\cos\,2\rho\tau\,d\tau\right), \tag{8.25}$$

where

$$V(x,\tau) = V_1(x,\tau) + V_1(x,-\tau),$$

$$V_1(x,\tau) = \begin{cases} 2(K(x,x-2\tau)+\widetilde{K}(x,x-2\tau)) + \displaystyle\int_{-x+2\tau}^{x} K(x,s)\widetilde{K}(x,s-2\tau)\,ds, & \tau > 0 \\ \displaystyle\int_{-x}^{x+2\tau} K(x,s)\widetilde{K}(x,s-2\tau)\,ds, & \tau < 0. \end{cases}$$

By Lemma 8.2, we have for $\Lambda < \delta_2$

$$|V(x,t)| < C. \tag{8.26}$$

Denote

$$g(x) = 2\Big(\widehat{q}(x) + \int_x^{\pi} V(x,t)\widehat{q}(x)\,dt\Big). \tag{8.27}$$

Substituting (8.25) into (8.24) we obtain for $\lambda = \lambda_{n1}$ and $\lambda = \widetilde{\lambda}_{n2}$

$$\int_0^{\pi} g(x)\cos 2\rho_{n1}x\,dx = \varphi'(\pi,\lambda_{n1})\widetilde{\varphi}(\pi,\lambda_{n1}),$$

$$\int_0^{\pi} g(x)\cos 2\widetilde{\rho}_{n2}x\,dx = \varphi'(\pi,\widetilde{\lambda}_{n2})\widetilde{\varphi}(\pi,\widetilde{\lambda}_{n2}).$$

We use Lemma 8.3 for $\rho_{2n+1} = 2\rho_{n,1}$, $\rho_{2n} = 2\widetilde{\rho}_{n,2}$, $\varepsilon_{2n+1} = \varphi'(\pi,\lambda_{n1})\widetilde{\varphi}(\pi,\lambda_{n1})$, $\varepsilon_{2n} = \varphi'(\pi,\widetilde{\lambda}_{n2})\widetilde{\varphi}(\pi,\widetilde{\lambda}_{n2})$. It follows from (8.5), (8.6), (8.18) and (8.19) that for $\Lambda < \delta_2$ $|g(x)| < C\Lambda$. Since $\widehat{q}(x)$ is the solution of the Volterra integral equation (8.27), it follows from (8.26) that $|\widehat{q}(x)| < C\Lambda$, and hence

$$|\widehat{h}| = \frac{1}{2}\left|\int_0^{\pi}\widehat{q}(x)\,dt\right| < C\Lambda.$$

Theorem 8.1 is proved.

8.2. Now we study stability of the solution of the IP from spectra for higher-order differential operators. Here we use another method connected with the development of the ideas of Levinson [3]. For this problem the method of § 1–7 is too rough. It is used here only to obtain auxiliary estimates. For simplicity, we confine ourselves to the study of self-adjoint fourth-order differential operators with symmetric coefficients. Analogous results are valid for differential operators of an arbitrary order.

Let $\{\lambda_n\}_{n\geq 1}$ and $\{\gamma_n\}_{n\geq 1}$ be eigenvalues of the boundary value problems Q_i, $i=1,2$ for the DE

$$\ell y \equiv y^{(4)} - (q_2(x)y')' + q_0(x)y = \lambda y = \rho^4 y, \qquad q_j(\pi - x) = q_j(x) \tag{8.28}$$

with the boundary conditions

$$y(0) = y''(0) = y(\pi) = y''(\pi) = 0, \qquad (\text{for } Q_1)$$

$$y(0) = y'(0) = y''(0) = y(\pi) = 0, \qquad (\text{for } Q_2)$$

respectively. Here $q_i(x)$ are real, and $q_i^{(\nu)}(x)$ are continuous on $[0,\pi]$ for $0 \leq i-\nu \leq 2$. We shall assume that the spectra of Q_1 and Q_2 are simple, and $\lambda_n \neq \gamma_k$ for all $n, k \geq 1$.

Theorem 8.2. *If $\lambda_n = \widetilde{\lambda}_n$, $\gamma_n = \widetilde{\gamma}_n$, $n \geq 1$, then $q_i(x) \equiv \widetilde{q}_i(x)$, $i = 0, 2$.*

Theorem 8.3. *There exists $\delta > 0$ (which depends on Q_i) such that if*

$$\Lambda \overset{df}{=} \sum_{n=1}^{\infty} (|\lambda_n - \widetilde{\lambda}_n| + |\gamma_n - \widetilde{\gamma}_n|) < \delta,$$

then

$$\max_{0\leq x\leq \pi} \left| \frac{d^\nu}{dx^\nu} \int_0^x (q_i(t) - \widetilde{q}_i(t))\, dt \right| < C\Lambda, \qquad 0 \leq i - \nu \leq 2.$$

Here and in the sequel, the symbol C denotes various positive constants in estimates which only depend on Q_i.

We first prove some auxiliary propositions. Observe that for any sufficiently smooth functions $y(x)$, $z(x)$ and any numbers a, $b \in [0,\pi]$ the following relations holds

$$\int_a^b (\ell y(x) z(x) - y(x) \ell z(x))\, dx = \Big|_a^b \langle y(x), z(x) \rangle, \tag{8.29}$$

where

$$\langle y, z \rangle \overset{df}{=} y'''z - y''z' + y'(z'' - q_2 z) - y(z''' - q_2 z').$$

In particular, if the functions $y(x,\lambda)$ and $z(x,\mu)$ are solutions of the equations $\ell y(x,\lambda) = \lambda y(x,\lambda)$ and $\ell z(x,\mu) = \mu z(x,\mu)$ respectively, then

$$(\lambda - \mu) \int_0^\pi y(x,\lambda) z(x,\mu)\, dx = \Big|_0^\pi \langle y(x,\lambda), z(x,\mu) \rangle. \tag{8.30}$$

Let $a_i(x,\lambda)$, $b_i(x,\lambda)$, $i = 1,2$ be solutions of (8.28) under the initial conditions

$$a_i^{(\nu-1)}(0,\lambda)=b_i^{(\nu-1)}(\pi,\lambda)=\delta_{2i,\nu}, \qquad \nu=\overline{1,4},$$

and let

$$a_3(x,\lambda)\overset{df}{=}a_1(x,\lambda)a_2(\pi,\lambda)-a_2(x,\lambda)a_1(\pi,\lambda),$$

$$b_3(x,\lambda)\overset{df}{=}b_1(x,\lambda)b_2(0,\lambda)-b_2(x,\lambda)b_1(0,\lambda).$$

Since $q_i(\pi-x)=q_j(x)$, we get

$$\begin{gathered} b_1(x,\lambda)\equiv -a_1(\pi-x,\lambda), \qquad b_2(x,\lambda)\equiv -a_2(\pi-x,\lambda), \\ b_3(x,\lambda)\equiv a_3(\pi-x,\lambda). \end{gathered} \tag{8.31}$$

Denote $G_\varepsilon=\{\lambda:|\lambda-\lambda_n|\geq\varepsilon,|\lambda-\gamma_n|\geq\varepsilon,n\geq 1\}$, $\Delta(\lambda)=a_3''(\pi,\lambda)$, $\delta(\lambda)=a_2(\pi,\lambda)$, $d(\lambda)=-a_3'(\pi,\lambda)$. The eigenvalues $\{\lambda_n\}$ and $\{\gamma_n\}$ of Q_1 and Q_2 coincide with the zeros of the entire functions $\Delta(\lambda)$ and $\delta(\lambda)$ respectively, and

$$\left.\begin{aligned} \sqrt[4]{\lambda_n}&=n+\sum_{\mu=1}^{3}\frac{\theta_{1\mu}}{n^\mu}+o\left(\frac{1}{n^3}\right), && n\to\infty, \\ \sqrt[4]{\gamma_n}&=(1+i)\left(n+\frac{1}{4}\right)+\sum_{\mu=1}^{3}\frac{\theta_{2\mu}}{(n+\frac14)^\mu}+o\left(\frac{1}{n^3}\right), && n\to\infty. \end{aligned}\right\} \tag{8.32}$$

Lemma 8.4. (1) *For* $n\to\infty$

$$\left.\begin{aligned} \delta(\lambda_n)&=\frac{1}{4n^3}\exp(n\pi)\left(1+O\left(\frac1n\right)\right), \\ \Delta(\gamma_n)&=-\frac{1}{8(n+\frac14)^2}\exp\left(2\left(n+\frac14\right)\pi\right)\left(1+O\left(\frac1n\right)\right), \\ \dot\delta(\gamma_n)&=\frac{(-1)^{n+1}\pi}{\sqrt2(2(n+\frac14))^6}\exp\left(\left(n+\frac14\right)\pi\right)\left(1+O\left(\frac1n\right)\right), \\ \dot\Delta(\lambda_n)&=\frac{(-1)^{n+1}\pi}{8n^5}\exp(n\pi)\left(1+O\left(\frac1n\right)\right), \\ d(\lambda_n)&=\frac{(-1)^{n+1}}{4n^3}\exp(n\pi)\left(1+O\left(\frac1n\right)\right), \\ a_1(\pi,\gamma_n)&=\frac{(-1)^n}{2\sqrt2(n+\frac14)}\exp\left(\left(n+\frac14\right)\pi\right)\left(1+O\left(\frac1n\right)\right). \end{aligned}\right\} \tag{8.33}$$

(2) *If* $|\rho| \to \infty$, $\arg \rho = \theta \in (0, \frac{\pi}{4})$, *then*

$$\left.\begin{aligned}
\delta(\rho) &= \frac{1}{4\rho^3} \exp(\rho\pi) \left(1 + O\left(\frac{1}{\rho}\right)\right), \\
\Delta(\rho) &= \frac{1}{4i\rho^2} \exp\left((1-i)\rho\pi\right) \left(1 + O\left(\frac{1}{\rho}\right)\right), \\
d(\lambda) &= -\frac{(1+i)}{8i\rho^3} \exp\left((1-i)\rho\pi\right) \left(1 + O\left(\frac{1}{\rho}\right)\right), \\
a_1(\pi, \lambda) &= \frac{1}{4\rho} \exp\left(\rho\pi\right) \left(1 + O\left(\frac{1}{\rho}\right)\right).
\end{aligned}\right\} \tag{8.34}$$

(3) *Let* S *be a fixed sector with property* (1.3). *Then for* $\rho \in \bar{S}$

$$\left.\begin{aligned}
&|a_2^{(\nu)}(x, \lambda)| < C\, |\rho^{\nu-3} \exp(\rho R_4 x)|, \\
&|a_3^{(\nu)}(x, \lambda)| < C\, |\rho^{\nu-4} \exp\left(\rho(R_4\pi - R_3 x)\right)|, \\
&|\delta(\lambda)| > C\, |\rho^{-3} \exp\left(\rho R_4 \pi\right)|, \qquad \rho \in G_\varepsilon, \\
&|\Delta(\lambda)| > C\, |\rho^{-2} \exp\left(\rho(R_3 + R_4)\pi\right)|, \qquad \rho \in G_\varepsilon.
\end{aligned}\right\} \tag{8.35}$$

(4) *For* $n^{-3}|\lambda - \gamma_n| < C_0$

$$|a_j^{(\nu)}(x, \lambda) < C\, n^{\nu - 2j + 1} \exp\left(nx\right), \qquad j = 1, 2, \tag{8.36}$$

and for $n^{-3}|\lambda - \lambda_n| < C_0$

$$|a_3^{(\nu)}(x, \lambda)| < C\, n^{\nu-4} \exp\left(n\pi\right). \tag{8.37}$$

Proof. Take the FSS B_0 for the DE (8.28). Then

$$a_1(x, \lambda) = \sum_{k=1}^{4} A_k(\rho) y_k(x, \rho).$$

From the initial conditions $a_1^{(\nu-1)}(0, \lambda) = \delta_{2\nu}$, $\nu = \overline{1,4}$ we calculate

$$A_k(\rho) = \frac{1}{4(\rho R_k)} \left(1 + O\left(\frac{1}{\rho}\right)\right).$$

Consequently, for $|\lambda| \to \infty$ uniformly in $x \in [0, \pi]$ we have

$$a_1^{(\nu)}(x, \lambda) = \frac{1}{4} \sum_{k=1}^{4} (\rho R_k)^{\nu - 1} \exp\left(\rho R_k x\right) \left(1 + O\left(\frac{1}{\rho}\right)\right). \tag{8.38}$$

Analogously we obtain the asymptotics for the functions $a_2^{(\nu)}(x,\lambda)$ and $a_3^{(\nu)}(x,\lambda)$:

$$\left.\begin{aligned} a_2^{(\nu)}(x,\lambda) &= \frac{1}{4}\sum_{k=1}^{4}(\rho R_k)^{\nu-3}\exp(\rho R_k x)\left(1+O\left(\frac{1}{\rho}\right)\right), \\ a_3^{(\nu)}(x,\lambda) &= \sum_{k=1}^{4}(\rho R_k)^{\nu-4}\sum_{\substack{j=1\\ j\neq k}}^{4} C_{jk}\exp(\rho(R_j\pi - R_k x))\left(1+O\left(\frac{1}{\rho}\right)\right), \end{aligned}\right\} \tag{8.39}$$

where C_{jk} are constants, and also for the functions $\frac{d}{d\lambda}a_i^{(\nu)}(x,\lambda)$. From this we derive all the assertions of Lemma 8.4.

Denote

$$\left.\begin{aligned} \alpha_{n1} &= \frac{1}{\Delta(\gamma_n)}\int_0^\pi a_2(x,\gamma_n)b_3(x,\gamma_n)\,dx, \\ \alpha_{n2} &= \frac{1}{\delta^2(\lambda_n)}\int_0^\pi a_3^2(x,\lambda_n)\,dx. \end{aligned}\right\} \tag{8.40}$$

Lemma 8.5.

$$\Delta(\gamma_n) = -a_1^2(\pi,\gamma_n), \tag{8.41}$$

$$b_3(x,\lambda_n) = k_n a_3(x,\lambda_n), \qquad k_n^2 = 1, \tag{8.42}$$

$$k_n = (-1)^{n+1}\operatorname{sign}\delta(\lambda_n), \tag{8.43}$$

$$\alpha_{n1} = \frac{\dot\delta(\gamma_n)}{a_1(\pi,\gamma_n)}, \qquad \alpha_{n2} = (-1)^{n+1}\frac{\dot\Delta(\lambda_n)}{|\delta(\lambda_n)|}. \tag{8.44}$$

Proof. We consider the boundary value problems W_1 and W_2 for (8.28) with the boundary conditions $y(0) = y'(0) = y''(0) = y''(\pi) = 0$ (for W_1) and $y(0) = y''(0) = y'''(0) - q_2(0)y'(0) = y(\pi) = 0$ (for W_2), respectively. We observe that the adjoint problem W_1^* has the boundary conditions $y(0) = y(\pi) = y''(\pi) = y'''(\pi) - q_2(\pi)y'(\pi) = 0$. Since the coefficients of (8.28) are symmetric, the eigenvalues of W_1 coincide with the eigenvalues of W_2, i.e. sp $W_1 =$ sp W_2 (sp. = spectrum). On the other hand, the eigenvalues of W_1 and W_2 coincide with the zeros of the entire functions of the order 1/4 $w_1(\lambda) = a_2''(\pi,\lambda)$ and $w_2(\lambda) = a_1(\pi,\lambda) + q_2(0)a_2(\pi,\lambda)$, respectively. It follows from (8.38) and (8.39) that $\lim (w_2(\lambda))^{-1}w_1(\lambda) = 1$ as $|\lambda|\to\infty$, $\arg\lambda = \theta \neq 0,\pi$. Consequently, $w_1(\lambda) = w_2(\lambda)$, i.e.

$$a_2''(\pi,\lambda) \equiv a_1(\pi,\lambda) + q_2(0)a_2(\pi,\lambda). \tag{8.45}$$

Further, since $\Delta(\lambda) = a_1''(\pi,\lambda)a_2(\pi,\lambda) - a_2''(\pi,\lambda)a_1(\pi,\lambda)$, we have $\Delta(\gamma_n) = -a_2''(\pi,\gamma_n)a_1(\pi,\gamma_n)$. From this we obtain (8.41), in view of (8.45). The functions $b_3(x,\lambda_n)$ and $a_3(x,\lambda_n)$ are the eigenfunctions of Q_1, and consequently $b_3(x,\lambda_n) = k_n a_3(x,\lambda_n)$. Using (8.31), we get

$$b_3(x,\lambda_n) = k_n a_3(x,\lambda_n) = k_n b_3(\pi - x,\lambda_n) = k_n^2 a_3(\pi - x,\lambda_n) = k_n^2 b_3(x,\lambda_n).$$

Hence $k_n^2 = 1$. Thus, (8.42) is proved.

It follows from the condition $\operatorname{sp} Q_1 \cap \operatorname{sp} Q_2 = \oslash$ that $d(\lambda_n) \neq 0$. Indeed, if we suppose that $\Delta(\lambda^*) = d(\lambda^*) = 0$ for a certain λ^*, then $\delta(\lambda^*) \neq 0$. Since

$$b_3(0,\lambda) = b_3(\pi,\lambda) = b_3''(\pi,\lambda) = 0,$$

$$b_3'(0,\lambda) = d(\lambda), \qquad b_3''(0,\lambda) = \Delta(\lambda), \qquad b_3'(\pi,\lambda) = -\delta(\lambda),$$

then the function $b_3(x,\lambda^*) \not\equiv 0$ is the eigenfunction and λ^* is the eigenvalue of Q_2. This contradicts the relation $a_2(\pi,\lambda^*) = \delta(\lambda^*) \neq 0$.

Further, using (8.30), one gets

$$\begin{aligned}(\lambda - \mu)\int_0^\pi b_3(x,\lambda) b_3(x,\mu)\,dx &= \Big|_0^\pi \langle b_3(x,\lambda), b_3(x,\mu)\rangle \\ &= b_3''(0,\lambda)b_3'(0,\mu) - b_3''(0,\mu)b_3'(0,\lambda)\end{aligned}$$

or

$$\int_0^\pi b_3(x,\lambda)b_3(x,\mu)\,dx = \frac{\Delta(\lambda)d(\mu) - \Delta(\mu)d(\lambda)}{\lambda - \mu}.$$

Hence we have

$$\int_0^\pi b_3^2(x,\lambda)\,dx = \dot\Delta(\lambda)d(\lambda) - \dot d(\lambda)\Delta(\lambda),$$

or

$$d^2(\lambda)\int_0^\pi b_3^2(x,\lambda)\,dx = \frac{d}{d\lambda}\left(\frac{\Delta(\lambda)}{d(\lambda)}\right).$$

Therefore for real λ the real-valued function $(d(\lambda))^{-1}\Delta(\lambda)$ is monotonically increasing. Thus, the zeros of $d(\lambda)$ and $\Delta(\lambda)$ are real and alternate. Taking (8.33) into account, we have

$$\operatorname{sign} d(\lambda_n) = (-1)^{n+1}. \tag{8.46}$$

Differentiating (8.42) and setting $x = 0$, we get $d(\lambda_n) = k_n\delta(\lambda_n)$. Then from (8.46), in view of the relation $k_n^2 = 1$, we obtain (8.43).

It follows from (8.30) that

$$(\lambda - \gamma_n) \int_0^\pi a_3(x,\lambda)b_3(x,\gamma_n)\,dx = \Big|_0^\pi \langle a_3(x,\lambda), b_3(x,\gamma_n)\rangle$$

$$= -a_3'(0,\lambda)b_3''(0,\gamma_n) = -\delta(\lambda)\Delta(\gamma_n).$$

Hence

$$\int_0^\pi a_3(x,\gamma_n)b_3(x,\gamma_n)\,dx = -\dot{\delta}(\gamma_n)\Delta(\gamma_n). \tag{8.47}$$

Since $a_3(x,\gamma_n) = -a_1(\pi,\gamma_n)a_2(x,\gamma_n)$, it follows from (8.40) and (8.47) that $\alpha_{n1} = (a_1(\pi,\gamma_n))^{-1}\dot{\delta}(\gamma_n)$. In the same way we calculate

$$(\lambda - \lambda_n) \int_0^\pi b_3(x,\lambda)a_3(x,\lambda_n)\,dx = \Delta(\lambda)\delta(\lambda_n),$$

and consequently

$$\int_0^\pi b_3(x,\lambda)a_3(x,\lambda_n)\,dx = \dot{\Delta}(\lambda_n)\delta(\lambda_n).$$

Using (8.42), (8.43) and (8.40), we obtain

$$\alpha_{n2} = \frac{\dot{\Delta}(\lambda_n)}{k_n\delta(\lambda_n)} = (-1)^{n+1}\frac{\dot{\Delta}(\lambda_n)}{|\delta(\lambda_n)|}.$$

Lemma 8.5 is proved.

Let

$$\mathfrak{M}_{m,m+1}(\lambda) = \frac{\Delta_{m,m+1}(\lambda)}{\Delta_{mm}(\lambda)}, \qquad m = \overline{1,3}$$

be the WF's for the DE (8.28) and LF

$$U_{1a}(y) = y(a), \quad U_{2a}(y) = y''(a), \quad U_{3a}(y) = y'(a), \quad U_{4a}(y) = y'''(a), \quad a = 0, T.$$

Denote by $\{\lambda_{\ell m}\}_{\ell \ge 1}$ the zeros of $\Delta_{mm}(\lambda)$, and

$$\beta_{\ell m} = (\dot{\Delta}_{mm}(\lambda_{\ell m}))^{-1}\,\Delta_{m,m+1}(\lambda_{\ell m}).$$

It is obvious that $\lambda_{\ell 1} = \lambda_{\ell 3} = \gamma_\ell$, $\lambda_{\ell 2} = \lambda_\ell$. Comparing (8.34) and (1.18), we calculate $\Delta(\lambda) \equiv -\Delta_{22}(\lambda)$, $\delta(\lambda) \equiv \Delta_{11}(\lambda) \equiv \Delta_{33}(\lambda)$. Similarly, $d(\lambda) \equiv -\Delta_{23}(\lambda)$, $a_1(\pi,\lambda) \equiv -\Delta_{34}(\lambda)$, $a_2''(\pi,\lambda) \equiv \Delta_{12}(\lambda)$. Moreover, (8.45) becomes $\Delta_{12}(\lambda) = -\Delta_{34}(\lambda) + q_2(0)\Delta_{11}(\lambda)$. Consequently, in view of Lemma 8.5, we have $\beta_{\ell 1} = -\beta_{\ell 3} = (\alpha_{\ell 1})^{-1}$, $\beta_{\ell 2} = (\alpha_{\ell 2})^{-1}$. Thus, the specification of the numbers $\{\lambda_n, \gamma_n, \alpha_{n1}, \alpha_{n2}\}_{n\geq 1}$ determines the WF's $\{\mathfrak{M}_{m,m+1}(\lambda)\}_{m=\overline{1,3}}$ uniquely and, by virtue of the results of §7, determines the coeffitients $q_0(x)$ and $q_2(x)$ of (8.28) uniquely.

Now let us prove Theorem 8.2. Let $\lambda_n = \widetilde{\lambda}_n$, $\gamma_n = \widetilde{\gamma}_n$ for all $n \geq 1$. The numbers $\{\lambda_n\}_{n\geq 1}$ and $\{\gamma_n\}_{n\geq 1}$ coincide with the zeros of the entire functions $\Delta(\lambda)$ and $\delta(\lambda)$, respectively. Using (8.34) we get $\Delta(\lambda) \equiv \widetilde{\Delta}(\lambda)$, $\delta(\lambda \equiv \widetilde{\delta}(\lambda)$. It follows from (8.41) and (8.44) that $\alpha_{n2} = \widetilde{\alpha}_{n2}$, $\alpha_{n1} = \varepsilon_n \widetilde{\alpha}_{n1}$, $\varepsilon_n = \pm 1$. Denote $\Omega_0 = \{n : \varepsilon_n = -1\}$. By virtue of the asymptotic formulae (8.33), we have $\varepsilon_n = 1$ for large n. Thus, Ω_0 is a finite set.

Let us show that $\Omega_0 = \oslash$. Indeed, let us assume that $\Omega_0 \neq 0$. Since Ω_0 is a finite set, then the main equation of the IP (4.41) is a linear algebraic system with the determinant

$$\mathcal{D}(x) = \prod_{n\in\Omega_0} \mathcal{D}_n^2(x),$$

where

$$\mathcal{D}_n(x) = 1 - \frac{2}{\alpha_{n1}} \int_0^x a_2(t,\gamma_n) b_3(t,\gamma_n)\, dt \, \frac{1}{\Delta(\gamma_n)}. \tag{8.48}$$

It follows from the solvability of the main equation that

$$\mathcal{D}(x) \neq 0, \qquad x \in [0,\pi].$$

On the other hand, since $\mathcal{D}_n(0) = 1$, $\mathcal{D}_n(\pi) = -1$, there exists $x_0 \in (0,\pi)$ such that $\mathcal{D}(x_0) = 0$. This contradiction means that $\Omega_0 \neq \oslash$. Thus, $\varepsilon_n = 1$ for all $n \geq 1$ and consequently $\alpha_{ni} = \widetilde{\alpha}_{ni}$, $n \geq 1$, $i = 1,2$. Hence $q_j(x) \equiv \widetilde{q}_j(x)$, $j = 0,2$. Theorem 8.2 is proved.

Using (8.36), (8.37) and Schwartz's lemma, we obtain for $|\gamma_n - \widetilde{\gamma}_n| n^{-3} < C_0$, $|\lambda_n - \widetilde{\lambda}_n| n^{-3} < C_0$ the following estimates

$$|a_j^{(\nu)}(x,\gamma_n) - a_j^{(\nu)}(x,\widetilde{\gamma}_n)| < C n^{\nu-2j-2} |\gamma_n - \widetilde{\gamma}_n| \exp(nx), \qquad j = 1,2, \tag{8.49}$$

$$|a_3^{(\nu)}(x,\lambda_n) - a_3^{(\nu)}(x,\widetilde{\lambda}_n)| < C n^{\nu-7} |\lambda_n - \widetilde{\lambda}_n| \exp(n\pi), \tag{8.50}$$

$$|a_j^{(\nu)}(x,\widetilde{\gamma}_n)| < C n^{\nu-2j+1} \exp(nx), \qquad j = 1,2, \tag{8.51}$$

$$|a_3^{(\nu)}(x,\widetilde{\lambda}_n)| < C n^{\nu-4} \exp(n\pi), \tag{8.52}$$

$$|\delta(\widetilde{\gamma}_n)| < C n^{-6} |\gamma_n - \widetilde{\gamma}_n| \exp(n\pi). \tag{8.53}$$

Denote

$$\left.\begin{aligned}\omega_{kn}^{(1)} &= \frac{\lambda_k - \widetilde{\lambda}_k}{\lambda_k - \lambda_n} + \frac{\widetilde{\lambda}_n - \lambda_n}{\lambda_k - \lambda_n}, & \omega_{kn}^{(2)} &= \frac{\gamma_k - \widetilde{\gamma}_k}{\gamma_k - \gamma_n} + \frac{\widetilde{\gamma}_n - \gamma_n}{\gamma_k - \gamma_n}, \\ \omega_{kn}^{(3)} &= \frac{\lambda_k - \widetilde{\lambda}_k}{\lambda_k - \gamma_n} + \frac{\widetilde{\gamma}_n - \gamma_n}{\lambda_k - \gamma_n}, & \omega_{kn}^{(4)} &= \frac{\gamma_k - \widetilde{\gamma}_k}{\gamma_k - \lambda_n} + \frac{\widetilde{\lambda}_n - \lambda_n}{\gamma_k - \lambda_n},\end{aligned}\right\} \tag{8.54}$$

and

$$\omega_n = \sum_{i=1}^{2} \sum_{\substack{k=1 \\ k \neq n}}^{\infty} |\omega_{kn}^{(i)}| + \sum_{i=3}^{4} \sum_{k=1}^{\infty} |\omega_{kn}^{(i)}|.$$

Lemma 8.6. (1) *The following relation holds*

$$\sum_{n=1}^{\infty} n^2 \omega_n < C\Lambda. \tag{8.55}$$

(2) *If* $\omega_n < \frac{1}{8}$ *then*

$$|(\Delta(\gamma_n))^{-1} \widetilde{\Delta}(\widetilde{\gamma}_n) - 1| < 4\omega_n, \tag{8.56}$$

$$|(\dot{\Delta}(\lambda_n))^{-1} \dot{\widetilde{\Delta}}(\widetilde{\lambda}_n) - 1| < 4\omega_n, \tag{8.57}$$

$$|(\delta(\lambda_n))^{-1} \widetilde{\delta}(\widetilde{\lambda}_n) - 1| < 4\omega_n, \tag{8.58}$$

$$||(\delta(\lambda_n))^{-1} \widetilde{\delta}(\widetilde{\lambda}_n)| - 1| < 4\omega_n, \tag{8.59}$$

$$|(\dot{\delta}(\gamma_n))^{-1} \dot{\widetilde{\delta}}(\widetilde{\gamma}_n) - 1| < 4\omega_n, \tag{8.60}$$

$$\frac{1}{2} \leq |(\Delta(\gamma_n))^{-1} \widetilde{\Delta}(\widetilde{\gamma}_n)| \leq 2, \tag{8.61}$$

$$\frac{1}{2} \leq |(\dot{\Delta}(\lambda_n))^{-1} \dot{\widetilde{\Delta}}(\widetilde{\lambda}_n)| \leq 2, \tag{8.62}$$

$$\frac{1}{2} \leq |(\delta(\lambda_n))^{-1} \widetilde{\delta}(\widetilde{\lambda}_n)| \leq 2, \tag{8.63}$$

$$\frac{1}{2} \leq |(\dot{\delta}(\gamma_n))^{-1} \dot{\widetilde{\delta}}(\widetilde{\gamma}_n)| \leq 2. \tag{8.64}$$

Proof. By virtue of (8.32), we get

$$\frac{1}{|\lambda_k - \lambda_n|} < \frac{C}{|k^4 - n^4|}, \qquad k \neq n.$$

In view of (8.13), we have

$$\sum_{\substack{k=1\\k\neq n}}^{\infty}\frac{n^2}{|\lambda_k-\lambda_n|}<C\sum_{\substack{k=1\\k\neq n}}^{\infty}\frac{n^2}{|k^4-n^4|}<C\sum_{\substack{k=1\\k\neq n}}^{\infty}\frac{1}{|k^2-n^2|}<C,$$

$$\sum_{\substack{n=1\\n\neq k}}^{\infty}\frac{n^2}{|\lambda_k-\lambda_n|}<C\sum_{\substack{n=1\\n\neq k}}^{\infty}\frac{n^2}{|k^4-n^4|}<C\sum_{\substack{n=1\\n\neq k}}^{\infty}\frac{1}{|k^2-n^2|}<C.$$

Hence

$$\begin{aligned}\sum_{n=1}^{\infty}n^2\sum_{\substack{k=1\\k\neq n}}^{\infty}|\omega_{kn}^{(1)}| &< \sum_{n=1}^{\infty}|\tilde\lambda_n-\lambda_n|\sum_{\substack{k=1\\k\neq n}}^{\infty}\frac{n^2}{|\lambda_k-\lambda_n|}\\ &\quad+\sum_{k=1}^{\infty}|\lambda_k-\tilde\lambda_k|\sum_{\substack{n=1\\n\neq k}}^{\infty}\frac{n^2}{|\lambda_k-\lambda_n|}<C\Lambda.\end{aligned}$$

Similarly,

$$\sum_{n=1}^{\infty}n^2\sum_{\substack{k=1\\k\neq n}}^{\infty}|\omega_{kn}^{(2)}|<C\Lambda,\qquad \sum_{k=1}^{\infty}n^2\sum_{k=1}^{\infty}|\omega_{kn}^{(i)}|<C\Lambda,\qquad i=3,4.$$

Thus, (8.55) is proved.

Further, the entire function $\Delta(\lambda)$ of the order 1/4 is determined to within a multiplicative constant by its zeros $\{\lambda_k\}_{k\geq 1}$, so that

$$\Delta(\lambda)=B_1\prod_{k=1}^{\infty}\left(1-\frac{\lambda}{\lambda_k}\right), \tag{8.65}$$

(the case when $\lambda=0$ is an eigenvalue of Q_1 requires obvious changes). Then

$$\frac{\tilde\Delta(\lambda)}{\Delta(\lambda)}=\frac{\tilde B_1}{B_1}\prod_{k=1}^{\infty}\frac{\lambda_k}{\tilde\lambda_k}\prod_{k=1}^{\infty}\left(1+\frac{\tilde\lambda_k-\lambda_k}{\lambda_k-\lambda}\right).$$

In view of (8.32) and (8.34), we obtain

$$\lim\frac{\tilde\Delta(\lambda)}{\Delta(\lambda)}=1,\qquad \lim\prod_{k=1}^{\infty}\left(1+\frac{\tilde\lambda_k-\lambda_k}{\lambda_k-\lambda}\right)=1,\qquad |\lambda|\to\infty,\qquad \arg\lambda=\frac{\pi}{2},$$

and consequently

$$\frac{\tilde B_1}{B_1}\prod_{k=1}^{\infty}\frac{\lambda_k}{\tilde\lambda_k}=1. \tag{8.66}$$

Using (8.65) and (8.66), we have

$$\frac{\widetilde{\Delta}(\widetilde{\gamma}_n)}{\Delta(\gamma_n)} = \prod_{k=1}^{\infty}\left(\frac{\widetilde{\lambda}_k - \widetilde{\gamma}_n}{\lambda_k - \gamma_n}\right) = \prod_{k=1}^{\infty}(1 - \omega_{kn}^{(3)}).$$

Let $\omega_n < \frac{1}{8}$. Then

$$\sum_{k=1}^{\infty} |\ln(1 - \omega_{kn}^{(3)})| \le \sum_{k=1}^{\infty}\sum_{\nu=1}^{\infty} \frac{|\omega_{kn}^{(3)}|^{\nu}}{\nu} < 2\sum_{k=1}^{\infty} |\omega_{kn}^{(3)}| < 2\omega_n,$$

and hence $|\ln((\Delta(\gamma_n))^{-1}\widetilde{\Delta}(\widetilde{\gamma}_n))| < 2\omega_n$. It follows from this that $|(\Delta(\gamma_n))^{-1}\widetilde{\Delta}(\widetilde{\gamma}_n)| < 4\omega_n$. Analogously we obtain (8.57)–(8.60). Relations (8.61)–(8.64) evidently follow from (8.56)–(8.60).

Lemma 8.7. *There exists $\delta_1 > 0$ such that if $\Lambda < \delta_1$, then*

$$\left.\begin{array}{ll} |(\alpha_{n1})^{-1}\widetilde{\alpha}_{n1} - 1| < C_1\omega_n, & |(\alpha_{n2})^{-1}\widetilde{\alpha}_{n2} - 1| < C_1\omega_n, \\ |(a_1(\pi,\gamma_n))^{-1}\widetilde{a}_1(\pi,\widetilde{\gamma}_n) - 1| < C_1\omega_n, & C_1 = 200. \end{array}\right\} \tag{8.67}$$

Proof. Choose $\delta_1^0 > 0$ such that $\omega_n < 0,005$ for $\Lambda < \delta_1^0$. It follows from Lemmas 8.5 and 8.6 that

$$\left|\frac{\widetilde{\alpha}_{n2}}{\alpha_{n2}} - 1\right| = \left|\left|\frac{\delta(\lambda_n)}{\widetilde{\delta}(\widetilde{\lambda}_n)}\right| \frac{\dot{\widetilde{\Delta}}(\widetilde{\lambda}_n)}{\dot{\Delta}(\lambda_n)} - 1\right|$$

$$\le \left|\frac{\delta(\lambda_n)}{\widetilde{\delta}(\widetilde{\lambda}_n)}\right| \left|\frac{\dot{\widetilde{\Delta}}(\widetilde{\lambda}_n)}{\dot{\Delta}(\lambda_n)} - 1\right| + \left|\left|\frac{\delta(\lambda_n)}{\widetilde{\delta}(\widetilde{\lambda}_n)}\right| - 1\right| < 16\,\omega_n,$$

$$\left|\frac{\widetilde{\alpha}_{n1}^2}{\alpha_{n1}^2} - 1\right| = \left|\left(\frac{\dot{\widetilde{\delta}}(\widetilde{\gamma}_n)}{\dot{\delta}(\gamma_n)}\right)^2 \left(\frac{a_1(\pi,\gamma_n)}{\widetilde{a}_1(\pi,\widetilde{\gamma}_n)}\right)^2 - 1\right|$$

$$= \left|\left(\frac{\dot{\widetilde{\delta}}(\widetilde{\gamma}_n)}{\dot{\delta}(\gamma_n)}\right)^2 \frac{\Delta(\gamma_n)}{\widetilde{\Delta}(\widetilde{\gamma}_n)} - 1\right| \le \left|\frac{\dot{\widetilde{\delta}}(\widetilde{\gamma}_n)}{\dot{\delta}(\gamma_n)}\right|^2 \left|\frac{\Delta(\gamma_n)}{\widetilde{\Delta}(\widetilde{\gamma}_n)} - 1\right| + \left|\left(\frac{\dot{\widetilde{\delta}}(\widetilde{\gamma}_n)}{\dot{\delta}(\gamma_n)}\right)^2 - 1\right| < 44\,\omega_n.$$

Therefore there exist numbers $\varepsilon_n \pm 1$ such that

$$|(\alpha_{n1})^{-1}\widetilde{\alpha}_{n1}\varepsilon_n - 1| < 44\,\omega_n.$$

Denote $\Omega_0 = \{n : \varepsilon_n = -1\}$. It follows from (8.33) and (8.44) that Ω_0 is a finite set, because $\varepsilon_n = 1$ for large n. As in the proof of Theorem 8.2, it can be shown that $\Omega_0 = \oslash$. Indeed, suppose that $\Omega_0 \neq \oslash$. The solvability condition of the main equation of the IP in this case takes the form

$$\mathcal{D}^+(x) \stackrel{df}{=} \prod_{n\in\Omega_0} \mathcal{D}_n^2(x) + \mathcal{D}^0(x) \neq 0, \qquad x \in [0,\pi],$$

where the functions $\mathcal{D}_n(x)$ are defined via (8.48), $\mathcal{D}^0(x)$ is a continuous function, and $|\mathcal{D}^0(x)| < C\Lambda$. Since $\mathcal{D}_n(0) = 1$, $\mathcal{D}_n(\pi) = -1$, there exists $x_0 \in (0,\pi)$ such that

$$\prod_{n\in\Omega_0} \mathcal{D}_n^2(x_0) = 0.$$

Therefore, for a sufficiently small Λ there exists $x^* \in (0,\pi)$ such that $\mathcal{D}^+(x^*) = 0$. This contradiction means that $\Omega_0 = \oslash$, i.e. $\varepsilon_n = 1$ for all $n \geq 1$. Thus, there exists $\delta_1 > 0$ such that for $\Lambda < \delta_1$ the estimate $|(\alpha_{n1})^{-1}\tilde{\alpha}_{n1} - 1| < 44\,\omega_n$ is valid.

Using (8.44) and Lemma 8.6, we obtain

$$\left|\frac{\tilde{a}_1(\pi,\tilde{\gamma}_n)}{a_1(\pi,\gamma_n)} - 1\right| = \left|\frac{\dot{\tilde{\delta}}(\tilde{\gamma}_n)}{\dot{\delta}(\gamma_n)}\frac{\alpha_{n1}}{\tilde{\alpha}_{n1}} - 1\right| \leq \left|\frac{\dot{\tilde{\delta}}(\tilde{\gamma}_n)}{\dot{\delta}(\gamma_n)}\right|\left|\frac{\alpha_{n1}}{\tilde{\alpha}_{n1}} - 1\right| + \left|\frac{\dot{\tilde{\delta}}(\tilde{\gamma}_n)}{\dot{\delta}(\gamma_n)} - 1\right| < 200\,\omega_n.$$

Lemma 8.7 is proved.

Corollary 8.1. *If $\Lambda < \delta_1$, then*

$$\sum_{n=1}^{\infty} n^2 |(\alpha_{ni})^{-1}\tilde{\alpha}_{ni} - 1| < C\Lambda, \qquad i = 1,2$$

$$|\tilde{a}_1(\pi,\tilde{\gamma}_n) - a_1(\pi,\tilde{\gamma}_n)| < \frac{C}{n}\exp(n\pi)\left(\frac{|\gamma_n - \tilde{\gamma}_n|}{n^3} + \omega_n\right). \tag{8.69}$$

Indeed, (8.68) follows from (8.55) and (8.67). Further,

$$|\tilde{a}_1(\pi,\tilde{\gamma}_n) - a_1(\pi,\tilde{\gamma}_n)| < |\tilde{a}_1(\pi,\tilde{\gamma}_n) - a_1(\pi,\gamma_n)| + |a_1(\pi,\gamma_n) - a_1(\pi,\tilde{\gamma}_n)|.$$

From (8.49), (8.36) and (8.67) we then obtain (8.69).

Denote

$$\xi_\ell = \frac{|\lambda_\ell - \tilde{\lambda}_\ell|}{\ell^3} + \frac{|\gamma_\ell - \tilde{\gamma}_\ell|}{\ell^3} + \left|\frac{\tilde{\alpha}_{\ell 1}}{\alpha_{\ell 1}} - 1\right| + \left|\frac{\tilde{\alpha}_{\ell 2}}{\alpha_{\ell 2}} - 1\right|.$$

According to (8.68)

$$\sum_{\ell=1}^{\infty} \xi_\ell \ell^2 < C\Lambda,$$

for $\Lambda < \delta_1$. It follows from Lemma 4.11 that there exists $\delta_2 > 0$ ($\delta_2 < \delta_1$) such that if $\Lambda < \delta_2$, then

$$\left.\begin{aligned}
&|\widetilde{a}_2^{(\nu)}(x,\widetilde{\gamma}_n)| < Cn^{\nu-3}\exp(nx),\\
&|\widetilde{a}_3^{(\nu)}(x,\widetilde{\lambda}_n)| < Cn^{\nu-4}\exp(n\pi)\\
&|\widetilde{a}_2^{(\nu)}(x,\widetilde{\gamma}_n) - a_2^{(\nu)}(x,\widetilde{\gamma}_n)| < Cn^{\nu-4}\exp(nx),\\
&|\widetilde{a}_3^{(\nu)}(x,\widetilde{\lambda}_n) - a_3^{(\nu)}(x,\widetilde{\lambda}_n)| < Cn^{\nu-5}\exp(n\pi).
\end{aligned}\right\} \tag{8.70}$$

Denote

$$\left.\begin{aligned}
&\Phi_{n1} = \frac{\widetilde{a}_1(\pi,\widetilde{\gamma}_n)}{\dot{\widetilde{\delta}}(\widetilde{\gamma}_n)\widetilde{\Delta}(\widetilde{\gamma}_n)}(a_1(\pi,\widetilde{\gamma}_n) - \widetilde{a}_1(\pi,\widetilde{\gamma}_n)), \quad \Phi_{n2} = \frac{\widetilde{a}_1(\pi,\widetilde{\gamma}_n)\delta(\widetilde{\gamma}_n)}{\dot{\widetilde{\delta}}(\widetilde{\gamma}_n)\widetilde{\Delta}(\widetilde{\gamma}_n)},\\
&r_n(x) = \Phi_{n1}b_2(x,\widetilde{\gamma}_n) + \Phi_{n2}b_1(x,\widetilde{\gamma}_n),\\
&w_n(x) = \Phi_{n1}a_2(x,\widetilde{\gamma}_n) + \Phi_{n2}a_1(x,\widetilde{\gamma}_n),\\
&y_n(x) = \frac{1}{\dot{\widetilde{\Delta}}(\widetilde{\lambda}_n)\widetilde{\delta}(\widetilde{\lambda}_n)}[(b_3(x,\widetilde{\lambda}_n) - b_3(x,\lambda_n)) - k_n(a_3(x,\widetilde{\lambda}_n) - a_3(x,\lambda_n))].
\end{aligned}\right\} \tag{8.71}$$

Using (8.31), (8.33), (8.41), (8.43), (8.50)–(8.53), (8.61)–(8.64), (8.69) and (8.70), we obtain the following estimates

$$\left.\begin{aligned}
&|\Phi_{n1}| < Cn^6\exp(-n\pi)\left(\frac{|\gamma_n - \widetilde{\gamma}_n|}{n^3} + \omega_n\right),\\
&|\Phi_{n2}| < Cn\exp(-n\pi)|\widetilde{\gamma}_n - \widetilde{\gamma}_n|,\\
&|y_n^{(\mu)}(x)\widetilde{a}_3^{(\nu)}(x,\widetilde{\lambda}_n)| < Cn^{\nu+\mu-3}|\lambda_n - \widetilde{\lambda}_n|,\\
&|r_n^{(\mu)}(x)\widetilde{a}_2^{(\nu)}(x,\widetilde{\gamma}_n)| < Cn^{\nu+\mu}\left(\frac{|\gamma_n - \widetilde{\gamma}_n|}{n^3} + \omega_n\right),\\
&|w_n^{(\mu)}(x)\widetilde{b}_2^{(\nu)}(x,\widetilde{\gamma}_n)| < Cn^{\nu+\mu}\left(\frac{|\gamma_n - \widetilde{\gamma}_n|}{n^3} + \omega_n\right),\\
&|r_n^{(\mu)}(x)(\widetilde{a}_2^{(\nu)}(x,\widetilde{\gamma}_n) - a_2^{(\nu)}(x,\widetilde{\gamma}_n))| < Cn^{\nu+\mu-1}\left(\frac{|\gamma_n - \widetilde{\gamma}_n|}{n^3} + \omega_n\right),\\
&|w_n^{(\mu)}(x)(\widetilde{b}_2^{(\nu)}(x,\widetilde{\gamma}_n) - b_2^{(\nu)}(x,\widetilde{\gamma}_n))| < Cn^{\nu+\mu-1}\left(\frac{|\gamma_n - \widetilde{\gamma}_n|}{n^3} + \omega_n\right).
\end{aligned}\right\} \tag{8.72}$$

Now let us prove Theorem 8.3. We consider the function

$$G^0(x,t,\lambda)=\begin{cases}\dfrac{1}{\widetilde{\delta}(\lambda)}b_2(x,\lambda)\widetilde{a}_2(t,\lambda)-\dfrac{1}{\widetilde{\Delta}(\lambda)\widetilde{\delta}(\lambda)}b_3(x,t)\widetilde{a}_3(t,\lambda), & x>t\\[2ex] \dfrac{1}{\widetilde{\delta}(\lambda)}a_2(x,\lambda)\widetilde{b}_2(t,\lambda)-\dfrac{1}{\widetilde{\Delta}(\lambda)\widetilde{\delta}(\lambda)}a_3(x,t)\widetilde{b}_3(t,\lambda), & x<t.\end{cases} \tag{8.73}$$

We observe that for $\widetilde{\ell}=\ell$ the function $G^0(x,t,\lambda)$ coincides with the Green's function of Q_1:

$$G(x,t,\lambda)=\begin{cases}\dfrac{1}{\delta(\lambda)}b_2(x,\lambda)a_2(t,\lambda)-\dfrac{1}{\Delta(\lambda)\delta(\lambda)}b_3(x,\lambda)a_3(t,\lambda), & x>t,\\[2ex] \dfrac{1}{\delta(\lambda)}a_2(x,\lambda)b_2(t,\lambda)-\dfrac{1}{\Delta(\lambda)\delta(\lambda)}a_3(x,\lambda)b_3(t,\lambda), & x<t,\end{cases}$$

and the following equality holds

$$\begin{aligned}&\frac{1}{\delta(\lambda)}(b_2(x,\lambda)a_2'''(x,\lambda)-a_2(x,\lambda)b_2'''(x,\lambda))\\ &\qquad+\frac{1}{\Delta(\lambda)\delta(\lambda)}(a_3(x,\lambda)b_3'''(x,\lambda)-b_3(x,\lambda)a_3'''(x,\lambda))\equiv-1.\end{aligned} \tag{8.74}$$

Let $f(x)\in C^2[0,\pi]$, $f(0)=f(\pi)=0$, and let $\Gamma_N=\{\lambda:|\lambda|=R_N\}$ be the circumference in the λ-plane with radius $R_N\to\infty$ such that Γ_N are $\varepsilon>0$ distant from the spectra $\{\widetilde{\lambda}_n\}_{n\ge1}$ and $\{\widetilde{\gamma}_n\}_{n\ge1}$. We consider the function

$$\begin{aligned}y(x,\lambda)&=\int_0^\pi G^0(x,t,\lambda)f(t)\,dt=\frac{1}{\widetilde{\delta}(\lambda)}\Big(b_2(x,\lambda)\int_0^x\widetilde{a}_2(t,\lambda)f(t)\,dt\\ &+a_2(x,\lambda)\int_x^\pi\widetilde{b}_2(t,\lambda)f(t)\,dt-\frac{1}{\widetilde{\Delta}(\lambda)\widetilde{\delta}(\lambda)}(b_3(x,\lambda)\int_0^x\widetilde{a}_3(t,\lambda)f(t)\,dt\\ &+a_3(x,\lambda)\int_x^\pi\widetilde{b}_3(t,\lambda)f(t)\,dt\Big).\end{aligned} \tag{8.75}$$

Since $\widetilde{\ell}\widetilde{a}_j(x,\lambda)=\lambda\widetilde{a}_j(x,\lambda)$, $\widetilde{\ell}\widetilde{b}_j(x,\lambda)=\lambda\widetilde{b}_j(x,\lambda)$, we can rewrite (8.75) as

$$\begin{aligned}y(x)&=\frac{1}{\lambda\widetilde{\delta}(\lambda)}\Big(b_2(x,\lambda)\int_0^x\widetilde{\ell}\widetilde{a}_2(t,\lambda)f(t)\,dt+a_2(x,\lambda)\int_x^\pi\widetilde{\ell}\,\widetilde{b}_2(t,\lambda)f(t)\,dt\Big)\\ &-\frac{1}{\lambda\widetilde{\Delta}(\lambda)\widetilde{\delta}(\lambda)}\Big(b_3(x,\lambda)\int_0^x\widetilde{\ell}\widetilde{a}_3(t,\lambda)f(t)\,dt+a_3(x,\lambda)\int_x^\pi\widetilde{\ell}\,\widetilde{b}_3(t,\lambda)f(t)\,dt\Big).\end{aligned}$$

Integration by parts yields

$$y(x,\lambda)=\frac{f(x)}{\lambda\widetilde{\delta}(\lambda)}(b_2(x,\lambda)\widetilde{a}_2'''(x,\lambda)-a_2(x,\lambda)\widetilde{b}_2'''(x,\lambda))$$
$$+\frac{f(x)}{\lambda\widetilde{\Delta}(\lambda)\widetilde{\delta}(\lambda)}(a_3(x,\lambda)\widetilde{b}_3'''(x,\lambda)-b_3(x,\lambda)\widetilde{a}_3'''(x,\lambda))+y_0(x,\lambda), \tag{8.76}$$

where

$$y_0(x,\lambda)=-\frac{f'(x)}{\lambda\widetilde{\delta}(\lambda)}(b_2(x,\lambda)\widetilde{a}_2''(x,\lambda)-a_2(x,\lambda)\widetilde{b}_2''(x,\lambda))$$
$$-\frac{f'(x)}{\lambda\widetilde{\Delta}(\lambda)\widetilde{\delta}(\lambda)}(a_3(x,\lambda)\widetilde{b}_3''(x,\lambda)-b_3(x,\lambda)\widetilde{a}_3''(x,\lambda))$$
$$+\frac{1}{\lambda\widetilde{\delta}(\lambda)}\left(b_2(x,\lambda)\int_0^x\widetilde{a}_2''(t,\lambda)f''(t)\,dt+a_2(x,\lambda)\int_x^\pi\widetilde{b}_2''(t,\lambda)f''(t)\,dt\right)$$
$$-\frac{1}{\lambda\widetilde{\Delta}(\lambda)\widetilde{\delta}(\lambda)}\left(b_3(x,\lambda)\int_0^x\widetilde{a}_3''(t,\lambda)f''(t)\,dt+a_3(x,\lambda)\int_x^\pi\widetilde{b}_3''(t,\lambda)f''(t)\,dt\right)$$
$$+\frac{1}{\lambda\widetilde{\delta}(\lambda)}\left(b_2(x,\lambda)\int_0^x\widetilde{\ell}_1\widetilde{a}_2(t,\lambda)f(t)\,dt+a_2(x,\lambda)\int_x^\pi\widetilde{\ell}_1\widetilde{b}_2(t,\lambda)f(t)\,dt\right)$$
$$-\frac{1}{\lambda\widetilde{\Delta}(\lambda)\widetilde{\delta}(\lambda)}\left(b_3(x,\lambda)\int_0^x\widetilde{\ell}_1\widetilde{a}_3(t,\lambda)f(t)\,dt+a_3(x,\lambda)\int_x^\pi\widetilde{\ell}_1\widetilde{b}_3(t,\lambda)f(t)\,dt\right),$$

$$\ell_1 y=-(q_2(x)y')'+q_0(x)y.$$

Taking (8.74) into account, we transform (8.76) to the form

$$y(x,\lambda)=\frac{f(x)}{\lambda}+y_1(x,\lambda), \tag{8.77}$$

where

$$y_1(x,\lambda)=\frac{f(x)}{\lambda\widetilde{\delta}(\lambda)}((b_2(x,\lambda)-\widetilde{b}_2(x,\lambda))\widetilde{a}_2'''(x,\lambda)-(a_2(x,\lambda)-\widetilde{a}_2(x,\lambda))\widetilde{b}_2'''(x,\lambda))$$
$$+\frac{f(x)}{\lambda\widetilde{\Delta}(\lambda)\widetilde{\delta}(\lambda)}\Big((a_3(x,\lambda)-\widetilde{a}_3(x,\lambda))\widetilde{b}_3'''(x,\lambda)$$
$$-(b_3(x,\lambda)-\widetilde{b}_3(x,\lambda))\widetilde{a}_3'''(x,\lambda)\Big)+y_0(x,\lambda).$$

Using (8.35), (8.38) and (8.39), we obtain

$$|y_1(x,\lambda)| < C|\rho|^{-5}, \qquad \lambda \in \Gamma_N, \qquad x \in [0,\pi].$$

Hence, by virtue of (8.77), we conclude

$$\lim_{N\to\infty} \max_{0\le x\le\pi} \left| f(x) - \frac{1}{2\pi i} \int\limits_{\Gamma_N} y(x,\lambda)\, d\lambda \right| = 0. \tag{8.78}$$

Now we calculate the residues of $G^0(x,t,\lambda)$:

$$\operatorname{res}_{\lambda=\tilde\lambda_n} G^0(x,t,\lambda) = -\frac{1}{\dot{\tilde\Delta}(\tilde\lambda_n)\tilde\delta(\tilde\lambda_n)} \begin{cases} b_3(x,\tilde\lambda_n)\tilde a_3(t,\tilde\lambda_n), & x > t, \\ a_3(x,\tilde\lambda_n)\tilde b_3(t,\tilde\lambda_n), & x < t, \end{cases} \tag{8.79}$$

$$\operatorname{res}_{\lambda=\tilde\gamma_n} G^0(x,t,\lambda)$$

$$= \frac{1}{\tilde\Delta(\tilde\gamma_n)\dot{\tilde\delta}(\tilde\gamma_n)} \begin{cases} \tilde\Delta(\tilde\gamma_n) b_2(x,\tilde\gamma_n)\tilde a_2(t,\tilde\gamma_n) - b_3(x,\tilde\gamma_n)\tilde a_3(t,\tilde\gamma_n), & x > t, \\ \tilde\Delta(\tilde\gamma_n) a_2(x,\tilde\gamma_n)\tilde b_2(t,\tilde\gamma_n) - a_3(x,\tilde\gamma_n)\tilde b_3(t,\tilde\gamma_n), & x < t, \end{cases} \tag{8.80}$$

Since

$$\tilde a_3(t,\tilde\gamma_n) = -\tilde a_1(\pi,\tilde\gamma_n)\tilde a_2(t,\tilde\gamma), \qquad \tilde\Delta(\tilde\gamma_n) = -\tilde a_1^2(\pi,\tilde\gamma_n),$$

$$\tilde a_1(\pi,\tilde\gamma_n) = -\tilde b_1(0,\tilde\gamma_n), \qquad \tilde b_3(t,\tilde\gamma_n) = -\tilde b_1(0,\tilde\gamma_n)\tilde b_2(t,\tilde\gamma_n)$$

we get

$$\tilde\Delta(\tilde\gamma_n) b_2(x,\tilde\gamma_n)\tilde a_2(t,\tilde\gamma_n) - b_3(x,\tilde\gamma_n)\tilde a_3(t,\tilde\gamma_n)$$

$$= (\tilde a_1(\pi,\tilde\gamma_n)((a_1(\pi,\tilde\gamma_n) - \tilde a_1(\pi,\tilde\gamma_n))b_2(x,\tilde\gamma_n) + \delta(\tilde\gamma_n)b_1(x,\tilde\gamma_n)))\tilde a_2(t,\tilde\gamma_n).$$

Similarly,

$$\tilde\Delta(\tilde\gamma_n) a_2(x,\tilde\gamma_n)\tilde b_2(t,\tilde\gamma_n) - a_3(x,\tilde\gamma_n)\tilde b_3(t,\tilde\gamma_n)$$

$$= (\tilde a_1(\pi,\tilde\gamma_n)((a_1(\pi,\tilde\gamma_n) - \tilde a_1(\pi,\tilde\gamma_n))a_2(x,\tilde\gamma_n) + \delta(\tilde\gamma_n)a_1(x,\tilde\gamma_n)))\tilde b_2(t,\tilde\gamma_n).$$

Thus, (8.80) becomes

$$\operatorname{res}_{\lambda=\widetilde{\gamma}_n} G^0(x,t,\lambda) = \begin{cases} r_n(x)\widetilde{a}_2(t,\widetilde{\gamma}_n), & x > t, \\ w_n(x)\widetilde{b}_2(t,\widetilde{\gamma}_n), & x < t, \end{cases} \tag{8.81}$$

where the functions $r_n(x)$ and $w_n(x)$ are define via (8.71).

It follows from (8.75), (8.78), (8.79) and (8.81) that

$$f(x) = \sum_{n=1}^{\infty}\Bigg[\frac{1}{\dot{\widetilde{\Delta}}(\widetilde{\lambda}_n)\widetilde{\delta}(\widetilde{\lambda}_n)}\Bigg(b_3(x,\widetilde{\lambda}_n)\int_0^x \widetilde{a}_3(t,\widetilde{\lambda}_n)f(t)\,dt$$

$$+a_3(x,\widetilde{\lambda}_n)\int_x^{\pi}\widetilde{b}_3(t,\widetilde{\lambda}_n)f(t)\,dt\Bigg) - r_n(x)\int_0^x \widetilde{a}_2(t,\widetilde{\gamma}_n)f(t)\,dt$$

$$-w_n(x)\int_x^{\pi}\widetilde{b}_2(t,\widetilde{\gamma}_n)f(t)\,dt\Bigg].$$

Adding and subtracting the function

$$a_3(x,\widetilde{\lambda}_n)\int_0^x \widetilde{b}_3(t,\widetilde{\lambda}_n)f(t)\,dt$$

in the right-hand side of the last equality and using the relation $\widetilde{b}_3(t,\lambda_n) = \widetilde{k}_n\widetilde{a}_3(t,\widetilde{\lambda}_n)$, we obtain

$$f(x) = \sum_{n=1}^{\infty}\Bigg[\frac{\widetilde{k}_n}{\dot{\widetilde{\Delta}}(\widetilde{\lambda}_n)\widetilde{\delta}(\widetilde{\lambda}_n)}a_3(x,\widetilde{\lambda}_n)\int_0^{\pi}\widetilde{a}_3(t,\widetilde{\lambda}_n)f(t)\,dt$$

$$+\frac{1}{\dot{\widetilde{\Delta}}(\widetilde{\lambda}_n)\widetilde{\delta}(\widetilde{\lambda}_n)}(b_3(x,\widetilde{\lambda}_n) - \widetilde{k}_n a_3(x,\widetilde{\lambda}_n))\int_0^x \widetilde{a}_3(t,\widetilde{\lambda}_n)f(t)\,dt$$

$$-r_n(x)\int_0^x \widetilde{a}_2(t,\widetilde{\gamma}_n)f(t)\,dt - w_n(x)\int_x^{\pi}\widetilde{b}_2(t,\widetilde{\gamma}_n)f(t)\,dt\Bigg].$$

Because of (8.43) and (8.59), we can choose $\delta > 0$ so that $k_n = \widetilde{k}_n$ for $\Lambda < \delta$. Then

$$\begin{aligned} f(x) &= \sum_{n=1}^{\infty}\Bigg[\frac{\widetilde{k}_n}{\dot{\widetilde{\Delta}}(\widetilde{\lambda}_n)\widetilde{\delta}(\widetilde{\lambda}_n)} a_3(x,\widetilde{\lambda}_n)\int_0^\pi \widetilde{a}_3(t,\widetilde{\lambda}_n)f(t)\,dt \\ &+ y_n(x)\int_0^x \widetilde{a}_3(t,\widetilde{\lambda}_n)f(t)\,dt - r_n(x)\int_0^x \widetilde{a}_2(t,\widetilde{\gamma}_n)f(t)\,dt \\ &- w_n(x)\int_x^\pi \widetilde{b}_2(t,\widetilde{\gamma}_n)f(t)\,dt\Bigg]. \end{aligned} \tag{8.82}$$

In particular, for $\widetilde{\ell} = \ell$ we have $r_n(x) = w_n(x) = y_n(x) = 0$. Thus, we obtain the well-known expansion theorem

$$f(x) = \sum_{n=1}^{\infty} A_n a_3(x,\lambda_n), \tag{8.83}$$

where

$$A_n = \frac{k_n}{\dot{\Delta}(\lambda_n)\delta(\lambda_n)}\int_0^\pi f(t)a_3(t,\lambda_n)\,dt = \Big(\int_0^\pi a_3^2(t,\lambda_n)\,dt\Big)^{-1}\int_0^\pi f(t)a_3(t,\lambda_n)\,dt.$$

Comparing (8.82) with (8.83), we obtain

$$\begin{aligned} \sum_{n=1}^{\infty}\widetilde{A}_n\widetilde{a}_3(x,\widetilde{\lambda}_n) &= \sum_{n=1}^{\infty}\Bigg[\widetilde{A}_n a_3(x,\widetilde{\lambda}_n) + y_n(x)\int_0^x \widetilde{a}_3(t,\widetilde{\lambda}_n)f(t)\,dt \\ &- r_n(t)\int_0^x \widetilde{a}_2(t,\widetilde{\gamma}_n)f(t)\,dt - w_n(x)\int_x^\pi \widetilde{b}_2(t,\widetilde{\gamma}_n)f(t)\,dt\Bigg], \end{aligned} \tag{8.84}$$

where

$$\widetilde{A}_n = \frac{\widetilde{k}_n}{\dot{\widetilde{\Delta}}(\widetilde{\lambda}_n)\widetilde{\delta}(\widetilde{\lambda}_n)}\int_0^\pi f(t)\widetilde{a}_3(t,\widetilde{\lambda}_n)\,dt,$$

and the series converges absolutely and uniformly for $x \in [0,\pi]$.

Now let $f(x)$ satisfy the conditions $f(x) \in C^4[0,\pi]$, $\ell f(x) \in C^2[0,\pi]$; $f^{(\nu)}(0) = f^{(\nu)}(\pi) = 0$, $\nu = 0,2$; $\ell f(0) = \ell f(\pi) = 0$. We put $\widetilde{\ell} f(x)$ in (8.84) instead of $f(x)$. Using (8.29) we compute

$$\int_0^\pi \widetilde{\ell} f(t)\widetilde{a}_3(t,\widetilde{\lambda}_n)\,dt = \int_0^\pi f(t)\widetilde{\ell}\widetilde{a}_3(t,\widetilde{\lambda}_n)\,dt = \widetilde{\lambda}_n\int_0^\pi f(t)\widetilde{a}_3(t,\widetilde{\lambda}_n)\,dt.$$

From (8.84), in view of (8.29), we obtain

$$\sum_{n=1}^{\infty}\widetilde{\lambda}_n\widetilde{A}_n\widetilde{a}_3(x,\widetilde{\lambda}_n)=\sum_{n=1}^{\infty}\Big[\widetilde{\lambda}_n\widetilde{A}_n a_3(x,\widetilde{\lambda}_n)+\widetilde{\lambda}_n y_n(x)\int_0^x \widetilde{a}_3(t,\widetilde{\lambda}_n)f(t)\,dt$$

$$-\widetilde{\gamma}_n r_n(x)\int_0^x \widetilde{a}_2(t,\widetilde{\gamma}_n)f(t)\,dt-\widetilde{\gamma}_n w_n(x)\int_x^\pi \widetilde{b}_2(t,\widetilde{\gamma}_n)f(t)\,dt$$

$$+y_n(x)(f'''(x)\widetilde{a}_3(x,\widetilde{\lambda}_n)-f''(x)\widetilde{a}_3'(x,\widetilde{\lambda}_n)+f'(x)(\widetilde{a}_3''(x,\widetilde{\lambda}_n)$$

$$-\widetilde{q}_2(x)\widetilde{a}_3(x,\widetilde{\lambda}_n))-f(x)(\widetilde{a}_3''(x,\widetilde{\lambda}_n)-\widetilde{q}_2(x)\widetilde{a}_3'(x,\widetilde{\lambda}_n)))$$

$$-r_n(x)(f'''(x)\widetilde{a}_2(x,\widetilde{\gamma}_n)-f''(x)\widetilde{a}_2'(x,\widetilde{\gamma}_n)+f'(x)(\widetilde{a}_2''(x,\widetilde{\gamma}_n) \tag{8.85}$$

$$-\widetilde{q}_2(x)\widetilde{a}_2(x,\widetilde{\gamma}_n))-f(x)(\widetilde{a}_2''(x,\widetilde{\gamma}_n)-\widetilde{q}_2(x)\widetilde{a}_2'(x,\widetilde{\gamma}_n)))$$

$$-w_n(x)(f'''(x)\widetilde{b}_2(x,\widetilde{\gamma}_n)-f''(x)\widetilde{b}_2'(x,\widetilde{\gamma}_n)+f'(x)(\widetilde{b}_2''(x,\widetilde{\gamma}_n)$$

$$-\widetilde{q}_2(x)\widetilde{b}_2(x,\widetilde{\gamma}_n))-f(x)(\widetilde{b}_2'''(x,\widetilde{\gamma}_n)-\widetilde{q}_2(x)\widetilde{b}_2'(x,\widetilde{\gamma}_n)))\Big].$$

On the other hand, applying the operator ℓ, we calculate

$$\ell\left(\sum_{n=1}^{\infty}\widetilde{A}_n a_3(x,\widetilde{\lambda}_n)\right)=\sum_{n=1}^{\infty}\widetilde{\lambda}_n\widetilde{A}_n a_3(x,\widetilde{\lambda}_n),$$

$$\begin{aligned}\ell\left(\sum_{n=1}^{\infty}\widetilde{A}_n\widetilde{a}_3(x,\widetilde{\lambda}_n)\right) &= \widetilde{\ell}\left(\sum_{n=1}^{\infty}\widetilde{A}_n\widetilde{a}_3(x,\widetilde{\lambda}_n)\right)+\widehat{\ell}\left(\sum_{n=1}^{\infty}\widetilde{A}_n\widetilde{a}_3(x,\widetilde{\lambda}_n)\right)\\ &=\sum_{n=1}^{\infty}\widetilde{\lambda}_n\widetilde{A}_n\widetilde{a}_3(x,\widetilde{\lambda}_n)+\widehat{\ell}f(x)\\ &=\sum_{n=1}^{\infty}\widetilde{\lambda}_n\widetilde{A}_n\widetilde{a}_3(x,\widetilde{\lambda}_n)-(\widehat{q}_2(x)f'(x))'+\widehat{g}_0(x)f(x).\end{aligned}$$

It follows from (8.84) that

$$\sum_{n=1}^{\infty}\widetilde{\lambda}_n\widetilde{A}_n\widetilde{a}_3(x,\widetilde{\lambda}_n)-(\widehat{q}_2(x)f'(x))'+\widehat{q}_0(x)f(x)$$

$$=\sum_{n=1}^{\infty}\widetilde{\lambda}_n\widetilde{A}_n\widetilde{a}_3(x,\widetilde{\lambda}_n)+\ell\Big[\sum_{n=1}^{\infty}\Big(y_n(x)\int_0^x\widetilde{a}_3(t,\widetilde{\lambda}_n)f(t)\,dt \tag{8.86}$$

$$-r_n(x)\int_0^x\widetilde{a}_2(t,\widetilde{\gamma}_n)f(t)\,dt-w_n(x)\int_x^\pi\widetilde{b}_2(t,\widetilde{\gamma}_n)f(t)\,dt\Big)\Big].$$

Since

$$\ell(yz)=z\ell y+4y'''z'+6y''z''+4y'z'''+yz^{(4)}-q_2y'z'-(q_2yz')',$$

we get

$$\ell\Big[y_n(x)\int_0^x\widetilde{a}_3(t,\widetilde{\lambda}_n)f(t)\,dt\Big]=\widetilde{\lambda}_ny_n(x)\int_0^x\widetilde{a}_3(t,\widetilde{\lambda}_n)f(t)\,dt$$

$$+4y_n'''(x)\widetilde{a}_3(x,\widetilde{\lambda}_n)f(x)+6y_n''(x)(\widetilde{a}_3(x,\widetilde{\lambda}_n)f(x))'+4y_n'(x)(\widetilde{a}_3(x,\widetilde{\lambda}_n)f(x))''$$

$$+y_n(x)(\widetilde{a}_3(x,\widetilde{\lambda}_n)f(x))'''-q_2(x)y_n'(x)\widetilde{a}_3(x,\widetilde{\lambda}_n)f(x)-(q_2(x)y_n(x)\widetilde{a}_3(x,\widetilde{\lambda}_n)f(x))',$$

$$\ell\Big[r_n(x)\int_0^x\widetilde{a}_2(t,\widetilde{\gamma}_n)f(t)\,dt\Big]=\widetilde{\gamma}_nr_n(x)\int_0^x\widetilde{a}_2(t,\widetilde{\gamma}_n)f(t)\,dt$$

$$+4r_n'''(x)\widetilde{a}_2(x,\widetilde{\gamma}_n)f(x)+6r_n''(x)(\widetilde{a}_2(x,\widetilde{\gamma}_n)f(x))'+4r_n'(x)(\widetilde{a}_2(x,\widetilde{\gamma}_n)f(x))''$$

$$+r_n(x)(\widetilde{a}_2(x,\widetilde{\gamma}_n)f(x))'''-q_2(x)r_n'(x)\widetilde{a}_2(x,\widetilde{\gamma}_n)f(x)-(q_2(x)r_n(x)\widetilde{a}_2(x,\widetilde{\gamma}_n)f(x))',$$

$$\ell\Big[w_n(x)\int_x^\pi\widetilde{b}_2(t,\widetilde{\gamma}_n)f(t)\,dt\Big]=\widetilde{\gamma}_nw_n(x)\int_x^\pi\widetilde{b}_2(t,\widetilde{\gamma}_n)f(t)\,dt$$

$$-4w_n'''(x)\widetilde{b}_2(x,\widetilde{\gamma}_n)f(x)-6w_n''(x)(\widetilde{b}_2(x,\widetilde{\gamma}_n)f(x))'-4w_n'(x)(\widetilde{b}_2(x,\widetilde{\gamma}_n)f(x))''$$

$$-w_n(x)(\widetilde{b}_2(x,\widetilde{\gamma}_n)f(x))'''+q_2(x)w_n'(x)\widetilde{b}_2(x,\widetilde{\gamma}_n)f(x)+(q_2(x)w_n(x)\widetilde{b}_2(x,\widetilde{\gamma}_n)f(x))'.$$

Comparing (8.85) with (8.86) and using the arbitrariness of $f(x)$, we obtain

$$\int_0^x \widehat{q}_{2-j}(t)\,dt = \sum_{n=1}^{\infty}\int_0^x (F_j[y_n(t),\widetilde{a}_3(t,\widetilde{\lambda}_n)] - F_j[r_n(t),\widetilde{a}_2(t,\widetilde{\gamma}_n)] + F_j[w_n(t),\widetilde{b}_2(t,\widetilde{\gamma}_n)])\,dt, \qquad j=0,2; \tag{8.87}$$

$$\frac{d^{\mu+1}}{dx^{\mu+1}}\int_0^x \widehat{q}_2(t)\,dt = \sum_{n=1}^{\infty}(F_\mu[y_n(x),\widetilde{a}_3(x,\widetilde{\lambda}_n)] - F_\mu[r_n(x),\widetilde{a}_2(x,\widetilde{\gamma}_n)] + F_\mu[w_n(x),\widetilde{b}_2(x,\widetilde{\gamma}_n)]), \qquad \mu=0,1; \tag{8.88}$$

where

$$F_0[y,z] = -4(yz)', \qquad F_1[y,z] = -6y''z - 8y'z' - 2yz'' + \widehat{q}_2 yz,$$

$$F_2[y,z] = 4y'''z + 6y''z' + 4y'z'' + 2yz''' - 2q_2y'z - q_2'yz - (q_2+\widetilde{q}_2)yz'.$$

The proof of Theorem 8.3 for $i=2$, $\nu=0,1,2$ follows from (8.87), (8.88) and (8.72). The case $i=\nu=0$ is more difficult. We transform (8.87):

$$\int_0^x \widehat{q}_0(t)\,dt = \sum_{n=1}^{\infty}\int_0^x (F_2[y_n(t),\widetilde{a}_3(t,\widetilde{\lambda}_n)] - F_2[r_n(t),\widetilde{a}_2(t,\widetilde{\gamma}_n) - a_2(t,\widetilde{\gamma}_n)] + F_2[w_n(t),\widetilde{b}_2(t,\widetilde{\gamma}_n) - b_2(t,\widetilde{\gamma}_n)])\,dt$$
$$+\sum_{n=1}^{\infty}\int_0^x (-F_2[r_n(t),a_2(t,\widetilde{\gamma}_n)] + F_2[w_n(t),b_2(t,\widetilde{\gamma}_n)])\,dt.$$

Substituting $r_n(x)$ and $w_n(x)$ from (8.71) into the second integral we arrive at

$$\int_0^x \widehat{q}_0(t)\,dt = s_0(x) + s_1(x),$$

where

$$s_0(x) = \sum_{n=1}^{\infty}\Phi_{n1}\int_0^x (F_2[a_2(t,\widetilde{\gamma}_n),b_2(t,\widetilde{\gamma}_n)] - F_2[b_2(t,\widetilde{\gamma}_n),a_2(t,\widetilde{\gamma}_n)])\,dt,$$

$$s_1(x) = \sum_{n=1}^{\infty} \Phi_{n2} \int_0^x (F_2[a_1(t,\widetilde{\gamma}_n), b_2(t,\widetilde{\gamma}_n)] - F_2[b_1(t,\widetilde{\gamma}_n), a_2(t,\widetilde{\gamma}_n)])\, dt$$

$$+ \sum_{n=1}^{\infty} \int_0^x (F_2[y_n(t), \widetilde{a}_3(t,\widetilde{\lambda}_n)] - F_2[r_n(t), \widetilde{a}_2(t,\widetilde{\gamma}_n) - a_2(t,\widetilde{\gamma}_n)]$$

$$+ F_2[w_n(t), \widetilde{b}_2(t,\widetilde{\gamma}_n) - b_2(t,\widetilde{\gamma}_n]) \, dt.$$

Using (8.31), (8.51) and (8.72),we get the estimate $|s_1(x)| < C\Lambda$. To obtain the estimate for $s_0(x)$ we essentially use the cancellation of the main terms of the asymptotics. Let us denote

$$I_n(x) = a_2'''(x,\widetilde{\gamma}_n) b_2(x,\widetilde{\gamma}_n) + a_2''(x,\widetilde{\gamma}_n) b_2'(x,\widetilde{\gamma}_n) - a_2'(x,\widetilde{\gamma}_n) b_2''(x,\widetilde{\gamma}_n) - a_2(x,\widetilde{\gamma}_n) b_2'''(x,\widetilde{\gamma}_n).$$

Taking (8.32) and (8.39) into account, we calculate the asymptotics of $I_n(x)$ for $n \to \infty$:

$$I_n(x) = -\frac{1}{4(n+\frac{1}{4})^3} \exp\left(\left(n+\frac{1}{4}\right)\pi\right) \cos\left(n+\frac{1}{4}\right)(\pi - 2x)\left(1 + O\left(\frac{1}{n}\right)\right). \tag{8.89}$$

It is clear that

$$F_2[a_2(t,\widetilde{\gamma}_n), b_2(t,\widetilde{\gamma}_n)] - F_2[b_2(t,\widetilde{\gamma}_n), a_2(t,\widetilde{\gamma}_n)]$$

$$= 2I_n(x) + F_2^0[a_2(t,\widetilde{\gamma}_n), b_2(t,\widetilde{\gamma}_n)] - F_2^0[b_2(t,\widetilde{\gamma}_n), a_2(t,\widetilde{\gamma}_n)],$$

where

$$F_2^0[y,z] = -2q_2 y' z - q_2' yz - (q_2 + \widetilde{q}_2) yz'.$$

Using (8.31), (8.51) and (8.89) we obtain the estimate $|s_0(x)| < C\Lambda$. Thus, the assertion of Theorem 8.3 for $i = \nu = 0$ is proved. This completes the proof of Theorem 8.3.

9. Method of standard models. Information condition

We consider the DE and LF $L = (\ell, U)$ of the form (1.1)–(1.2) on the half-line or on the finite interval ($T \le \infty$) and study the IP of recovering the N ($1 \le N \le n-1$) coefficients of the DE from given N Weyl functions.

9.1. Let sets of positive integers $\kappa = \{\kappa_j\}_{j=\overline{1,N}}$, $I = \{(k_i,\gamma_i), i = \overline{1,N}\}$, $2 \le \kappa_1 < \ldots < \kappa_N \le n$, $1 \le k_i < \gamma_i \le n$ be given. The IP is formulated as follows.

Problem 9.1. Given the WF's $\{\mathfrak{M}_{k_i,\gamma_i}(\lambda)\}_{i=\overline{1,N}}$ and the coefficients $p_\nu(x)$, $n-\nu\notin\kappa$, construct the functions $\{p_{n-\kappa_j}(x)\}_{j=\overline{1,N}}$.

We study this problem under the *apriori* condition that the functions $p_{n-\kappa_j}(x)$, $j=\overline{1,N}$ are piecewise-analytic (the rest of the coefficients $p_\nu(x)$, $n-\nu\notin\kappa$ are known integrable functions).

For convenience, we number the given WF's in a different way. Let

$$I=\{(m_\tau,\gamma_{\tau\eta}),\tau=\overline{1,\theta},\eta=\overline{1,N_\tau}\},\qquad 1\le m_\tau<\gamma_{\tau 1}<\ldots<\gamma_{\tau,N_\tau}\le n,$$

$$N_1+\ldots+N_\theta=N,\qquad m_\tau\ne m_{\tau'}\qquad(\tau\ne\tau').$$

Denote $\mathfrak{M}_s(\lambda)=\mathfrak{M}_{m_\tau,\gamma_{\tau\eta}}(\lambda)$, $s=\overline{1,N}$. Here and below, the positive integer s has a unique representation $s=N_1+\ldots+N_{\tau-1}+\eta$, $1\le\eta\le N_\tau$. Then our IP can be written as follows.

Problem 9.2. Given the WF's $\{\mathfrak{M}_s(\lambda)\}_{s=\overline{1,N}}$ and the coefficients $p_\nu(x)$, $n-\nu\notin\kappa$, construct the functions $\{p_{n-\kappa_j}(x)\}_{j=\overline{1,N}}$.

We note that using the given WF's we can find not only the DE, but also the coefficients of the LF. However, we assume for brevity that the LF are known. We also assume that the enumeration of the R_k in (1.3) is chosen for the sector $S_0=\{\rho:\arg\rho\in(0,\frac{\pi}{n})\}$. Denote

$$\omega_\xi^*(R)=(-R)^{\sigma_{\xi 0}^*},\qquad \Omega^*(j_1,\ldots,j_p)=\det[\omega_{j_\nu}^*(R_k^*)]_{\nu,k=\overline{1,p}},$$

$$\Omega_\mu^*(j_1,\ldots,j_p)=\det[\omega_{j_\nu}^*(R_k^*)]_{\nu=\overline{1,p};,k=\overline{1,p+1}\setminus\mu}.$$

Let us give a classification of the IP. For this we consider the matrices $A_\ell=[A_{\ell sj}]_{s=\overline{1,N};j=\overline{1,p}}$, $\ell\ge 1$, where p is such that $\ell\in[\kappa_p,\kappa_{p+1})$, $\kappa_0=1$, $\kappa_{N+1}=\infty$ (if $\ell\ge\kappa_N$ then A_ℓ is a square matrix); here

$$\begin{aligned}A_{\ell sj}&=\frac{(-1)^{n+m_\tau+\gamma_{\tau\eta}+1}}{\Omega(\overline{1,m_\tau})\Omega^*(\overline{1,n-m_\tau})}\\&\quad\times\sum_{\mu=1}^{m_\tau}\sum_{\nu=1}^{n-m_\tau}\frac{(-1)^{\mu+\nu}\Omega_\mu(\overline{1,m_\tau-1})\Omega_\nu^*(\overline{1,n-m_\tau}\setminus n-\gamma_{\tau\eta}+1)}{R_\mu^{\kappa_j}(-R_\mu-R_\nu^*)^{\ell-\kappa_j+1}}.\end{aligned}$$

Definition 9.1. If $A_\ell\#0$ for all $\ell\ge 1$, then the set $\{\mathfrak{M}_s(\lambda)\}_{s=\overline{1,N}}$ of the WF's is said to be a P_κ-system.

This definition distinguishes classes of WF's which have sufficient information for solution of the IP. We therefore call $A_\ell\#0$, $\ell\ge 1$ an information condition. It is easily shown that this condition is independent of the choice of a sector (see below).

If the information condition is not satisfied, then the IP does not have a unique solution.

9.2. Let us study the IP for P_κ-systems. To solve the IP we use the so-called method of standard models in which we construct a sequence of model differential operators of the form (1.1) "approaching" the unknown operator. The method allows us to obtain an algorithm of the solution of the IP. We first prove some auxiliary propositions.

Lemma 9.1. *Let*

$$r(x) = (\alpha!)^{-1} x^\alpha (h + p(x)), \qquad \alpha \geq 0, \qquad p(x) \in C[0, b], \qquad p(0) = 0,$$

$$H(x, z) = \exp(-zx)(1 + z^{-1}\xi(x, z)),$$

where the function $\xi(x, z)$ is continuous and bounded for

$$x \in [0, b], \qquad |z| \geq z_0, \qquad z \in Q \stackrel{df}{=} \left\{ z : \arg z \in [-\frac{\pi}{2} + \delta_0, \frac{\pi}{2} - \delta_0], \delta_0 > 0 \right\}.$$

Then for $|z| \to \infty$, $z \in Q$ we have

$$\int_0^b r(x) H(x, z)\, dx = z^{-\alpha - 1}(h + o(1)).$$

Proof. Consider the integral

$$z^{\alpha+1} \int_0^b r(x) H(x, z)\, dx = h z^{\alpha+1} \int_0^b \frac{x^\alpha}{\alpha!} \exp(-zx)\, dx + z^{\alpha+1} \int_0^b \frac{x^\alpha}{\alpha!} p(x) \exp(-zx)\, dx$$

$$+ z^\alpha \int_0^b r(x) \xi(x, z) \exp(-zx)\, dx = J_1(z) + J_2(z) + J_3(z).$$

In the domain Q we get $\operatorname{Re} z \geq \varepsilon_0 |z|$, $\varepsilon_0 > 0$. Since

$$\int_0^\infty \frac{x^\alpha}{\alpha!} \exp(-zx)\, dx = \frac{1}{z^{\alpha+1}},$$

it follows that

$$J_1(z) = h - hz^{\alpha+1}\int_b^{\infty}\frac{x^\alpha}{\alpha!}\exp(-zx)\,dx.$$

Hence $J_1(z) - h \to 0$ for $z \to \infty$, $z \in Q$.

Let $\varepsilon > 0$. Choose $\delta = \delta(\varepsilon)$ such that $|p(x)| < \frac{\varepsilon}{2}\varepsilon_0^{\alpha+1}$ for $x \in [0, \delta]$. Then

$$|J_2(z)| < \frac{\varepsilon}{2}(|z|\varepsilon_0)^{\alpha+1}\int_0^{\delta}\frac{x^\alpha}{\alpha!}\exp(-\varepsilon_0|z|x)\,dx + |z|^{\alpha+1}\int_\delta^b\frac{x^\alpha}{\alpha!}|p(x)|\exp(-\varepsilon_0|z|x)\,dx$$

$$< \frac{\varepsilon}{2} + |z|^{\alpha+1}\exp(-\varepsilon_0|z|\delta)\int_0^{b-\delta}\frac{(x+\delta)^\alpha}{\alpha!}|p(x+\delta)|\exp(-\varepsilon_0|z|x)\,dx.$$

For $|z| \to \infty$, $z \in Q$ the second term can be made less than $\frac{\varepsilon}{2}$. In virtue of the arbitrariness of ε, we have $J_2(z) \to 0$ for $|z| \to \infty$, $z \in Q$.

Since $|h + p(x)|\,|\xi(x,z)| < C$, it follows that for $z \in Q$

$$|J_3(z)| < C|z|^\alpha\int_0^b\frac{x^\alpha}{\alpha!}\exp(-\varepsilon_0|z|x)\,dx < \frac{C}{|z|\varepsilon_0^{\alpha+1}},$$

i.e. $J_3(z) \to 0$ for $|z| \to \infty$, $z \in Q$. Lemma 9.1 is proved.

Denote

$$\psi_\tau(x,\lambda) = \Phi_{m_\tau}(x,\lambda), \qquad \beta_\tau = m_\tau + N_\tau, \qquad \gamma_{\tau 0} = m_\tau,$$

$$\mathcal{M}_\tau = \{k : k = \overline{1, m_\tau}; \gamma_{\tau 1}, \ldots, \gamma_{\tau, N_\tau}\}, \qquad \mathcal{M}_{\tau\eta} = \{k : k \in \mathcal{M}_\tau, k \neq \gamma_{\tau\eta}\},$$

$$q_\tau = \det[R_k^\mu]_{k=\overline{\beta_\tau+1,n};\,\mu=\overline{\beta_\tau,n-1}}, \qquad Q = [q_{\xi\nu}^\tau]_{\xi=\overline{1,n};\,\nu=\overline{0,n-1}},$$

where

$$q_{\xi\nu}^\tau = (-1)^{\nu-\beta_\tau+1}(q_\tau)^{-1}\det[R_k^\mu]_{k=\overline{\beta_\tau+1,n};\,\mu=\xi-1,\overline{\beta_\tau,n-1}\setminus\nu}$$

for $\xi = \overline{1,\beta_\tau}$, $\nu = \overline{\beta_\tau, n-1}$, and $q_{\xi\nu}^\tau = \delta_{\nu,\xi-1}$ otherwise.

The following lemma allows us to solve the IP in steps.

Lemma 9.2. *For a fixed $a \in (0,T)$, the WS's $\psi_\tau(x,\lambda)$, $\tau = \overline{1,\theta}$ satisfy the boundary conditions*

$$U^\tau_{\xi a}(\psi_\tau) = \mathfrak{N}^\tau_\xi(\lambda, a), \qquad \xi = \overline{1,\beta_\tau}, \tag{9.1}$$

where

$$U^\tau_{\xi a}(y) = \sum_{\nu=0}^{n-1} Q^\tau_{\xi\nu}(\lambda, a) y^{(\nu)}(a),$$

$$\mathfrak{N}^\tau_\xi(\lambda, a) = J^\tau_{\xi 0}(\lambda, a) + \sum_{\eta=1}^{N_\tau} J^\tau_{\xi\eta}(\lambda, a)\mathfrak{M}_s(\lambda),$$

$$J^\tau_{\xi\eta}(\lambda, a) = \frac{(-1)^{n-N_\tau+\eta-1}}{\Delta_\tau(\lambda, a)} \det\,[z_k^{(\mu)}(a,\lambda), U_{j0}(z_k)]_{k=\overline{1,n};\mu=\xi-1,\overline{\beta_\tau,n-1};j\in M_{\tau\eta}},$$

$$Q^\tau_{\xi\nu}(\lambda, a) = \frac{(-1)^{n-\beta_\tau+1}}{\Delta_\tau(\lambda, a)} \det\,[z_k^{(\mu)}(a,\lambda), U_{j0}(z_k)]_{k=\overline{1,n};\mu=\xi-1,\overline{\beta_\tau,n-1}\setminus\nu;j\in M_\tau},$$

$$(\nu = \overline{\beta_\tau, n-1}),$$

$$Q^\tau_{\xi\nu}(\lambda, a) = \delta_{\xi,\nu-1}, \qquad (\nu = \overline{0, \beta_\tau - 1}),$$

$$\Delta_\tau(\lambda, a) = \det\,[z_k^{(\mu)}(a,\lambda), U_{j0}(z_k)]_{k=\overline{1,n};\mu=\overline{\beta_\tau,n-1};j\in M_\tau}\,.$$

Here $\{z_k(x,\lambda)\}_{k=\overline{1,n}}$, $x \in [0,a]$ is a certain FSS of the DE (1.1). For $|\lambda| \to \infty$, $\arg\lambda = \varphi \in (0,\pi)$, we have

$$Q^\tau_{\xi\nu}(\lambda, a) = \rho^{\xi-1-\nu} q^\tau_{\xi\nu}(1 + O(\rho^{-1})).$$

Indeed, from the relation

$$\psi_\tau(x,\lambda) = \sum_{\mu=1}^{n} b_{\tau\mu}(\lambda) z_\mu(x,\lambda)$$

for $x = 0$ and $x = a$ we get

$$\sum_{\mu=1}^{n} b_{\tau\mu}(\lambda) z_\mu^{(\nu)}(a,\lambda) = \psi_\tau^{(\nu)}(a,\lambda); \qquad \sum_{\mu=1}^{n} b_{\tau\mu}(\lambda) U_{j0}(z_\mu(x,\lambda)) = \delta_{j,m_\tau},$$

$$\sum_{\mu=1}^{n} b_{\tau\mu}(\lambda) U_{\gamma_{\tau\eta},0}(z_\mu(x,\lambda)) = \mathfrak{M}_s(\lambda), \qquad \nu = \overline{0,n-1}, \qquad j = \overline{1,m_\tau}, \qquad \eta = \overline{1,N_\tau}.$$

Solving this algebraic system for each $\tau = \overline{1,\theta}$ with respect to $\{b_{\tau\mu}(\lambda)\}_{\mu=\overline{1,N}}$, $\{\psi_\tau^{(\nu)}(a,\lambda)\}_{\nu=\overline{0,\beta_\tau-1}}$, we obtain (9.1). Choosing $z_k(x,\lambda) = y_k(x,\lambda)$, where $\{y_k(x,\rho)\}_{k=\overline{1,N}}$ is the FSS B_0 and using the asymptotic properties (1.5) of the functions $y_k(x,\rho)$, we obtain the asymptotic formula for $Q^\tau_{\xi\nu}(\lambda,a)$. We observe that the functions $\mathfrak{N}^\tau_\xi(\lambda,a)$ and $Q^\tau_{\xi\nu}(\lambda,a)$ are defined from L for $x\in(0,a)$ and from the WF's $\{\mathfrak{M}_s(\lambda)\}_{s=\overline{1,N}}$.

We define $U^\tau_{\xi a}(y) = y^{(\xi-1)}(a)$ for $\xi = \overline{\beta_\tau+1,n}$ and denote

$$\omega^1_{\xi\tau}(R) = \sum_{\nu=0}^{n-1} q^\tau_{\xi\nu} R^\nu, \qquad \xi = \overline{1,n}.$$

The functions $\omega^1_{\xi\tau}(R)$ are the characteristic polynomials for the LF $U^\tau_{\xi a}$ (they do not depend on $a\in(0,T)$. Define LF $U^{\tau,*}_{\xi a}$, $\xi=\overline{1,n}$ from the relation

$$\langle y,z\rangle\Big|_{x=a} = \sum_{\xi=1}^n U^\tau_{n-\xi+1,a}(y) U^{\tau,*}_{\xi a}(z).$$

Denote

$$\omega^{1,*}_{\xi\tau}(R) = \sum_{\nu=0}^{n-1} q^{\tau,*}_{\xi\nu} R^\nu$$

the characteristic polynomials for the LF $U^{\tau,*}_{\xi a}$. It is clear that $q^{\tau,*}_{\xi\nu} = (-1)^\nu \mathcal{D}^\tau_{n+1-\xi,n-1-\nu}$, where $\mathcal{D}_\tau = [\mathcal{D}^\tau_{\mu\nu}]_{\mu=\overline{1,n};\nu=\overline{0,n-1}}$ is the matrix of algebraic minors of Q_τ. For Ω^1_τ, $\Omega^1_{\tau\mu}$, $\Omega^{1,*}_\tau$ and $\Omega^{1,*}_{\tau\mu}$ the same formulas are used as for Ω, Ω_μ, Ω^* and Ω^*_μ with $\omega^1_{\xi\tau}(R)$ and $\omega^{1,*}_{\xi\tau}(R)$ replacing $\omega_\xi(R)$ and $\omega^*_\xi(R)$.

Let us show that

$$\operatorname{rank}\,[\omega^1_{\xi\tau}(R_k)]_{\xi=\overline{1,\beta_\tau};k=\overline{1,m_\tau}} = m_\tau. \tag{9.2}$$

Indeed,

$$[\omega^1_{\xi\tau}(R)]_{\xi=\overline{1,n}} = Q_\tau[R^{\xi-1}]_{\xi=\overline{1,n}}.$$

Since $\det Q_\tau = 1$, $\det[R_k^{\xi-1}]_{k,\xi=\overline{1,n}} \neq 0$ it follows that $\Omega_\tau(R_1,\dots,R_n)\neq 0$. However

$$\omega^1_{\xi\tau}(R) = (q_\tau)^{-1}\det[R^\nu, R^\nu_{\beta_\tau+1},\dots,R^\nu_n]_{\nu=\xi-1,\overline{\beta_\tau+1,n}}, \qquad (\xi=\overline{1,\beta_\tau}).$$

Then $\omega^1_{\xi\tau}(R_k)=0$, $\xi=\overline{1,\beta_\tau}$, $k=\overline{\beta_\tau+1,n}$. Hence $\Omega^1_\tau(R_1,\dots,R_{\beta_\tau})\neq 0$, i.e. (9.2) is valid.

Let $\{\varepsilon_{ji}\}_{j=\overline{1,\beta_\tau}}$ denote a permutation of the numbers $\overline{1,\beta_\tau}$ such that $\Omega^1_\tau(\varepsilon_{1\tau},\dots,\varepsilon_{m_\tau,\tau})\neq 0$. By (9.2), such a permutation exists. Let the functions $\psi^*_s(x,\lambda,a)$, $s=\overline{1,N}$; $a\geq 0$, $x\in(a,T)$ be solutions of the DE (1.24) with the following conditions:

$$U_{\xi 0}^{*}(\psi_{s}^{*}) = \delta_{\xi, n-\gamma_{r\eta}+1}, \quad (\xi = \overline{1, n-m_{\tau}}) \quad \text{for} \quad a = 0;$$

$$U_{\xi a}^{\tau,*}(\psi_{s}^{*}) = \delta_{\xi, n-\varepsilon_{m_{\tau}+\eta,\tau}+1}, \quad (\xi = \overline{1, n-\beta_{\tau}, n-\varepsilon_{m_{\tau}+1,\tau}+1, \ldots, n-\varepsilon_{\beta_{\tau},\tau}+1})$$

for $a > 0$, and also $U_{\xi\tau}^{*}(\psi_s^*) = 0$, $\xi = \overline{1, m_\tau}$ for $T < \infty$, and $\psi_s^*(x, \lambda, a) = O(\exp(\rho R_{n-m_\tau}^* x))$, $x \to \infty$ for $T = \infty$ ($a \geq 0$). Denote

$$\Lambda_{nj}^{\tau}(\lambda, a) = -U_{n-\varepsilon_{jr}+1, a}(\psi_s^*), \quad a > 0.$$

For $\ell \geq 1$, $k = 1, 2$ we consider the matrices $A_\ell^k = [A_{\ell s j}^k]$ $s = \overline{1, N}$, $j = \overline{1, p}$, where

$$A_{\ell sj}^1 = \frac{\Omega(\overline{1, m_\tau - 1})}{R_{m_\tau}^{\kappa_j} \Omega(\overline{1, m_\tau})} \sum_{\nu=1}^{n-m_\tau} \frac{(-1)^{n-\beta_\tau+\eta+\nu}}{(-R_{m_\tau} - R_\nu^*)^{\ell+1-\kappa_j}}$$

$$\times \frac{\Omega_{\tau\nu}^{1,*}(\overline{1, n-\beta_\tau}; n-\varepsilon_{m_\tau+1,\tau}+1, \ldots, n-\varepsilon_{\beta_\tau,\tau}+1 \setminus n-\varepsilon_{m_\tau+\eta,\tau}+1)}{\Omega_\tau^{1,*}(\overline{1, n-\beta_\tau}; n-\varepsilon_{m_\tau+1,\tau}+1, \ldots, n-\varepsilon_{\beta_\tau,\tau}+1)},$$

$$A_{\ell sj}^2 = \frac{\Omega(\overline{1, m_\tau - 1})(-1)^{m_\tau+\gamma_{r\eta}+1}\Omega^*(\overline{1, n-m_\tau} \setminus n-\gamma_{r\eta}+1)}{R_{m_\tau}^{\kappa_j}\Omega(\overline{1, m_\tau})(R_{m_\tau+1}-R_{m_\tau})^{\ell-\kappa_j+1}\Omega^*(\overline{1, n-m_\tau})}.$$

We shall write $A[a, b]$ ($PA[a, b]$) for the set of functions analytic (piecewise-analytic) on $[a, b]$. Let $p_{n-\kappa_j}(x) \in PA[0, T]$, $r_{\ell j}^a = p_{n-\kappa_j}^{(\ell-\kappa_j)}(a+0)$, $r_\ell^a = [r_{\ell j}^a]_{j=\overline{1,p}}$. Denote by $\mathcal{P}_\kappa^a$ the set of $\tilde{L}$ such that $p_k(x) = \tilde{p}_k(x)$, $x > 0$, $k \neq n - \kappa_j$, $j = \overline{1, N}$, and $L = \tilde{L}$ for $x \in (0, a)$.

Lemma 9.3. *Let $p_{n-\kappa_j}(x)$, $\tilde{p}_{n-\kappa_j}(x) \in PA[0, T]$, $\tilde{L} \in \mathcal{P}_\kappa^a$, $r_\mu^a = \tilde{r}_\mu^a$, $\mu = \overline{1, \ell-1}$. Then for $|\lambda| \to \infty$, $\arg \lambda = \varphi \in (0, \pi)$, $k = 1, 2$ there exist finite limits*

$$X_{\ell s}^k(a) = \lim \mathcal{P}_{sk}(\rho, a)\rho^\ell,$$

where

$$\mathcal{P}_{s1}(\rho, a) = B_s(\lambda, a)\rho^{\upsilon_{s1}(a)} \exp(-\rho R_{m_\tau} a),$$

$$\mathcal{P}_{s2}(\rho, a) = B_s(\lambda, 0)\rho^{\upsilon_{s2}} \exp(\rho(R_{m_\tau+1} - R_{m_\tau})a), \quad B_s(\lambda, 0) = \widehat{\mathfrak{M}}_s(\lambda),$$

$$B_s(\lambda, a) = \mathfrak{N}_{\varepsilon_{m_\tau+\eta,\tau}}^\tau(\lambda, a) - \sum_{j=1}^{m_\tau} \tilde{\Lambda}_{\eta j}^\tau(\lambda, a)\mathfrak{N}_{\varepsilon_{jr}}^\tau(\lambda, a), \quad (a > 0),$$

$$\upsilon_{s1}(0) = \upsilon_{s2} = \sigma_{m_\tau, 0} - \sigma_{\gamma_{r\eta}, 0}, \quad \upsilon_{s1}(a) = \sigma_{m_\tau, 0} - \varepsilon_{m_\tau+\eta,\tau} + 1, \quad (a > 0).$$

Moreover

$$\left.\begin{array}{ll} A_\ell(r_\ell^0-\widetilde{r}_\ell^0)=[X_{\ell s}]_{s=\overline{1,N}}, & X_{\ell s}=X_{\ell s}^k(0); \\ A_\ell^k(r_\ell^a-\widetilde{r}_\ell^a)=[X_{\ell s}^k(a)]_{s=\overline{1,N}}, & a>0. \end{array}\right\} \tag{9.3}$$

Proof. Let us show that if $\widetilde{L}\in\mathcal{P}_\kappa^a$ then

$$\sum_{j=1}^N\int_a^T\widehat{p}_{n-\kappa_j}(x)\psi_\tau^{(n-\kappa_j)}(x,\lambda)\widetilde{\psi}_s^*(x,\lambda,a)\,dx=B_s(\lambda,a). \tag{9.4}$$

Indeed, let $\widetilde{L}\in\mathcal{P}_\kappa^0$. Then

$$\sum_{j=1}^N\int_0^T\widehat{p}_{n-\kappa_j}(x)\psi_\tau^{(n-\kappa_j)}(x,\lambda)\widetilde{\psi}_s^*(x,\lambda,0)\,dx=\int_0^T(\ell\psi_\tau(x,\lambda)-\widetilde{\ell}\psi_\tau(x,\lambda))\widetilde{\psi}_s^*(x,\lambda,0)\,dx.$$

It follows from (1.26) that

$$\int_0^T\widetilde{\ell}\psi_\tau(x,\lambda)\widetilde{\psi}_s^*(x,\lambda,0)\,dx=\Big|_0^T\langle\psi_\tau(x,\lambda),\widetilde{\psi}_s^*(x,\lambda,0)\rangle_{\widetilde{\ell}}+\int_0^T\psi_\tau(x,\lambda)\widetilde{\ell}^*\widetilde{\psi}_s^*(x,\lambda,0)\,dx.$$

Since

$$\ell\psi_\tau(x,\lambda)=\lambda\psi_\tau(x,\lambda),\qquad \widetilde{\ell}^*\widetilde{\psi}_s^*(x,\lambda,0)=\lambda\widetilde{\psi}_s^*(x,\lambda,0),$$

$$\begin{aligned}\Big|_0^T\langle\psi_\tau(x,\lambda),\widetilde{\psi}_s^*(x,\lambda,0)\rangle_{\widetilde{\ell}} &= \sum_{\xi=1}^n U_{\xi0}(\psi_\tau(x,\lambda))U_{n-\xi+1,0}^*(\widetilde{\psi}_s^*(x,\lambda,0))\\ &= U_{n-m_\tau+1,0}^*(\widetilde{\psi}_s^*(x,\lambda,0))+\mathfrak{M}_s(\lambda),\end{aligned}$$

we obtain

$$\sum_{j=1}^N\int_0^T\widehat{p}_{n-\kappa_j}(x)\psi_\tau^{(n-\kappa_j)}(x,\lambda)\widetilde{\psi}_s^*(x,\lambda,0)\,dx=U_{n-m_\tau+1,0}^*(\widetilde{\psi}_s^*(x,\lambda,0))+\mathfrak{M}_s(\lambda).$$

From this for $\ell=\widetilde{\ell}$ we have $\widetilde{\mathfrak{M}}_s(\lambda)=-U_{n-m_\tau+1,0}^*(\widetilde{\psi}_s^*(x,\lambda,0))$, and hence

$$\sum_{j=1}^{N}\int_0^T \widehat{p}_{n-\kappa_j}(x)\psi_\tau^{(n-\kappa_j)}(x,\lambda)\widetilde{\psi}_s^*(x,\lambda,0)\,dx = \widehat{\mathfrak{M}}_s(\lambda).$$

Thus, (9.4) for $a=0$ is proved. For $a>0$ the proof is completely analogous.

For definiteness, let $T=\infty$. Let $\{\mathcal{Y}_\mu(x,\rho)\}_{\mu=\overline{1,n}}$ be the FSS B_0. It follows from (1.8) and (1.10) that

$$\psi_\tau(x,\lambda) = \sum_{\mu=1}^{m_\tau} C_{\tau\mu}(\rho)\mathcal{Y}_\mu(x,\rho), \tag{9.5}$$

$$C_{\tau\mu}(\rho) = \frac{1}{\rho^{\sigma_{m_\tau,0}}}\left(\frac{(-1)^{m_\tau+\mu}}{\Omega(\overline{1,m_\tau})}\Omega_\mu(\overline{1,m_\tau-1}) + O\left(\frac{1}{\rho}\right)\right), \qquad |\rho|\to\infty. \tag{9.6}$$

Similarly,

$$\widetilde{\psi}_s^*(x,\lambda,a) = \sum_{\nu=1}^{n-m_\tau} \widetilde{C}_{s\nu}^*(\rho,a)\widetilde{\mathcal{Y}}^*(x,\rho), \qquad a\ge 0, \tag{9.7}$$

$$\widetilde{C}_{s\nu}^*(\rho,0) = \rho^{t_{\tau\eta}}\left(\frac{(-1)^{n-\gamma_{\tau\eta}+\nu+1}}{\Omega^*(\overline{1,n-m_\tau})}\Omega_\nu^*(1,n-m_\tau\backslash n-\gamma_{\tau\eta}+1) + O\left(\frac{1}{\rho}\right)\right),$$

$$\widetilde{C}_{s\nu}^*(\rho,a) = \rho^{t^1_{\tau\eta}}\exp(-\rho R_\nu^* a)\Bigg((-1)^{n-\beta_\tau+\eta+\nu}$$

$$\times\frac{\Omega_{\tau\nu}^{1,*}(\overline{1,n-\beta_\tau};n-\varepsilon_{m_\tau+1,\tau}+1,\dots,n-\varepsilon_{\beta_\tau,\tau}+1\backslash n-\varepsilon_{m_\tau+\eta,\tau}+1)}{\Omega_\tau^{1,*}(\overline{1,n-\beta_\tau};n-\varepsilon_{m_\tau+1,\tau}+1,\dots,n-\varepsilon_{\beta_\tau,\tau}+1)} + O\left(\frac{1}{\rho}\right)\Bigg),$$

$$a>0. \tag{9.8}$$

where $t_{\tau\eta}=\sigma_{\gamma_{\tau\eta},0}+1-n$, $t^1_{\tau\eta}=\varepsilon_{m_\tau+\eta,\tau}-n$ and $\widetilde{\mathcal{Y}}_\nu^*(x,\rho)=\exp(\rho R_\nu^* x)(1+O(\rho^{-1}))$ is the FSS B_0 for $\widetilde{\ell}^*$. Since $r_\mu^a=\widetilde{r}_\mu^a$, $\mu=\overline{1,\ell-1}$ it follows that

$$\widehat{p}_{n-\kappa_j}(x) = \frac{(x-a)^{\ell-\kappa_j}}{(\ell-\kappa_j)!}(r_{\ell j}^a-\widetilde{r}_{\ell j}^a+o(1)), \qquad x\to a+0.$$

By Lemma 9.1 we get

$$\int_0^T \widehat{p}_{n-\kappa_j}(x)\mathcal{Y}_\mu^{(n-\kappa_j)}(x,\rho)\widetilde{\mathcal{Y}}_\nu^*(x,\rho)\,dx$$

$$= \frac{(r_{\ell j}^a-\widetilde{r}_{\ell j}^a)\exp(\rho(R_\mu+R_\nu^*)a)}{\rho^{\ell+1-n}R_\mu^{\kappa_j}(-R_\mu-R_\nu^*)^{\ell+1-\kappa_j}}(1+o(1)), \qquad |\rho|\to\infty. \tag{9.9}$$

Substituting (9.5) and (9.7) in (9.4) we obtain

$$\sum_{\mu=1}^{m_\tau}\sum_{\nu=1}^{n-m_\tau} C_{\tau\mu}(\rho)\widetilde{C}^*_{s\nu}(\rho,a)\sum_{j=1}^{N}\int_a^T \widehat{p}_{n-\kappa_j}(x)\mathcal{Y}_\mu^{(n-\kappa_j)}(x,\rho)\widetilde{\mathcal{Y}}_\nu^*(x,\rho)\,dx \tag{9.10}$$
$$= B_s(\lambda,a), \qquad a\geq 0, \qquad \widetilde{L}\in\mathcal{P}_\kappa^a.$$

In particular,

$$\sum_{\mu=1}^{m_\tau}\sum_{\nu=1}^{n-m_\tau} C_{\tau\mu}(\rho)\widetilde{C}^*_{s\nu}(\rho,0)\sum_{j=1}^{N}\int_a^T \widehat{p}_{n-\kappa_j}(x)\mathcal{Y}_\mu^{(n-\kappa_j)}(x,\rho)\widetilde{\mathcal{Y}}_\nu^*(x,\rho)\,dx \tag{9.11}$$
$$= B_s(\lambda,0), \qquad a\geq 0, \qquad \widetilde{L}\in\mathcal{P}_\kappa^a.$$

From (9.10) and (9.11) we obtain the assertion of Lemma 9.3, in view of the asymptotic formulas (9.6) and (9.8).

Now let us show that the information condition $A_\ell\#0$ does not depend on the choice of a sector. We consider the sectors $S_\nu = \left\{\rho : \arg\rho\in\left(\frac{\nu\pi}{n},\frac{(\nu+1)\pi}{n}\right)\right\}$. It is sufficient to consider the information condition in two sectors S_0 and S_1. We observe that A_ℓ does not depend on the coefficients $p_k(x)$ and $u_{\xi\nu a}$ of the DE and LF.

Choose the functions $p_k(x)$ such that they are regular at $x=0$. Then in each sector S_ν we have the asymptotic formulas

$$\mathfrak{M}_{m\mu}(\lambda) = \rho^{\sigma_{\mu 0}-\sigma_{m0}}\sum_{\ell=0}^{\infty}\frac{1}{\rho^\ell}\mathfrak{M}^\ell_{m\mu}, \qquad \rho\in S_\nu, \qquad |\rho|\to\infty,$$

and $\mathfrak{M}^\ell_{m\mu} = \mathfrak{M}^\ell_{m\mu}(S_\nu)$ depend on S_ν.

Denote $\varepsilon_1 = \exp\left(\frac{2\pi i}{n}\right)$. If $\rho\in S_{-1}$ then $\rho\varepsilon_1\in S_1$ and hence

$$\mathfrak{M}^\ell_{m\mu}(S_1) = \mathfrak{M}^\ell_{m\mu}(S_{-1})\varepsilon_1^{\ell+\sigma_{m0}+\sigma_{\mu 0}}.$$

When $n-m$ is odd we get

$$\mathfrak{M}^\ell_{m\mu}(S_0) = \mathfrak{M}^\ell_{m\mu}(S_{-1}), \qquad \text{i.e.} \qquad \mathfrak{M}^\ell_{m\mu}(S_1) = \mathfrak{M}^\ell_{m\mu}(S_0)\varepsilon_1^{\ell+\sigma_{m0}-\sigma_{\mu 0}}.$$

When $n-m$ is even we have $\mathfrak{M}^\ell_{m\mu}(S_1) = \mathfrak{M}^\ell_{m\mu}(S_0)$. Hence $\mathfrak{M}^\ell_s(S_1) = \mathfrak{M}^\ell_s(S_0)B_{s\ell}$, where $B_{s\ell}=1$ when $n-m_\tau$ is even, and $B_{s\ell}=\varepsilon_1^{\ell+\sigma_{m_\tau,0}-\sigma_{\gamma_{\tau\eta},0}}$ when $n-m_\tau$ is odd.

Let $\widetilde{p}_k(x)\equiv 0$, $\widetilde{u}_{\xi\nu a}=0$, and $L\in\mathcal{P}^0_\kappa$, $r_1^0=\ldots=r^0_{\ell-1}=0$. Then, by Lemma 9.3 we have

$$A_\ell(S_0)r_\ell^0 = [\mathfrak{M}^\ell_s(S_0)]_{s=\overline{1,N}}, \qquad A_\ell(S_1)r_\ell^0 = [\mathfrak{M}^\ell_s(S_0)B_{s\ell}]_{s=\overline{1,N}}.$$

So the change from S_0 to S_1 leads to multiplication of the rows of A_ℓ by non-zero numbers. Hence the information condition $A_\ell \# 0$ does not depend on the choice of a sector.

Lemma 9.4. *Let $\widetilde{L} \in \mathcal{P}^0_\kappa$; $k = 1 \vee 2$; $\alpha \in (0,T)$, and let $A_\ell \# 0$, $A^k_\ell \# 0$, $\ell \geq 1$. If for $|\lambda| \to \infty$, $\arg \lambda = \varphi \in (0,\pi)$, $\lim \mathcal{P}_{sk}(\rho, a)\rho^\ell = 0$ for all $\ell \geq 1$, $a \in [0,\alpha)$, $s = \overline{1,N}$, then $\widetilde{L} \in \mathcal{P}^\alpha_\kappa$.*

Indeed, if there exist $a < \alpha$ and $\ell \geq 1$ such that $\widetilde{L} \in \mathcal{P}^a_\kappa$, $r^a_\mu = \widetilde{r}^a_\mu$, $\mu = \overline{1, \ell - 1}$; $r^a_\ell \neq \widetilde{r}^a_\ell$, then it follows from (9.3) and the conditions of the lemma that $X^k_{\ell, s_0}(a) \neq 0$ for a certain s_0. This contradiction proves the lemma.

From the propositions which were proved above we obtain the following theorem.

Theorem 9.1. *Let $A_\ell \# 0$, $\ell \geq 1$. Then Problem 9.2 has a unique solution in the class $p_{n-\kappa_j}(x) \in A[0,T)$, $j = \overline{1,N}$. If, further, $A^1_\ell \# 0$, $\ell \geq 1$ then Problem 9.2 has a unique solution in the class $p_{n-\kappa_j}(x) \in PA[0,T)$, $j = \overline{1,N}$. The solution of Problem 9.2 can be found by applying the following algorithm.*

Algorithm 9.1.

(1) Take $a = 0$.

(2) Compute $\{r^a_\ell\}_{\ell \geq 1}$. For this we do the following operations successively for $\ell = 1, 2, \ldots$: construct $\widetilde{L} \in \mathcal{P}^a_\kappa$ such that $\widetilde{r}^a_\mu = r^a_\mu$, $\mu = \overline{1, \ell - 1}$ and find r^a_ℓ from (9.3) for $k = 1$.

(3) Construct L for $x \in (a, \alpha)$ by the formula

$$p_{n-\kappa_j}(x) = \sum_{\ell=0}^{\infty} r^a_{\ell+\kappa_j, j} \frac{(x-a)^\ell}{\ell!}, \qquad j = \overline{1,N}. \tag{9.12}$$

(4) If $\alpha < T$, then we put $a := \alpha$ and go on to the step 2.

Remark 1. In Algorithm 9.1, the solution of the IP is sought in steps, whose lengths are determined by Lemma 9.4 as follows. Suppose that L for $x \in (0,a)$ and $\{r^a_\ell\}_{\ell \geq 1}$ have been found. Construct $\widetilde{L} \in \mathcal{P}^a_\kappa$ so that $\widetilde{r}^a_\ell = r^a_\ell$, $\ell \geq 1$. Put $\alpha = \sup\{b > a : \lim \mathcal{P}_{s1}(\rho, b)\rho^\ell = 0, \ell \geq 1, s = \overline{1,N}\}$. Then Lemma 9.4 implies that $\widetilde{L} \in \mathcal{P}^\alpha_\kappa$ i.e. we found L for $x \in (a, \alpha)$.

Remark 2. If the conditions of Theorem 9.1 are not satisfied, then a solution of the IP will not be unique. Indeed, let $T < \infty$, $n = 3$, $\sigma_{\xi 0} = \sigma_{\xi T} = 3 - \xi$. Consider the following IP: given the WF's $\{\mathfrak{M}_{12}(\lambda), \mathfrak{M}_{23}(\lambda)\}$, construct the coefficients $p_0(x)$ and $p_1(x)$. Thus, $N = 2$, $\kappa = \{2, 3\}$ and $I = \{(1,2), (2,3)\}$. It is easy to see that in this case the information condition is not satisfied, i.e. the set $\{\mathfrak{M}_{12}(\lambda), \mathfrak{M}_{23}(\lambda)\}$ of the WF's is not the P_κ-system. It was shown in 4.7 that a solution of this IP is not unique even in the class of analytic coefficients.

Remark 3. Theorem 9.1 remains in force when the condition $A_\ell^1 \# 0$ is replaced by the condition $A_\ell^2 \# 0$. When this is done, Algorithm 9.1 can be replaced by the simpler Algorithm 9.2.

Algorithm 9.2.

(1) Take $a = 0$.

(2) Compute $\{r_\ell^a\}_{\ell \geq 1}$. For this we do the following operations successively for $\ell = 1, 2, \ldots$: construct $\widetilde{L} \in \mathcal{P}_\kappa^a$ such that $\widetilde{r}_\mu^a = r_\mu^a$, $\mu = \overline{1, \ell - 1}$ and find r_ℓ^a from (9.3) for $k = 2$.

(3) Construct L for $x \in (a, \alpha)$ by (9.12)

(4) If $\alpha < T$ then we put $a := \alpha$ and go on to the step 2.

Algorithm 9.2 is simpler than Algorithm 9.1, since it does not require the computation of the functions $\mathfrak{N}_\xi^\tau(\lambda, a)$ and $Q_{\xi\nu}^\tau(\lambda, a)$ in each step.

Remark 4. Theorem 9.1 remains in force when the condition of piecewise analyticity is replaced by a more general condition ensuring that an asymptotics for the integral (9.9) exists.

9.3. Here we consider some special cases.

Case 1. We study the IP of recovering a single coefficient of the DE (1.1) from one WF. For definiteness, let $n = 2q$, $\sigma_{\xi 0} = \xi - 1$. Take a fixed integer κ $(2 \leq \kappa \leq n)$.

Problem 9.3. To find the function $p_{n-\kappa}(x)$ from the WF $\mathfrak{M}_{12}(\lambda)$ and the coefficients $p_\nu(x)$, $\nu \neq n - \kappa$.

In this case $N = 1$. Let us check the information condition for Problem 9.3. Denote

$$b = \det [R_k^j]_{k=\overline{2,n}; j=\overline{0,n-2}}, \qquad b_\nu = \det [R_k^j]_{k=\overline{2,n}\setminus\nu; j=\overline{0,n-3}}\,.$$

Lemma 9.5.

$$b^{-1} b_\nu = \frac{1}{n}(-1)^{n-\nu} R_\nu (R_\nu - R_1), \qquad \nu = \overline{2, n}. \tag{9.13}$$

Proof. Since b and b_ν are the Vandermonde determinant, it follows that

$$b = \prod_{2 \leq \xi < j \leq n} (R_j - R_\xi), \qquad b_\nu = \prod_{\substack{2 \leq \xi < j \leq n \\ \xi, j \neq \nu}} (R_j - R_\xi).$$

Then

$$b^{-1} b_\nu = \prod_{j=\nu+1}^{n} (R_j - R_\nu) \prod_{\xi=2}^{\nu-1} (R_\nu - R_\xi). \tag{9.14}$$

Denote

$$F(R) = R^n - 1 = \prod_{j=1}^{n} (R - R_j).$$

Compute

$$F'(R_\nu) = \prod_{\substack{j=1 \\ j\neq\nu}}^{n} (R_\nu - R_j).$$

Comparing with (9.14) we obtain

$$b^{-1} b_\nu = \frac{(-1)^{n-\nu}(R_\nu - R_1)}{F'(R_\nu)}. \tag{9.15}$$

Since $F'(R) = nR^{n-1}$, then $F'(R_\nu) = nR_\nu^{n-1} = nR_\nu^{-1}$. Hence, (9.13) follows from (9.15).

For Problem 9.3 we have

$$A_{\alpha+\kappa} = \frac{(-1)^n}{R_1^{\sigma_{10}} \Omega^*(\overline{1, n-1})} \sum_{\nu=1}^{n-1} \frac{(-1)^{\nu-1} \Omega_\nu^*(\overline{1, n-2})}{(-R_1 - R_\nu^*)^{\alpha+1}}, \qquad \alpha \geq 0.$$

Since

$$\Omega^*(\overline{1, n-1}) = (-1)^{(n-1)(n-2)/2} b, \qquad \Omega_\nu^*(\overline{1, n-2}) = (-1)^{(n-2)(n-3)/2} b_{n-\nu+1},$$

it follows that

$$A_{\alpha+\kappa} = \frac{(-1)^n}{R_1^{\sigma_{10}+\kappa} b} \sum_{\nu=2}^{n} \frac{(-1)^\nu b_\nu}{(R_\nu - R_1)^{\alpha+1}}, \qquad \alpha \geq 0.$$

Using (9.13) we compute

$$A_{\alpha+\kappa} = \frac{1}{nR_1^\kappa} \sum_{\nu=2}^{n} \frac{R_\nu}{(R_\nu - R_1)^\alpha} = \frac{(-R_1)^{1-\alpha}}{nR_1^\kappa} \sum_{\nu=-(q-1)}^{q-1} \frac{\varepsilon_\nu}{(1+\varepsilon_\nu)^\alpha},$$

where $\varepsilon_\nu = \exp\left(\frac{2\nu\pi i}{n}\right)$. Since $1 + \varepsilon_\nu = 2\cos\frac{\nu\pi}{n} \exp\left(\frac{\nu\pi i}{n}\right)$, we get

$$A_{\alpha+\kappa} = \frac{(-R_1)^{1-\alpha}}{n 2^\alpha R_1^\kappa} a_\alpha, \qquad \alpha \geq 0. \tag{9.16}$$

where

$$a_\alpha = 1 + 2\sum_{\nu=1}^{q-1} \left(\cos\frac{\nu\pi}{n}\right)^{-\alpha} \cos\frac{\nu\pi(\alpha-2)}{n}, \tag{9.17}$$

Denote

$$B_k = \sin\frac{\pi}{n}\left(\sin\frac{k\pi}{n}\right)^{-1}, \qquad \theta_p = \sum_{k=2}^{q}(B_k)^p.$$

Lemma 9.6.

$$\theta_p < 2\left(\frac{\pi}{3\sqrt{3}}\right)^p, \qquad p \geq 4, \qquad q \geq 3. \tag{9.18}$$

Proof. For $q = 3$ and $q = 4$, (9.18) is obvious. Since

$$\sum_{k=\mu+1}^{\infty}\frac{1}{k^p} < \int_{\mu}^{\infty}\frac{dx}{x^p} = \frac{1}{p-1}\frac{1}{\mu^{p-1}},$$

we have for $q \geq 5$

$$\begin{aligned}\theta_p &= \sum_{k=2}^{[2q/3]}(B_k)^p + \sum_{k=[2q/3]+1}^{q}(B_k)^p < \left(\frac{2\pi}{3\sqrt{3}}\right)^p\sum_{k=2}^{[2q/3]}\frac{1}{k^p} + \left(\frac{\pi}{2}\right)^p\sum_{k=[2q/3]+1}^{q}\frac{1}{k^p}\\ &\leq \left(\frac{\pi}{3\sqrt{3}}\right)^p + \left(\frac{2\pi}{3\sqrt{3}}\right)^p\sum_{k=3}^{\infty}\frac{1}{k^p} + \left(\frac{\pi}{2}\right)^p\sum_{k=[2q/3]+1}^{q}\frac{1}{k^p}\\ &\leq \left(\frac{\pi}{3\sqrt{3}}\right)^p\left(\frac{5}{3} + \left(\frac{3\sqrt{3}}{2}\right)^p\sum_{k=[2q/3]+1}^{q}\frac{1}{k^p}\right) < 2\left(\frac{\pi}{3\sqrt{3}}\right)^p.\end{aligned}$$

Lemma 9.6 is proved.

Lemma 9.7. *For all* $\alpha \geq 0$ $a_\alpha \neq 0$.

Proof. For $q = 2$ the lemma is obvious. Therefore, we now consider that $q \geq 3$. Let us write a_α in the following form

$$a_\alpha = 2(-1)^{[\alpha/2]-1}\left(\sin\frac{\pi}{n}\right)^{-\alpha}h_{\alpha-2},$$

where

$$h_\alpha = \sum_{k=1}^{q}(B_k)^{\alpha+2}(1 - \frac{1}{2}\delta_{kn})F_{k,\alpha},$$

$$F_{k,2\beta} = \cos\frac{k\pi\beta}{q}, \qquad F_{k,2\beta+1} = \sin\frac{k\pi(2\beta+1)}{2q}.$$

It is obvious that $a_\alpha \neq 0$ for $\alpha = 0, 1, 2, 3$. We will show that $h_\alpha \neq 0$ for $\alpha = 2, 3, \ldots$.

Let us at first consider $h_{2\beta}$, $\beta \geq 1$. If $q = 2m$, $\beta = (2p+1)m$, $p = 0, 1, 2, \ldots$, then

$$h_{2\beta} = \sum_{k=1}^{m} (-1)^k (B_{2k})^{2\beta+2} (1 - \frac{1}{2}\delta_{km}) \neq 0.$$

But if $q = 2m+1$ or $q = 2m$, $\beta \neq (2p+1)m$, then $\left|\cos \frac{\pi\beta}{q}\right| > \frac{1}{q}$, and, taking into account (9.18), we obtain

$$\left| h_{2\beta} \left(\cos \frac{\pi\beta}{q}\right)^{-1} - 1 \right| < q\theta_{2\beta+2} < 2q \left(\frac{\pi}{3\sqrt{3}}\right)^{2\beta+2} < 1, \qquad (\beta \geq \ln q).$$

Consequently, $h_{2\beta} \neq 0$ for $\beta \geq \ln q$. Since $[q/3] + 1 > \ln q$, then the case $\beta \leq [q/3]$ remains to be considered. But then $\cos \frac{\pi\beta}{q} \geq \frac{1}{2}$ and, consequently

$$\left| h_{2\beta} \left(\cos \frac{\pi\beta}{q}\right)^{-1} - 1 \right| < 2\theta_{2\beta+2} < 4 \left(\frac{\pi}{3\sqrt{3}}\right)^{2\beta+2} < 1.$$

Thus, $h_{2\beta} \neq 0$ for all $\beta = 1, 2, \ldots$.

Let us now consider $h_{2\beta+1}$, $\beta \geq 1$. Since $\left|\sin \frac{(2\beta+1)\pi}{2q}\right| \geq \frac{1}{q}$, then, taking into account (9.18), we obtain

$$\left| h_{2\beta+1} \left(\sin \frac{(2\beta+1)\pi}{2q}\right)^{-1} - 1 \right| < q\theta_{2\beta+3} < 2q \left(\frac{\pi}{3\sqrt{3}}\right)^{2\beta+3} < 1, \qquad \left(\beta \geq \ln \frac{q}{2}\right).$$

Consequently, $h_{2\beta+1} \neq 0$ for $\beta \geq \ln \frac{q}{2}$. The case $\beta < \ln \frac{q}{2}$ $(q \geq 6)$ remains to be considered. We denote $r_\beta = \left[\frac{2q}{2\beta+1}\right]$. Then $\sin \frac{k\pi}{2q}(2\beta+1) \geq 0$ for $k = \overline{1, r_\beta}$ and

$$\left| \left(h_{2\beta+1} - \sum_{k=2}^{r_\beta} (B_k)^{2\beta+3} \sin\left((2\beta+1)\frac{k\pi}{2q}\right) \sin \frac{\pi}{2q}(2\beta+1) \right)^{-1} - 1 \right| < f_\beta,$$

where

$$f_\beta = \frac{q}{2\beta+1} \sum_{k=r_\beta+1}^{q} (B_k)^{2\beta+3}.$$

Let us show that $f_\beta < 1$ for $1 \leq \beta < \ln \frac{q}{2}$. Indeed, since $r_\beta \geq 3$, then

$$\begin{aligned} f_\beta &\leq \frac{q}{2\beta+1} \left(\frac{\pi}{2}\right)^{2\beta+3} \sum_{k=r_\beta+1}^{q} \frac{1}{k^{2\beta+3}} \leq \frac{q}{2\beta+1} \frac{\pi}{4(\beta+1)} \left(\frac{\pi}{2r_\beta}\right)^{2\beta+2} \\ &\leq \frac{\pi^2}{6(\beta+1)} \left(\frac{\pi}{2r_\beta}\right)^{2\beta+1} \leq \frac{\pi^2}{6(\beta+1)} \left(\frac{\pi}{6}\right)^{2\beta+1} < 1. \end{aligned}$$

Thus, $h_{2\beta+1} \neq 0$ for all $\beta = 1, 2, 3, \ldots$. Lemma 9.7 is proved.

It follows from (9.16) and Lemma 9.7 that $A_\ell \neq 0$ for all ℓ, i.e. the information condition for Problem 9.3 is satisfied. Furthermore, it is obvious that $A_\ell^2 \neq 0$, $\ell \geq 1$. Thus, the following theorem is proved.

Theorem 9.2. *Problem 9.3 has a unique solution in the class $p_{n-\kappa}(x) \in PA[0,T)$ (the rest of $p_k(x) \in \mathcal{L}(0,T)$, $k \neq n-\kappa$). The solution of Problem 9.3 can be found by applying Algorithm 9.2.*

Case 2. We consider the IP of recovering all the coefficients of the DE (1.1) from the first row of the WM.

Problem 9.4. To find the DE (1.1) from the WF's $\{\mathfrak{M}_{1k}(\lambda)\}_{k=\overline{2,n}}$.

Below, in §12, we will show that the set $\{\mathfrak{M}_{1k}(\lambda)\}_{k=\overline{2,n}}$ of the WF's is the P_κ-system for $\kappa = \{2, \ldots, n\}$, and $A_\ell \# 0$, $A_\ell^1 \# 0$, $\ell \geq 1$. Therefore, from Theorem 9.1 we obtain the following theorem.

Theorem 9.3. *Problem 9.4 has a unique solution in the class $p_k(x) \in PA[0,T)$, $k = \overline{0, n-2}$. This solution can be found by Algorithm 9.1.*

The counterexample from 4.7 shows that there are no P_κ-systems for $\kappa = \{2, 3, \ldots, n\}$ with the exception of the first rwo of the WM.

Remark. Let $T < \infty$, and let G_k, $k = \overline{0, n-1}$ denote the boundary value problems for (1.1) with the conditions $y^{(k)}(0) = y(T) = \ldots = y^{(n-2)}(T) = 0$. It was shown in §7 (see Lemma 7.4) that the specification of each WF $\mathfrak{M}_{1k}(\lambda)$ is equivalent to the specification of two spectra for the problems G_0 and G_{k-1}. Hence, for $T < \infty$ Problem 9.3 is to find one of the coefficients of the DE from the two spectra of G_0 and G_1, and Problem 9.4 is to find the DE (1.1) from the system of n spectra of G_k, $k = \overline{0, n-1}$. L. Sakhnovich was the first to investigate an IP of this type. In [46] he proved a uniqueness theorem for recovering the two-term operator $\ell_1 y = y^{(n)} + p_0(x) y$ from the system of n spectra of G_k, $k = \overline{0, n-1}$ in the class of entire functions. The same result is established in [57] in the class of piecewise analytic functions. The transformation operator method is used in [46] and [57]. Thus, Theorems 9.2 and 9.3 essentially strengthen the results from [46] and [57]. We note that an IP for the two-term operator ℓ_1 in another formulation was considered in [49].

Now we consider Problem 9.4 on the half-line ($T = \infty$) in the case when $p_k(x) \equiv 0$ ($x > a$), $p_k(x) \in \mathcal{L}(0, a)$, $k = \overline{0, n-2}$ for a certain $a > 0$. Let $U_{\xi 0}(y) = y^{(\xi-1)}(0)$, for definiteness. Denote by $e(x, \rho)$ the Jost solution of the DE (1.1) under the condition

$$\lim_{x \to \infty} \exp(-\rho x) e(x, \rho) = 1, \qquad \arg(-\rho) \in \left(-\frac{\pi}{n}, \frac{\pi}{n}\right).$$

The Jost solution satisfies the following integral equation

$$e(x,\rho) = \exp(\rho x) + \frac{1}{n\rho^{n-1}} \int_x^a \Big(\sum_{j=1}^{n} R_j \exp(\rho R_j (x-t)) \times \sum_{\mu=0}^{n-2} p_\mu(t) e^{(\mu)}(t,\rho) \Big) \, dt, \qquad \arg(-\rho) \in \left(-\frac{\pi}{n}, \frac{\pi}{n}\right).$$

Since $p_k(x) \equiv 0$, $x > a$, then for each fixed $x \geq 0$, $\nu = \overline{0, n-1}$ the functions $e^{(\nu)}(x,\rho)$ are entire in ρ, and $e(x,\rho) \equiv \exp(\rho x)$ for $x \geq a$. The functions $\{e(x, \rho R_k)\}_{k=\overline{1,n}}$ form a FSS of the DE (1.1). For $|\rho| \to \infty$, $\arg(-\rho) \in \left(-\frac{\pi}{n}, \frac{\pi}{n}\right)$ we have

$$e^{(\nu)}(0,\rho) = \rho^{\nu}(1 + O(\rho^{-1})). \tag{9.19}$$

We can write the WS $\Phi_m(x,\lambda)$ from the Jost solutions:

$$\Phi_m(x,\lambda) = \sum_{k=1}^{m} \alpha_{km}(\rho) e(x, \rho R_k), \qquad \rho \in S.$$

The coefficients $\alpha_{km}(\rho)$ can be found from the conditions $\Phi_m^{(\nu-1)}(0,\lambda) = \delta_{m\nu}$, $\nu = \overline{1,m}$. We obtain

$$\Phi_m(x,\lambda) = \frac{\det\left[e(0,\rho R_k), \ldots, e^{(m-2)}(0,\rho R_k), e(x,\rho R_k)\right]_{k=\overline{1,m}}}{\det\left[e(0,\rho R_k), \ldots, e^{(m-2)}(0,\rho R_k), e^{(m-1)}(0,\rho R_k)\right]_{k=\overline{1,m}}}, \qquad \rho \in S$$

and hence

$$\mathfrak{M}_{m\xi}(\lambda) = \frac{\det\left[e^{(\nu)}(0,\rho R_k)\right]_{k=\overline{1,m};\,\nu=\overline{0,m-2},\xi-1}}{\det\left[e^{(\nu)}(0,\rho R_k)\right]_{k=\overline{1,m};\,\nu=\overline{0,m-1}}}, \qquad \rho \in S, \qquad \xi > m. \tag{9.20}$$

In particular,

$$\mathfrak{M}_{1\xi}(\lambda) = \frac{e^{(\xi-1)}(0,\rho R_1)}{e(0,\rho R_1)}, \qquad \rho \in S. \tag{9.21}$$

Since the functions $e(0,\rho), \ldots, e^{(n-1)}(0,\rho)$ are entire in ρ and not equal to zero simultaneously, it follows from (9.19) and (9.21) that the WF's $\{\mathfrak{M}_{1\xi}(\lambda)\}_{\xi=\overline{2,n}}$ determine the functions $\{e^{(\nu)}(0,\rho)\}_{\nu=\overline{0,n-1}}$ uniquely, and consequently, by virtue of (9.20), all the WM $\mathfrak{M}(\lambda)$. Thus, the following theorem is proved.

Theorem 9.4. *Problem 9.4 has a unique solution in the class of finite integrable coefficients* $p_k(x)$, $k=\overline{0,n-2}$. *This solution can be found by the following algorithm:*

(1) *we construct the functions* $\{e^{(\nu)}(0,\rho)\}_{\nu=\overline{0,n-1}}$ *from the WF's* $\{\mathfrak{M}_{1\xi}(\lambda)\}_{\xi=\overline{2,n}}$,

(2) *compute* $\mathfrak{M}(\lambda)=[\mathfrak{M}_{m\xi}(\lambda)]_{m,\xi=\overline{1,n}}$ *by* (9.20) *and* $\mathfrak{M}_{m\xi}(\lambda)=\delta_{m\xi}$, $\xi\le m$;

(3) *we construct the DE from the WM* $\mathfrak{M}(\lambda)$ *by the method of Part I (see Theorem 2.3).*

Case 3. We consider the IP of recovering a self-adjoint DE from the spectral function. For $n=2$, this IP was studied by Marchenko [5], [6], Gel'fand and Levitan [10], and for higher-order DE by L. Sakhnovich [47], [48] and Kahchatryan [50]. In particular, the transformation operator method for $n>2$ is used in [50] to prove a uniqueness theorem in the class of functions which are analytic in a certain sector.

The IP of recovering the self-adjoint operator from the spectral function can be reduced to the IP from the WS's. Thus, we can obtain a uniqueness theorem and an algorithm for the solution of the IP from the spectral function in the class of piecewise-analytic coefficients. For brevity, we consider only the case in which $n=4$, $U_{\xi a}(y)=y^{(\xi-1)}(a)$, $a=0,T$.

Let $\psi_1(x,\lambda)$ and $\psi_2(x,\lambda)$ be solutions of the DE

$$\ell y = y^{(4)}+p_2(x)y''+p_1(x)y'+p_0(x)y=\lambda y, \qquad 0\le x\le T\le\infty, \tag{9.22}$$

with the conditions $\psi_k^{(\xi-1)}(0,\lambda)=\delta_{\xi,3-k}$, $k,\xi=1,2$, and also $\psi_k(T,\lambda)=\psi_k'(T,\lambda)=0$ for $T<\infty$ and $\psi_k(x,\lambda)=O(1)$, $x\to\infty$ for $T=\infty$. Denote $\mathcal{M}(\lambda)=[\mathcal{M}_{kr}(\lambda)]_{k,r=1,2}$, $\mathcal{M}_{kr}(\lambda)=\psi_k^{(r+1)}(0,\lambda)$. It is known [92] that if the DE (9.22) is self-adjoint, then the specification of $\mathcal{M}(\lambda)$ is equivalent to the specification of the spectral function $\sigma(\lambda)=[\sigma_{kr}(\lambda)]_{k,r=1,2}$ of the DE with the conditions $y(0)=y'(0)=0$ (and $y(T)=y'(T)=0$ for $T<\infty$). The IP is formulated as follows: given the matrix $\mathcal{M}(\lambda)$ construct the DE (9.22).

Denote $d(\alpha)=[d_{kr}(\alpha,0),d_{kr}(\alpha+1,1),d_{kr}(\alpha+2,2)]_{k,r=1,2}$, where $d_{kr}(\alpha,\nu)=1+i^{\alpha+3\nu+k+r-1}-(1+i)^{\alpha+1}(i^{r-1}-i^{k+3\nu-1})$. It is easy to see that the information condition for this IP has the form $d(\alpha)\#0$, $\alpha\ge1$, and is clearly satisfied. Therefore, applying the method of standard models to this IP we obtain that the specification of $\mathcal{M}(\lambda)$ uniquely determines the DE (9.22) in the class $p_k(x)\in PA[0,T)$. In particular, if the DE (9.22) is self-adjoint, then the specification of the spectral function $\sigma(\lambda)$ uniquely determines the DE in the class $PA[0,T)$.

10. An inverse problem of elasticity theory

The problem of determining the dimensions of the transverse cross-sections of a beam from the given frequencies of its natural vibrations is examined. Frequency spectra are indicated which determine the dimensions of the transverse cross-sections of the beam uniquely. An effective procedure is presented for solving the inverse problem, and a uniqueness theorem is proved. The method of standard models is used to solve the inverse problems.

Consider the differential equation describing beam vibrations in the form

$$(h^\mu(x)y'')'' = \lambda h(x)y, \qquad 0 \le x \le T. \tag{10.1}$$

Here $h(x)$ is a function characterizing the beam transverse section, and $\mu = 1, 2, 3$ is a fixed number. We will assume that the function $h(x)$ is absolutely continuous in the segment $[0, T]$ and $h(x) > 0$, $h(0) = 1$. The inverse problem for (10.1) in the case $\mu = 2$ (similar transverse sections) was investigated in [70]–[71] to determine small changes in the beam transverse sections for given small changes in a finite number of its natural vibration frequencies.

10.1. Let $\{\lambda_{kj}\}_{k\ge 1}$, $j = 1, 2$ be the eigenvalues of boundary value problems Q_j for (10.1) with the boundary conditions

$$y(0) = y^{(j)}(0) = y(T) = y'(T) = 0.$$

The inverse problem is formulated as follows.

Problem 10.1. Find the function $h(x)$, $x \in [0, T]$ from the given spectra $\{\lambda_{kj}\}_{k\ge 1, j=1,2}$.

Let us show that this IP can be reduced to the IP of recovering the DE (10.1) from the WF. Let $\Phi(x, \lambda)$ be a solution of (10.1) under the conditions $\Phi(0, \lambda) = \Phi(T, \lambda) = \Phi'(T, \lambda) = 0$, $\Phi'(0, \lambda) = 1$. We set $\mathfrak{M}(\lambda) = \Phi''(0, \lambda)$. The function $\mathfrak{M}(\lambda)$ is called the WF for (10.1). Let the functions $C_\nu(x, \lambda)$, $\nu = \overline{0, 3}$ be solutions of (10.1) under the initial conditions $C_\nu^{(\mu)}(0, \lambda) = \delta_{\nu\mu}$, $\nu, \mu = \overline{0, 3}$. Denote

$$\Delta_j(\lambda) = C_{3-j}(T, \lambda)C_3'(T, \lambda) - C_3(T, \lambda)C_{3-j}'(T, \lambda), \qquad j = 1, 2.$$

Then

$$\Phi(x, \lambda) = (\Delta_1(\lambda))^{-1} \det\,[C_\nu(x, \lambda), C_\nu(T, \lambda), C_\nu'(T, \lambda)]_{\nu=1,2,3},$$

and hence $\mathfrak{M}(\lambda) = -(\Delta_1(\lambda))^{-1}\Delta_2(\lambda)$.

The eigenvalues $\{\lambda_{kj}\}_{k\ge 1, j=1,2}$ of the boundary value problems Q_j coincide with the zeros of the entire functions $\Delta_j(\lambda)$. As in §7 (see 7.3) it is easy to see that the functions $\Delta_j(\lambda)$ are uniquely determined by their zeros. Hence the specification of the spectra $\{\lambda_{kj}\}_{k\ge 1, j=1,2}$ uniquely determines the WF $\mathfrak{M}(\lambda)$. Thus, Problem 10.1 is reduced to the following IP.

Problem 10.2. Given the WF $\mathfrak{M}(\lambda)$, find $h(x)$, $x \in [0, T]$.

10.2. We will solve Problem 10.2 by the method of standard models. Let $\lambda = \rho^4$. In the same way as in §1, one can obtain the asymptotic formulas for $|\lambda| \to \infty$, $\arg\lambda = \varphi \ne 0$, $\rho \in S$, $x \in [0, T]$

$$\Phi^{(\nu)}(x, \lambda) = \rho^{\nu-1} \sum_{\xi=1}^{2} (R_\xi \gamma'(x))^\nu g_{\xi 0}(x) \exp\,(\rho R_\xi \gamma(x))(1 + O(\rho^{-1})), \tag{10.2}$$

where

$$\gamma(x) = \int_0^x (h(t))^{(1-\mu)/4}\,dt.$$

The functions $g_{\xi 0}(x)$ are absolutely continuous, and $g_{\xi 0}(x) \neq 0$, $g_{10}(0) = -g_{20}(0) = (R_1 - R_2)^{-1}$. In particular, $\mathfrak{M}(\lambda) = \rho(R_1 + R_2)(1 + O(\rho^{-1}))$.

Lemma 10.1. *Let $p(x) = h^{\mu}(x)$. The following relationship holds*

$$\int_0^T \Big(\widehat{h}(x)\lambda\Phi(x,\lambda)\widetilde{\Phi}(x,\lambda) - \widehat{p}(x)\Phi''(x,\lambda)\widetilde{\Phi}''(x,\lambda)\Big)\,dx = \widehat{\mathfrak{M}}(\lambda). \tag{10.3}$$

Proof. Denote

$$\ell_\lambda y = (p(x)y'')'' - \lambda h(x)y, \qquad \mathcal{L}(y,z) = (py'')'z - py''z' + py'z'' - y(pz'')'.$$

Then

$$\int_0^T \ell_\lambda y(x)z(x)\,dx = \Big|_0^T \mathcal{L}(y(x),z(x)) + \int_0^T y(x)\ell_\lambda z(x)\,dx. \tag{10.4}$$

Using (10.4) and the equality $\ell_\lambda\Phi(x,\lambda) = \widetilde{\ell}_\lambda\widetilde{\Phi}(x,\lambda) = 0$, we obtain

$$\begin{aligned}
\int_0^T (\ell_\lambda - \widetilde{\ell}_\lambda)\Phi(x,\lambda)\widetilde{\Phi}(x,\lambda)\,dx &= -\int_0^T \widetilde{\ell}_\lambda\Phi(x,\lambda)\widetilde{\Phi}(x,\lambda)\,dx \\
&= -\Big|_0^T \widetilde{\mathcal{L}}(\Phi(x,\lambda),\widetilde{\Phi}(x,\lambda)) - \int_0^T \Phi(x,\lambda)\widetilde{\ell}_\lambda\widetilde{\Phi}(x,\lambda)\,dx \\
&= \Phi'(0,\lambda)\widetilde{\Phi}''(0,\lambda) - \Phi''(0,\lambda)\widetilde{\Phi}'(0,\lambda) = -\widehat{\mathfrak{M}}(\lambda).
\end{aligned}$$

On the other hand, integrating by parts, we have

$$\begin{aligned}
\int_0^T (\ell_\lambda - \widetilde{\ell}_\lambda)\Phi(x,\lambda)\widetilde{\Phi}(x,\lambda)\,dx &= \int_0^T ((\widehat{p}(x)\Phi''(x,\lambda))'' - \lambda\widehat{h}(x)\Phi(x,\lambda))\widetilde{\Phi}(x,\lambda)\,dx \\
&= \Big|_0^T ((\widehat{p}(x)\Phi''(x,\lambda))'\widetilde{\Phi}(x,\lambda) - \widehat{p}(x)\Phi''(x,\lambda)\widetilde{\Phi}'(x,\lambda) \\
&\quad + \int_0^T (\widehat{p}(x)\Phi''(x,\lambda)\widetilde{\Phi}''(x,\lambda) - \lambda\widehat{h}(x)\Phi(x,\lambda)\widetilde{\Phi}(x,\lambda))\,dx.
\end{aligned}$$

Since the substitution vanishes, we hence obtain (10.3).

Lemma 10.2. *Consider the integral*

$$J(z) = \int_0^T f(x)H(x,z)\,dx, \tag{10.5}$$

$$f(x) \in C[0,T], \quad f(x) \sim f_\alpha \frac{x^\alpha}{\alpha!}, \quad (x \to +0), \quad H(x,z) = \exp(-za(x))\left((1 + \frac{\xi(x,z)}{z}\right),$$

$$a(x) \in C^1[0,T], \qquad 0 < a(x_1) < a(x_2), \qquad (0 < x_1 < x_2),$$

$$a^{(\nu)}(x) \sim a_0 x^{1-\nu}, \qquad (x \to +0,\ \nu = 0,1), \qquad a'(x) > 0,$$

where the function $\xi(x,z)$ *is continuous and bounded for* $x \in [0,T]$,

$$z \in Q = \left\{z : arg\, z \in \left[-\frac{\pi}{2} + \delta_0, \frac{\pi}{2} - \delta_0\right],\ \delta_0 > 0\right\}.$$

Then as $z \to \infty$, $z \in Q$

$$J(z) \sim f_\alpha (a_0 z)^{-\alpha-1}. \tag{10.6}$$

Proof. The function $t = a(x)$ has the inverse $x = b(t)$, where $b(t) \in C^1[0,T_1]$, $T_1 = a(T)$, $b(t) > 0$ $(t > 0)$ and $b^{(\nu)}(t) = \frac{1}{a_0}t^{1-\nu}$, $\nu = 0,1$ as $t \to +0$. Let us make the change of variable $t = a(x)$ in the integral in (10.5). We obtain

$$J(z) = \int_0^{T_1} g(t)\exp(-zt)\,(1 + z^{-1}\xi(b(t),z))\,dt, \tag{10.7}$$

where $g(t) = f(b(t))b'(t)$. It is clear that for $t \to +0$

$$g(t) \sim f_\alpha (a_0)^{-\alpha-1}\frac{t^\alpha}{\alpha!}.$$

Applying Lemma 9.1 to (10.7), we obtain (10.6).

Denote

$$A_\alpha = \frac{1}{(R_1 - R_2)^2}\sum_{\xi,s=1}^{2}\frac{(-1)^{\xi+s}(1 - \mu R_\xi^2 R_s^2)}{(R_\xi + R_s)^{\alpha+1}}, \qquad \alpha \geq 1.$$

Let us show that $A_\alpha \neq 0$ for all $\alpha \geq 1$. For definiteness, we put $\arg\rho \in (0, \frac{\pi}{4})$ i.e. $\{R_1, R_2\} = \{-1, i\}$. Then

$$A_\alpha = -\frac{1}{(R_1 - R_2)^2}\frac{1}{(-2)^{\alpha+1}}a_\alpha,$$

where

$$a_\alpha = (\mu - 1)(1 + i^{\alpha+1}) + 2(\mu + 1)(1 + i)^{\alpha+1}.$$

Since $|1 + i^{\alpha+1}| \le 2$, $|1 + i|^{\alpha+1} = (\sqrt{2})^{\alpha+1}$, it follows that $a_\alpha \neq 0$, $\alpha \ge 1$. Hence $A_\alpha \neq 0$ for all $\alpha \ge 1$.

Lemma 10.3. *As $x \to 0$ let $\hat{h}(x) \sim \widehat{h}_\alpha(\alpha!)^{-1}x^\alpha$. Then as $|\lambda| \to \infty$, $\arg\lambda = \varphi \neq 0$ there exists a finite limit $I_\alpha = \lim \rho^{\alpha-1}\widehat{\mathfrak{M}}(\lambda)$, and*

$$A_\alpha \widehat{h}_\alpha = I_\alpha. \tag{10.8}$$

Proof. Since $p(x) = h^\mu(x)$, then by virtue of the conditions of the lemma we have $\widehat{p}(x) \sim \mu\widehat{h}_\alpha(\alpha!)^{-1}x^\alpha$ as $x \to +0$. Using the asymptotic formulas (10.2) and Lemma 10.2 we find as $\arg\lambda = \varphi \neq 0$, $\rho \in S$, $\lambda \to \infty$:

$$\int_0^T \widehat{h}(x)\lambda\Phi(x,\lambda)\widetilde{\Phi}(x,\lambda)\,dx \sim \frac{1}{\rho^{\alpha-1}}\frac{\widehat{h}_\alpha}{(R_1 - R_2)^2}\sum_{\xi,s=1}^{2}\frac{(-1)^{\xi+s}}{(R_\xi + R_s)^{\alpha+1}},$$

$$\int_0^T \widehat{p}(x)\Phi''(x,\lambda)\widetilde{\Phi}''(x,\lambda)\,dx \sim \frac{1}{\rho^{\alpha-1}}\frac{\mu\widehat{h}_\alpha}{(R_1 - R_2)^2}\sum_{\xi,s=1}^{2}\frac{(-1)^{\xi+s}R_\xi^2R_s^2}{(R_\xi + R_s)^{\alpha+1}}.$$

Substituting the expressions obtained in (10.3), we obtain the assertion of the lemma.

From the facts presented above we have the following theorem.

Theorem 10.1. *Problem 10.2 has a unique solution in the class $h(x) \in A[0,T]$. This solution can be found according to the following algorithm:*

1. *we calculate $h_\alpha = h^{(\alpha)}(0)$, $\alpha \ge 0$, $h_0 = 1$. For this we successively perform operations for $\alpha = 1, 2, \ldots$: we construct the function $\widetilde{h}(x) \in A[0,T]$, $\widetilde{h}(x) > 0$ such that $\widetilde{h}^{(\nu)}(0) = h_\nu$, $\nu = \overline{0, \alpha - 1}$, and arbitrary in the rest, and we calculate h_α from (10.8).*

2. *we construct the function $h(x)$ from the formula*

$$h(x) = \sum_{\alpha=0}^{\infty} h_\alpha \frac{x^\alpha}{\alpha!}, \qquad 0 < x < R,$$

where

$$R = \left(\overline{\lim_{\alpha\to\infty}} \sqrt[\alpha]{\frac{|h_\alpha|}{\alpha!}}\right)^{-1}.$$

If $R < T$, then for $R < x < T$ the function $h(x)$ is constructed by analytic continuation.

We note that the IP in the class of piecewise-analytic functions can be solved in an analogous manner.

11. Differential operators with locally integrable coefficients

Here we investigate the IP for the non-self-adjoint differential operator (1.1) on the half-line with locally intergable analytic coefficients from the so-called generalized Weyl functions. To solve the IP we use connections with an IP for partial differential equations, and also use the Riemann–Fage formula [76] for the solution of the Cauchy problem for higher-order partial differential equations.

11.1. Let us introduce the space of generalized functions (distributions) by analogy with [8] (ch. 2). Let $\mathcal{D}$ be the set of all integrable on the real axis entire functions of exponential type with ordinary operations of addition and multiplication by complex numbers and with the following convergence: $z_k(\mu)$ is said to converge to $z(\mu)$ if the types σ_k of the functions $z_k(\mu)$ are bounded ($\sup \sigma_k < \infty$) and $\|z_k(\mu) - z(\mu)\|_{\mathcal{L}(-\infty,\infty)} \to 0$ as $k \to \infty$. The linear manifold $\mathcal{D}$ with such a convergence is our space of test functions.

Definition 11.1. All additive, homogeneous and continuous functionals $\langle z(\mu), R\rangle$, defined on $\mathcal{D}$, are called generalized functions (GF). The set of GF is denoted by $\mathcal{D}'$. The sequence of GF $R_k \in \mathcal{D}'$ converges to $R \in \mathcal{D}'$, if $\lim \langle z(\mu), R_k\rangle = \langle z(\mu), R\rangle$, $k \to \infty$ for any $z(\mu) \in \mathcal{D}$. A GF $R \in \mathcal{D}'$ is called regular if it is determined by the following formula

$$\langle z(\mu), R\rangle = \int_{-\infty}^{\infty} z(\mu)R(\mu)\,d\mu, \qquad R(\mu) \in \mathcal{L}_\infty.$$

Definition 11.2. Let the function $f(t)$ be locally integrable for $t > 0$ (i.e., it is integrable on every finite segment $[0,\sigma]$). A GF $L_f(\mu) \in \mathcal{D}'$ defined by the equality

$$\langle z(\mu), L_f(\mu)\rangle \stackrel{df}{=} \int_0^{\infty} f(t)\,dt \int_{-\infty}^{\infty} z(\mu)\exp(i\mu t)\,d\mu, \qquad z(\mu) \in \mathcal{D} \tag{11.1}$$

is called a generalized Fourier–Laplace transformation for the function $f(t)$. As $z(\mu) \in \mathcal{D}$ in (11.1), so $z(\mu) \in \mathcal{L}_2(-\infty,\infty)$. Therefore, by virtue of the Paley–Wiener theorem, the function

$$\int_{-\infty}^{\infty} z(\mu)\exp(i\mu t)\,d\mu$$

is continuous and finite. Consequently, the integral in (11.1) exists. We note that $f(t) \in \mathcal{L}(0,\infty)$ implies

$$\langle z(\mu), L_f(\mu)\rangle = \int_{-\infty}^{\infty} z(\mu)\,d\mu \int_0^{\infty} f(t)\exp(i\mu t)\,dt.$$

In this case $L_f(\mu)$ is, consequently, a regular GF and coincides with the ordinary Fourier–Laplace transformation for the function $f(t)$. Since

$$\frac{1}{\pi}\int_{-\infty}^{\infty}\frac{1-\cos\mu x}{\mu^2}\exp(i\mu t)\,d\mu = \begin{cases} x-t, & t<x, \\ 0, & t>x, \end{cases}$$

the following inversion formula occurs:

$$\int_0^x (x-t)f(t)\,dt = \langle \frac{1}{\pi}\frac{1-\cos\mu x}{\mu^2}, L_f(\mu)\rangle. \tag{11.2}$$

11.2. In this item we give the formulation and solution of the inverse problem for differential operators of the third order, in order to simplify calculations. The general case of arbitrary order operators will be briefly described in item 11.3.

Let us consider the following differential equation

$$\ell y \equiv y''' + p_1(x)y' + p_0(x)y = \lambda y = (i\mu)^3 y, \qquad x>0. \tag{11.3}$$

Denote $q_0(x) = -p_1(x)$, $q_1(x) = p_0(x) - p_1'(x)$, $B = \{x : |\arg x| < \frac{\pi}{6}\}$, $R_k = \exp(2(k-1)\pi i/3)$, $k = \overline{1,3}$ and assume that the functions $q_\nu(x)$ are regular for $x \in B$, $x = 0$ and continuous in $\bar{B}$. Let us consider the following integral equation:

$$\begin{aligned} Q(x,s) &= Q_1(x,s) + \sum_{\nu=0}^{1}\Bigg(-\frac{1}{3}\int_0^s \frac{(s-u)^\nu}{\nu!}\,du\int_0^x q_\nu(t)Q(t,u)\,dt \\ &+\sum_{k=2}^{3}\frac{R_k^{2-\nu}}{3(1-R_k)}\int_0^s du\int_0^{s-u}\frac{(s-u-\xi)^\nu}{\nu!} \\ &\times\left(q_\nu\left(\frac{\xi}{1-R_k}+x\right)Q\left(\frac{\xi}{1-R_k}+x,u\right)\right. \\ &\left.-q_\nu\left(\frac{\xi}{1-R_k}\right)Q\left(\frac{\xi}{1-R_k},u\right)\right)d\xi\Bigg), \end{aligned} \tag{11.4}$$

where

$$\begin{aligned} Q_1(x,s) &= \sum_{\nu=0}^{1}\Bigg(-\frac{1}{3}\frac{s^\nu}{\nu!}\int_0^x q_\nu(t)\,dt + \sum_{k=2}^{3}\frac{R_k^{2-\nu}}{3(1-R_k)}\int_0^s\frac{(s-\xi)^\nu}{\nu!} \\ &\times\left(q_\nu\left(\frac{\xi}{1-R_k}+x\right) - q_\nu\left(\frac{\xi}{1-R_k}\right)\right)d\xi\Bigg). \end{aligned} \tag{11.5}$$

Lemma 11.1. *In the domain $s \geq 0$, $x \in \bar{B}$ the integral equation* (11.4) *has a unique solution $Q(x,s)$; moreover, the function $Q(x,s)$ is continuous and, for any fixed $s \geq 0$ is regular with respect to $x \in B$.*

Proof. We shall solve the equation (11.4) by means of the method of successive approximations. Let us define the functions $Q_j(x,s)$ with the help of the recurrence formula:

$$\begin{aligned} Q_{j+1}(x,s) &= \sum_{\nu=0}^{1}\Bigg(-\frac{1}{3}\int_0^s \frac{(s-u)^\nu}{\nu!}\,du\int_0^x q_\nu(t)Q_j(t,u)\,dt \\ &\quad+\sum_{k=2}^{3}\frac{R_k^{2-\nu}}{3(1-R_k)}\int_0^s du\int_0^{s-u}\frac{(s-u-\xi)^\nu}{\nu!} \\ &\quad\times\left(q_\nu\left(\frac{\xi}{1-R_k}+x\right)Q_j\left(\frac{\xi}{1-R_k}+x,u\right)\right. \\ &\quad\left.-q_\nu\left(\frac{\xi}{1-R_k}\right)Q_j\left(\frac{\xi}{1-R_k},u\right)\right)d\xi\Bigg), \qquad j=1,2,\dots, \end{aligned} \tag{11.6}$$

where the function $Q_1(x,s)$ is defined by (10.5). Denote

$$r(y)=\max_\nu \sup_{\operatorname{Re}x=y,x\in B}|q_\nu(x)|, \qquad \mathcal{P}_a=\int_0^a r(y)\,dy, \qquad \gamma_j(s)=\sum_{k=j-1}^{2j-1}\frac{s^k}{k!}.$$

Now we shall prove by induction that the estimate

$$|Q_j(x,\lambda)|\leq\left(4\mathcal{P}_{s/2+\operatorname{Re}x}\right)^j\gamma_j(s), \qquad s\geq 0, \qquad x\in\bar{B}, \qquad j\geq 1 \tag{11.7}$$

is valid.

1) It follows from (11.5) that

$$\begin{aligned} |Q_1(x,s)| &\leq \sum_{\nu=0}^{1}\Bigg(\frac{1}{3}\frac{s^\nu}{\nu!}\left|\int_0^x q_\nu(t)\,dt\right| \\ &\quad+\frac{1}{3\sqrt{3}}\frac{s^\nu}{\nu!}\sum_{k=2}^{3}\int_0^s\left(\left|q_\nu\left(\frac{\xi}{1-R_k}+x\right)\right|+\left|q_\nu\left(\frac{\xi}{1-R_k}\right)\right|\right)\Bigg)\,d\xi. \end{aligned} \tag{11.8}$$

Since $x = |x| \exp(i\alpha)$, $|\alpha| \leq \frac{\pi}{6}$, we get

$$\begin{aligned}\left|\int_0^x q_\nu(t)\,dt\right| &\leq \int_0^{|x|} |q_\nu(\eta \exp(i\alpha))|\,d\eta \leq \int_0^{|x|} r(\eta\cos\alpha)\,d\eta \\ &= \frac{1}{\cos\alpha}\int_0^{\operatorname{Re} x} r(\eta)\,d\eta \leq \frac{2}{\sqrt{3}}\mathcal{P}_{\operatorname{Re} x}.\end{aligned} \tag{11.9}$$

Since $\operatorname{Re}\frac{1}{1-R_k} = \frac{1}{2}$, $k = 2, 3$, we obtain

$$\begin{aligned}&\int_0^s \left|q_\nu\left(\frac{\xi}{1-R_k} + x\right)\right| d\xi \\ &\leq \int_0^s r\left(\frac{\xi}{2} + \operatorname{Re} x\right) d\xi = 2\int_{\operatorname{Re} x}^{s/2+\operatorname{Re} x} r(\eta)\,d\eta \leq 2\mathcal{P}_{s/2+\operatorname{Re} x}.\end{aligned} \tag{11.10}$$

It follows from (11.8), (11.9) and (11.10) that

$$|Q_1(x,s)| \leq \left(\frac{2}{3\sqrt{3}}\mathcal{P}_{\operatorname{Re} x} + \frac{4}{3\sqrt{3}}\mathcal{P}_{s/2+\operatorname{Re} x} + \frac{4}{3\sqrt{3}}\mathcal{P}_{s/2}\right)\sum_{\nu=0}^{1}\frac{s^\nu}{\nu!} \leq \frac{10}{3\sqrt{3}}\mathcal{P}_{s/2+\operatorname{Re} x}\gamma_1(s).$$

(2) From (11.6) we have

$$\begin{aligned}|Q_{j+1}(x,s)| &= \sum_{\nu=0}^{1}\Bigg(\frac{1}{3}\int_0^s \frac{(s-u)^\nu}{\nu!}\,du\left|\int_0^x q_\nu(t)Q_j(t,u)\,dt\right| \\ &+\frac{1}{3\sqrt{3}}\sum_{k=2}^{3}\int_0^s \frac{(s-u)^\nu}{\nu!}\,du \\ &\times\int_0^{s-u}\left(\left|q_\nu\left(\frac{\xi}{1-R_k}+x\right)Q_j\left(\frac{\xi}{1-R_k}+x,u\right)\right|\right. \\ &\left.+\left|q_\nu\left(\frac{\xi}{1-R_k}\right)Q_j\left(\frac{\xi}{1-R_k},u\right)\right|\right)d\xi\Bigg).\end{aligned} \tag{11.11}$$

Assume that (11.7) has already been obtained for the function $Q_j(x,s)$. Then

$$\begin{aligned}\left|\int_0^x q_\nu(t)Q_j(t,u)\,dt\right| &\le \int_0^{|x|} |q_\nu(\eta\exp(i\alpha))Q_j(\eta\exp(i\alpha),u)|\,d\eta \\ &\le \int_0^{|x|} r(\eta\cos\alpha)\left(4\mathcal{P}_{u/2+\eta\cos\alpha}\right)^j \gamma_j(u)\,d\eta \\ &\le \frac{2}{3}\int_0^{\operatorname{Re}x} r(\eta)(4\mathcal{P}_{u/2+\eta})^j\gamma_j(u)\,d\eta \\ &\le \frac{2}{\sqrt{3}}(4\mathcal{P}_{s/2+\operatorname{Re}x})^j\mathcal{P}_{\operatorname{Re}x}\gamma_j(u),\end{aligned} \tag{11.12}$$

$$\begin{aligned}&\int_0^s \left|q_\nu\left(\frac{\xi}{1-R_k}+x\right)Q_j\left(\frac{\xi}{1-R_k}+x,u\right)\right|d\xi \\ &\le \gamma_j(u)\int_0^{s-u} r\left(\frac{\xi}{2}+\operatorname{Re}x\right)\left(4\mathcal{P}_{u/2+\xi/2+\operatorname{Re}x}\right)^j d\xi \\ &\le \gamma_j(u)(4\mathcal{P}_{s/2+\operatorname{Re}x})^j\int_0^s r\left(\frac{\xi}{2}+\operatorname{Re}x\right)d\xi \le \gamma_j(u)(4\mathcal{P}_{s/2+\operatorname{Re}x})^{j+1}.\end{aligned} \tag{11.13}$$

Since

$$\sum_{\nu=0}^{1}\int_0^s \frac{(s-u)^\nu}{\nu!}\gamma_j(u)\,du \le 2\gamma_{j+1}(s),$$

it follows from (11.11), (11.12) and (11.13) that

$$\begin{aligned}|Q_{j+1}(x,s)| &\le \frac{10}{3\sqrt{3}}\mathcal{P}_{s/2+\operatorname{Re}x}(4\mathcal{P}_{s/2+\operatorname{Re}x})^j\sum_{\nu=0}^{1}\int_0^s\frac{(s-u)^\nu}{\nu!}\gamma_j(u)\,du \\ &\le (4\mathcal{P}_{s/2+\operatorname{Re}x})^{j+1}\gamma_{j+1}(s).\end{aligned}$$

Thus, (11.7) is proved.

Now let us construct the function $Q(x,s)$ via the formula

$$Q(x,s)=\sum_{j=1}^{\infty}Q_j(x,s). \tag{11.14}$$

Using (11.7) we obtain

$$\sum_{j=1}^{\infty} |Q_j(x,s)| \le \sum_{j=1}^{\infty} (4\mathcal{P}_{s/2+\mathrm{Re}x})^j \gamma_j(s)$$

$$\le \sum_{j=1}^{\infty} (4\mathcal{P}_{s/2+\mathrm{Re}x})^j \sum_{k=j-1}^{\infty} \frac{s^k}{k!} = \sum_{k=0}^{\infty} \frac{s^k}{k!} \sum_{j=1}^{k+1} (4\mathcal{P}_{s/2+\mathrm{Re}x})^j .$$

Thus, the series (11.14) converges absolutely and uniformly on compacts. It is obvious that the function $Q(x,s)$ is the desired solution of (11.4). Lemma 11.1 is proved.

Note that if $\mathcal{P}_\infty < \infty$, then we have $|Q(x,s)| < C \exp(Cs)$. The proof of the Lemma implies that if the functions $q_\nu(x)$ are regular for $|x| < \delta$, then the function $Q(x,s)$ is regular in the domain

$$\mathcal{F}_\delta = \left\{ (x,s) : |x| < \delta, |s| < \sqrt{3}\delta, \left| x + \frac{s}{1 - R_k} \right| < \delta, k = 2,3 \right\}.$$

Let the function $Q(x,s)$ be a solution of (11.4). Denote $u(x,t) = Q(x, t-x)$, $0 \le x \le t < \infty$; $u(x,t) = 0$, $t < x$, and consider the GF

$$\Phi(x,\mu) = \exp(i\mu x) + L_u(\mu), \tag{11.15}$$

i.e.,

$$\langle z(\mu), \Phi(x,\mu) \rangle = \int_{-\infty}^{\infty} z(\mu)\, e^{i\mu x}\, d\mu + \int_{x}^{\infty} u(x,t)\, dt \int_{-\infty}^{\infty} z(\mu)\, e^{i\mu t}\, d\mu, \tag{11.16}$$

$$z(\mu) \in \mathcal{D}.$$

Put also

$$\langle z(\mu), (i\mu)^3 \Phi(x,\mu) \rangle = \langle (i\mu)^3 z(\mu), \Phi(x,\mu) \rangle,$$

$$\langle z(\mu), \Phi^{(j)}(x,\mu) \rangle = \frac{d^j}{dx^j} \langle z(\mu), \Phi(x,\mu) \rangle, \qquad j = \overline{1,3}$$

for $z(\mu) \in \mathcal{D}$, $\mu^2 z(\mu) \in \mathcal{L}(-\infty,\infty)$.

Theorem 11.1. *The following relations*

$$\ell\Phi(x,\mu) - (i\mu)^3\Phi(x,\mu) = 0, \qquad \Phi(0,\mu) = 1,$$

are valid.

Proof. Equality (11.4) can be transformed to the following form

$$\begin{aligned}
Q(x,s) = \sum_{\nu=0}^{1}\sum_{k=2}^{3}\frac{R_k^{2-\nu}}{3}\Bigg[&\int\limits_0^{x+s/(1-R_k)} \frac{(s-(1-R_k)(\eta-x))^\nu}{\nu!}\, q_\nu(\eta)\,d\eta \\
&- \int\limits_0^{s/(1-R_k)} \frac{(s-(1-R_k)\eta)^\nu}{\nu!}\, q_\nu(\eta)\,d\eta \\
&+ \int\limits_0^{s} du \int\limits_0^{x+(s-u)/(1-R_k)} \frac{((s-u)-(1-R_k)(\eta-x))^\nu}{\nu!}\, q_\nu(\eta)Q(\eta,u)\,d\eta \\
&- \int\limits_0^{s} du \int\limits_0^{(s-u)/(1-R_k)} \frac{((s-u)-(1-R_k)\eta)^\nu}{\nu!}\, q_\nu(\eta)Q(\eta,u)\,d\eta\Bigg].
\end{aligned} \tag{11.17}$$

Indeed, we can first transform $Q_1(x,s)$. Via changes of variables in the right-hand side of (11.5) $\eta = \frac{\xi}{1-R_k} + x$ and $\eta = \frac{\xi}{1-R_k}$ respectively, we obtain

$$\begin{aligned}
Q_1(x,s) \;=\; \sum_{\nu=0}^{1}\Bigg[&-\frac{1}{3}\frac{s^\nu}{\nu!}\int\limits_0^{x} q_\nu(t)\,dt \\
&+ \sum_{k=2}^{3}\frac{R_k^{2-\nu}}{3}\int\limits_0^{x+s/(1-R_k)} \frac{(s-(1-R_k)(\eta-x))^\nu}{\nu!} q_\nu(\eta)\,d\eta \\
&- \int\limits_0^{s/(1-R_k)} \frac{(s-(1-R_k)\eta)^\nu}{\nu!} q_\nu(\eta)\,d\eta\Bigg].
\end{aligned} \tag{11.18}$$

Using the regularity of the integrand, we can make the change

$$\int_{x}^{x+s/(1-R_k)} = \int_{0}^{x+s/(1-R_k)} - \int_{0}^{x}.$$

Since

$$\sum_{k=1}^{3} R_k^j = 0, \qquad j = 1, 2,$$

we get that the integral $\int_0^x$ can be cancelled, and (11.18) has the form

$$\begin{aligned} Q_1(x,s) &= \sum_{\nu=0}^{1}\sum_{k=2}^{3} \frac{R_k^{2-\nu}}{3}\Bigg[\int_0^{x+s/(1-R_k)} \frac{(s-(1-R_k)(\eta-x))^\nu}{\nu!}\, q_\nu(\eta)\, d\eta \\ &\quad - \int_0^{s/(1-R_k)} \frac{(s-(1-R_k)\eta)^\nu}{\nu!}\, q_\nu(\eta)\, d\eta\Bigg). \end{aligned}$$

The rest of the terms in the right-hand side of (11.4) can be transformed in an analogous way.

Differentiating (11.17) we get

$$\begin{aligned} Q_x(x,s) &= \sum_{k=2}^{3}\Bigg[\frac{R_k^2}{3} q_0\left(x+\frac{s}{1-R_k}\right) + \frac{R_k(1-R_k)}{3}\int_0^{x+\frac{s}{1-R_k}} q_1(\eta)\, d\eta \\ &\quad + \frac{R_k^2}{3}\int_0^s q_0\left(x+\frac{s-u}{1-R_k}\right) Q\left(x+\frac{s-u}{1-R_k}, u\right) du \qquad (11.19) \\ &\quad + \frac{R_k(1-R_k)}{3}\int_0^s du \int_0^{x+\frac{s-u}{1-R_k}} q_1(\eta) Q(\eta, u)\, d\eta\Bigg], \end{aligned}$$

$$Q_s(x,s)=\sum_{k=2}^{3}\Bigg[\frac{R_k^2}{3(1-R_k)}q_0\left(x+\frac{s}{1-R_k}\right)$$

$$+\frac{R_k}{3}\int_0^{x+s/(1-R_k)} q_1(\eta)\,d\eta-\frac{R_k^2}{3(1-R_k)}q_0\left(\frac{s}{1-R_k}\right)$$

$$-\frac{R_k}{3}\int_0^{s/(1-R_k)} q_1(\eta)\,d\eta+\frac{R_k^2}{3(1-R_k)}\int_0^s q_0\left(x+\frac{s-u}{(1-R_k)}\right)Q\left(x+\frac{s-u}{1-R_k},u\right)du$$

$$+\frac{R_k^3}{3}\int_0^s du\int_0^{x+(s-u)/(1-R_k)} q_1(\eta)Q(\eta,u)\,d\eta$$

$$-\frac{R_k^2}{3(1-R_k)}\int_0^s q_0\left(\frac{s-u}{1-R_k}\right)Q\left(\frac{s-u}{1-R_k},u\right)du$$

$$-\frac{R_k^3}{3}\int_0^s du\int_0^{\frac{s-u}{1-R_k}} q_1(\eta)Q(\eta,u)\,d\eta+\frac{R_k^2}{3}\int_0^x q_0(\eta)Q(\eta,s)\,d\eta\Bigg], \tag{11.20}$$

$$Q_{xx}(x,s)=\sum_{k=2}^{3}\Bigg[\frac{R_k^2}{3}q_0'\left(x+\frac{s}{1-R_k}\right)+\frac{R_k(1-R_k)}{3}q_1\left(x+\frac{s}{1-R_k}\right)$$

$$+\frac{R_k^2}{3}\int_0^s\Bigg(q_0'\left(x+\frac{s-u}{1-R_k}\right)Q\left(x+\frac{s-u}{1-R_k},u\right)$$

$$+q_0\left(x+\frac{s-u}{1-R_k}\right)Q_x\left(x+\frac{s-u}{1-R_k},u\right)\Bigg)du \tag{11.21}$$

$$+\frac{R_k(1-R_k)}{3}\int_0^s q_1\left(x+\frac{s-u}{1-R_k}\right)Q\left(x+\frac{s-u}{1-R_k},u\right)du\Bigg],$$

$$Q_{xs}(x,s) = \sum_{k=2}^{3}\Bigg[\frac{R_k^2}{3(1-R_k)}q_0'\left(x+\frac{s}{1-R_k}\right)+\frac{R_k}{3}q_1\left(x+\frac{s}{1-R_k}\right)$$
$$+\frac{R_k^2}{3(1-R_k)}\int_0^s\left(q_0'\left(x+\frac{s-u}{1-R_k}\right)Q\left(x+\frac{s-u}{1-R_k},u\right)\right.$$
$$\left.+q_0\left(x+\frac{s-u}{1-R_k}\right)Q_x\left(x+\frac{s-u}{1-R_k},u\right)\right)du \tag{11.22}$$
$$+\frac{R_k}{3}\int_0^s q_1\left(x+\frac{s-u}{1-R_k}\right)Q\left(x+\frac{s-u}{1-R_k},u\right)du+\frac{R_k^2}{3}q_0(x)Q(x,s)\Bigg],$$

From this we get

$$\frac{\partial^3 Q(x,s)}{\partial x^3}-3\left(\frac{\partial^3 Q(x,s)}{\partial x^2\partial s}-\frac{\partial^3 Q(x,s)}{\partial x\partial s^2}\right)$$
$$= q_0'(x)Q(x,s)+q_0(x)Q_x(x,s)-q_1(x)Q(x,s)-q_0(x)Q_s(x,s)$$

or

$$\frac{\partial^3 Q(x,s)}{\partial x^3}-3\frac{\partial^3 Q(x,s)}{\partial x^2\partial s}+3\frac{\partial^3 Q(x,s)}{\partial x\partial s^2}$$
$$= -p_0(x)Q(x,s)+p_1(x)\left(\frac{\partial Q_x(x,s)}{\partial s}-\frac{\partial Q(x,s)}{\partial x}\right). \tag{11.23}$$

Moreover, it follows from (11.17) and (11.20) that

$$\left.\begin{aligned} &Q(0,s)=0, \quad Q(x,0)=\frac{1}{3}\int_0^x p_1(t)\,dt,\\ &Q_s(x,s)\big|_{s=0}=\frac{1}{3}\left(p_1(x)-p_1(0)-\int_0^x (p_0(t)-p_1(t)Q(t,0))\,dt\right), \end{aligned}\right\} \tag{11.24}$$

Since $u(x,t)=Q(x,t-x)$, $0\le x\le t$, then (11.23) and (11.24) imply that

$$\frac{\partial^3 u(x,t)}{\partial t^3}+\frac{\partial^3 u(x,t)}{\partial x^3}+p_1(x)\,\frac{\partial u(x,t)}{\partial x}+p_0(x)u(x,t)=0, \tag{11.25}$$

$$u(0,t)=0, \tag{11.26}$$

$$\left.\begin{aligned} u_x(x,t)\Big|_{t=x} &= \frac{1}{3}\left(p_1(0) + \int_0^x (p_0(\xi) - p_1(\xi)u(\xi,\xi))\,d\xi\right) \\ u(x,x) &= \frac{1}{3}\int_0^x p_1(t)\,dt. \end{aligned}\right\} \tag{11.27}$$

Consequently,

$$3\frac{d}{dx}u(x,x) = p_1(x), \qquad 3\frac{d}{dx}\left(u_x(x,t)\Big|_{t=x}\right) + p_1(x)u(x,x) = p_0(x). \tag{11.28}$$

Further, using (11.16), we calculate

$$\begin{aligned} \langle z(\mu), \ell\Phi(x,\mu)\rangle &= \int_{-\infty}^{\infty} z(\mu)\Big[(i\mu)^3 - (i\mu)^2 u(x,x) + (i\mu)(p_1(x) \\ &-2\frac{d}{dx}u(x,x) - u_x(x,t)\Big|_{t=x}) + (p_0(x) - p_1(x)u(x,x) - \frac{d^2}{dx^2}u(x,x) \\ &-\frac{d}{dx}(u_x(x,t)\Big|_{t=x}) - u_{xx}(x,t)\Big|_{t=x})\Big]\exp(i\mu x)\,d\mu \\ &+\int_x^{\infty}\left(\frac{\partial^3 u(x,t)}{\partial x^3} + p_1(x)\frac{\partial u(x,t)}{\partial x} + p_0(x)u(x,t)\right)dt\int_{-\infty}^{\infty} z(\mu)\exp(i\mu t)\,d\mu. \end{aligned} \tag{11.29}$$

On the other hand, integration by parts gives

$$\begin{aligned} \langle z(\mu), -(i\mu)^3\Phi(x,\mu)\rangle &= \int_{-\infty}^{\infty}(-i\mu)^3 z(\mu)\exp(i\mu x)\,d\mu \\ &+\int_x^{\infty} u(x,t)\,dt\int_{-\infty}^{\infty}(-i\mu)^3 z(\mu)\exp(i\mu t)\,d\mu = \int_{-\infty}^{\infty} z(\mu)\Big[-(i\mu)^3 \\ &+(i\mu)^2 u(x,x) - (i\mu)u_t(x,t)\Big|_{t=x} + u_{tt}(x,t)\Big|_{t=x}\Big]\exp(i\mu x)\,d\mu \\ &+\int_x^{\infty}\frac{\partial^3 u(x,t)}{\partial t^3}\,dt\int_{-\infty}^{\infty} z(\mu)\exp(i\mu t)\,d\mu. \end{aligned} \tag{11.30}$$

Since

$$u_x(x,t)\big|_{t=x}+u_t(x,t)\big|_{t=x}=\frac{d}{dx}u(x,x),$$

$$\frac{d^2}{dx^2}u(x,x)+u_{xx}(x,t)\big|_{t=x}-u_{tt}(x,t)\big|_{t=x}=2\frac{d}{dx}\left(u_x(x,t)\big|_{t=x}\right),$$

it follows from (11.29) and (11.30) that

$$\langle z(\mu),\ell\Phi(x,\mu)-(i\mu)^3\Phi(x,\mu)\rangle=\int_{-\infty}^{\infty}z(\mu)\Big[(i\mu)\left(p_1(x)-3\frac{d}{dx}u(x,x)\right)$$

$$+\left(p_0(x)-3\frac{d}{dx}\left(u_x(x,t)\big|_{t=x}\right)-p_1(x)u(x,x)\right)\Big]\exp(i\mu x)\,d\mu$$

$$+\int_x^{\infty}\left(\frac{\partial^3u(x,t)}{\partial t^3}+\frac{\partial^3u(x,t)}{\partial x^3}+p_1(x)\frac{\partial u(x,t)}{\partial x}+p_0(x)u(x,t)\right)\int_{-\infty}^{\infty}z(\mu)\exp(i\mu t)\,d\mu.$$

By virtue of (11.25) and (11.28) we have

$$\langle z(\mu),\ell\Phi(x,\mu)-(i\mu)^3\Phi(x,\mu)\rangle=0.$$

From (11.16) for $x=0$ and (11.26) we obtain

$$\langle z(\mu),\Phi(0,\mu)\rangle=\int_{-\infty}^{\infty}z(\mu)\,d\mu,$$

i.e., $\Phi(0,\mu)=1$. Theorem 11.1 is proved.

Definition 11.3. The GF $\Phi(x,\mu)$ is called the Weyl generalized solution of the DE (11.3), and the functions $\mathfrak{M}_\nu(\mu)=\Phi^{(\nu)}(0,\mu)$, $\nu=1,2$ are called the Weyl generalized functions (WGF).

Note that if $\mathcal{P}_\infty<\infty$, then $|u(x,t)|<C\exp(Ct)$, and the function

$$\Phi(x,\mu)=\exp(i\mu x)+\int_x^{\infty}u(x,t)\exp(i\mu t)\,dt,\qquad \arg\mu\in\left(\frac{\pi}{6},\frac{5\pi}{6}\right)$$

is the ordinary Weyl solution.

The IP for the DE (11.3) can be formulated as follows:

Problem 11.1. Given the WGF's $\{\mathfrak{M}_\nu(\mu)\}_{\nu=1,2}$, construct the functions $\{p_k(x)\}_{k=0,1}$.

For this IP let us prove the uniqueness theorem.

Theorem 11.2. *If* $\mathfrak{M}_\nu(\mu) = \widetilde{\mathfrak{M}}_\nu(\mu)$, $\nu = 1, 2$, *then* $p_k(x) = \widetilde{p}_k(x)$, $x \geq 0$, $k = 0, 1$.

Proof. We denote

$$h_\nu(t) = \frac{\partial^\nu}{\partial x^\nu} u(x,t)\big|_{x=0}, \qquad \nu = 1, 2, \qquad t \geq 0.$$

Taking into account (11.27), from (11.16) we deduce that

$$\langle z(\mu), \Phi'(0,\mu)\rangle = \int_{-\infty}^{\infty} z(\mu)(i\mu)\, d\mu + \int_0^\infty h_1(t)\, dt \int_{-\infty}^{\infty} z(\mu) \exp(i\mu t)\, d\mu,$$

$$\langle z(\mu), \Phi''(0,\mu)\rangle = \int_{-\infty}^{\infty} z(\mu)((i\mu)^2 - \frac{2}{3} p_1(0))\, d\mu + \int_0^\infty h_2(t)\, dt \int_{-\infty}^{\infty} z(\mu) \exp(i\mu t)\, d\mu$$

for $z(\mu) \in \mathcal{D}$ and $\mu^2 z(\mu) \in \mathcal{L}(-\infty, \infty)$. According to (11.27) we have $h_1(0) = \frac{1}{3} p_1(0)$. Hence

$$\mathfrak{M}_1(\mu) = (i\mu) + L_{h_1}(\mu), \qquad \mathfrak{M}_2(\mu) = (i\mu)^2 - 2h_1(0) + L_{h_2}(\mu).$$

Using the inversion formula (11.2), we calculate

$$\left.\begin{aligned} h_1(t) &= \frac{d^2}{dt^2}\langle \frac{1}{\pi}\frac{1 - \cos \mu t}{\mu^2}, \mathfrak{M}_1(\mu) - i\mu\rangle, \\ h_2(t) &= \frac{d^2}{dt^2}\langle \frac{1}{\pi}\frac{1 - \cos \mu t}{\mu^2}, \mathfrak{M}_2(\mu) - (i\mu)^2 + 2h_1(0)\rangle. \end{aligned}\right\} \qquad (11.31)$$

Under the conditions of the theorem being fulfilled we obtain from (11.31) that $h_\nu(t) = \widetilde{h}_\nu(t)$, $t \geq 0$, $\nu = 1, 2$. But (11.25) and (11.26) then imply

$$\frac{\partial^3 \widehat{u}(x,t)}{\partial t^3} + \frac{\partial^3 \widehat{u}(x,t)}{\partial x^3} + p_1(x)\frac{\partial \widehat{u}(x,t)}{\partial x} + p_0(x)\widehat{u}(x,t) + \widehat{p}_1(x)\frac{\partial \widetilde{u}(x,t)}{\partial x} + \widehat{p}_0(x)\widetilde{u}(x,t) = 0,$$

$$\left.\frac{\partial^\nu \widehat{u}(x,t)}{\partial x^\nu}\right|_{x=0} = 0, \qquad \nu = 0, 1, 2.$$

For this Cauchy problem we use the Riemann–Fage formula (see [76]) in the vicinity of the point $x = t = 0$, and obtain

$$\begin{aligned}\widehat{u}(x,t) &= -\int_0^x d\xi_1 \int_0^{\xi_1} d\xi_2 \int_0^{\xi_2} \Big(\sum_{\nu=0}^{1} \widehat{p}_\nu(\xi_3)\widetilde{u}_\nu(\xi_3, -t + x + (R_2 - R_1)\xi_1 \\ &\quad +(R_3 - R_2)\xi_2 - R_3\xi_3)\Big) V(0,0,\xi_3, x - \xi_1, \xi_1 - \xi_2, \xi_2)\, d\xi_3, \\ u_\nu &= \frac{\partial^\nu u}{\partial x^\nu},\end{aligned}$$

where V is the Riemann–Fage function. Changing the order of integration, we get

$$\widehat{u}(x,t) = \int_0^x \sum_{\nu=0}^{1} \widehat{p}_\nu(\xi) B_\nu(x,t,\xi)\, d\xi, \tag{11.32}$$

where

$$\begin{aligned}B_\nu(x,t,\xi) &= -\int_\xi^x d\xi_1 \int_\xi^{\xi_1} \widetilde{u}_\nu(\xi, -t + x + (R_2 - R_1)\xi_1 \\ &\quad +(R_3 - R_2)\xi_2 - R_3\xi) V(0,0,\xi, x - \xi_1, \xi_1 - \xi_2, \xi_2)\, d\xi_2, \\ &\quad \nu = 0,1.\end{aligned}$$

Since

$$\left.\frac{\partial^{i+j} B_\nu(x,t,\xi)}{\partial x^i \partial t^j}\right|_{\xi = x} = 0, \qquad i,j = 0,1,$$

it follows from (11.32) that

$$\frac{\partial^{i+j} \widehat{u}(x,t)}{\partial x^i \partial t^j} = \int_0^x \sum_{\nu=0}^{1} \widehat{p}_\nu(\xi) \frac{\partial^{i+j} B_\nu(x,t,\xi)}{\partial x^i \partial t^j}\, d\xi, \qquad i,j = 0,1,2. \tag{11.33}$$

Using (11.28), we get

$$\widehat{p}_1(x) = 3\frac{d}{dx}\widehat{u}(x,x),$$

$$\widehat{p}_0(x) = 3\frac{d}{dx}\big(\widehat{u}_x(x,t)\big|_{t=x}\big) + p_1(x)\widehat{u}(x,x) + 3\widetilde{u}(x,x)\frac{d}{dx}\widehat{u}(x,x).$$

From this and from (11.33) it follows that

$$\widehat{p}_k(x) = \int_0^x \sum_{\nu=0}^{1} A_{k\nu}(x,\xi)\widehat{p}_\nu(\xi)\,d\xi, \qquad k = 0,1,$$

where

$$A_{1\nu}(x,\xi) = 3\frac{d}{dx}B_\nu(x,x,\xi),$$

$$A_{0\nu}(x,\xi) = 3\frac{d}{dx}\left(\frac{\partial B_\nu(x,t,\xi)}{\partial x}\bigg|_{t=x}\right) + p_1(x)B_\nu(x,x,\xi) + 3\widetilde{u}(x,x)\frac{d}{dx}B_\nu(x,x,\xi).$$

Consequently, $\widehat{p}_k(x) = 0$, $k = 0,1$. Theorem 11.2 is proved.

The solution of the inverse problem. Let there be given the WGF $\mathfrak{M}_\nu(\mu)$, $\nu = 1,2$ of the DE (11.3). We construct the functions $h_\nu(t)$, $\nu = 1,2$ by (11.31). Since $u(x,t) = Q(x,t-x)$, $Q(0,s) = 0$, we have

$$h_1(s) = Q_x(0,s), \qquad h_2(s) = Q_{xx}(0,s) - 2Q_{xs}(0,s). \tag{11.34}$$

Note that the functions $h_\nu(s)$ are regular at the point $s = 0$. More precisely, if the functions $q_k(x)$ are regular for $|x| < \delta$, then the functions $h_\nu(s)$ are regular for $|s| < \sqrt{3}\delta$.

Denote $q_0^1(x) = q_0'(x)$, $Q^1(x,s) = Q_x(x,s)$. Using (11.34), (11.19), (11.21) and (11.22), we calculate

$$\begin{aligned} h_{2-\nu}^{(\nu)} &= \frac{1}{3}\sum_{k=2}^{3}\Bigg(\frac{R_k^{\nu+1}}{1-R_k}q_0^1\left(\frac{s}{1-R_k}\right) + R_k^\nu q_1\left(\frac{s}{1-R_k}\right) \\ &\quad + \frac{R_k^{\nu+1}}{1-R_k}\int_0^s \Bigg(q_0^1\left(\frac{s-u}{1-R_k}\right)Q\left(\frac{s-u}{1-R_k},u\right) \\ &\quad + q_0\left(\frac{s-u}{1-R_k}\right)Q^1\left(\frac{s-u}{1-R_k},u\right)\Bigg)\,du \\ &\quad + R_k^\nu\int_0^s q_1\left(\frac{s-u}{1-R_k}\right)Q\left(\frac{s-u}{1-R_k},u\right)du\Bigg), \qquad \nu = 0,1. \end{aligned} \tag{11.35}$$

Having solved (11.35) with respect to the functions $q_0^1(x)$ and $q_1(x)$, we obtain

$$q_1(x) - \sum_{k=2}^{3} R_k^2 \int_0^x (q_0^1(u)Q(u,(1-R_k)(x-u)) \\ +q_0(u)Q^1(u,(1-R_k)(x-u)))\,du = I_1(x), \tag{11.36}$$

$$q_0^1(x) + 3\sum_{k=2}^{3} R_k^2 \int_0^x q_1(u)Q(u,(1-R_k)(x-u))\,du = I_2(x), \tag{11.37}$$

where

$$I_1(x) = 3\sum_{k=2}^{3} \frac{1}{R_k^2 - R_k}(R_k h_1'((1-R_k)x) - h_2((1-R_k)x)),$$

$$I_2(x) = -3\sum_{k=2}^{3} R_k^2(R_k h_1'((1-R_k)x) - h_2((1-R_k)x)).$$

Since $q_0(0) = -p_1(0) = -3h_1(0)$, we get

$$q_0(x) = -3h_1(0) + \int_0^x q_0^1(u)\,du. \tag{11.38}$$

Equality (11.19) can be rewritten as follows

$$\begin{aligned} Q^1(x,s) \;=\; & \frac{1}{3}\Bigg(R_k^2\Bigg(-3h_1(0) + \int_0^{x+s/(1-R_k)} q_0^1(u)\,du\Bigg) \\ & + R_k(1-R_k)\int_0^{x+s/(1-R_k)} q_1(\eta)\,d\eta \\ & + R_k^2 \int_0^s q_0\left(x + \frac{s-u}{1-R_k}\right) Q\left(x+\frac{s-u}{1-R_k}, u\right) du \\ & + R_k(1-R_k)\int_0^s du \int_0^{x+(s-u)/(1-R_k)} q_1(\eta)Q(\eta,u)\,d\eta\Bigg). \end{aligned} \tag{11.39}$$

Let us consider the system of nonlinear integral equations (11.17) and (11.36)–(11.39) with respect to the functions $q_0(x)$, $q_0^1(x)$, $q_1(x)$, $Q(x,s)$ and $Q^1(x,s)$.

We solve this system by the method of successive approximations in a sufficiently small neighborhood of the point $x = s = 0$. Then we obtain the following obvious result.

Lemma 11.2. *Let the functions $h_\nu(s)$, $\nu = 1,2$ be regular for $s = 0$. Then there exists $\delta > 0$ such that the functions $h_\nu(s)$ are regular for $|s| < \sqrt{3}\delta$, and there exist the unique functions $q_0(x)$, $q_0^1(x)$, $q_1(x)$, which are regular for $|x| < \delta$, and the unique functions $Q(x,s)$, $Q^1(x,s)$ which are regular in $\mathcal{F}_\delta$, where all the functions satisfy the system* (11.17), (11.36)–(11.39), *i.e. the system* (11.17), (11.36)–(11.39) *is uniquely solvable in a neighborhood of the point $x = s = 0$. The solution of the system* (11.17), (11.36)–(11.39) *can be found by the method of successive approximations, and $q_0^1(x) = q_0'(x)$, $Q^1(x,s) = Q_x(x,s)$.*

Thus we can construct the solution of the IP via the following algorithm.

Algorithm 11.1. Given the WGF $\mathfrak{M}_\nu(\mu)$, $\nu = 1,2$ of the DE (11.3).

(1) We construct the functions $h_\nu(t)$, $\nu = 1,2$ by formulae (11.31) .

(2) Having solved the system (11.17), (11.36)–(11.39), we find the functions $q_0(x)$, $q_0^1(x)$, $q_1(x)$, $Q(x,s)$, $Q^1(x,s)$.

(3) We construct the functions $p_1(x) = -q_0(x)$, $p_0(x) = q_1(x) - q_0^1(x)$, $|x| < \delta$.

(4) By means of analytic continuation we obtain the functions $p_0(x)$ and $p_1(x)$ for $x > 0$.

11.3. Similar results are also valid for the DE of an arbitrary order

$$\ell y \equiv y^{(n)} + \sum_{k=0}^{n-2} p_k(x) y^{(k)} = \lambda y = (i\mu)^n y, \qquad x > 0. \tag{11.40}$$

In this case

$$q_\nu(x) = \sum_{j=0}^{\nu} (-1)^{j+1} C_{n-2-j}^{\nu-j} p_{n-2-j}^{(\nu-j)}(x), \qquad \nu = \overline{0, n-2},$$

$$B = \left\{ x : |\arg x| < \frac{\pi}{2} - \frac{\pi}{n} \right\}, \qquad R_k = \exp\left(\frac{2(k-1)\pi i}{n} \right), \qquad k = \overline{1,n}.$$

The function $Q(x,s)$ can be determined by the integral equation

$$Q(x,s) = Q_1(x,s) + \sum_{\nu=0}^{n-2}\Bigg(-\frac{1}{n}\int_0^s \frac{(s-u)^\nu}{\nu!}\,du \int_0^x q_\nu(t)Q(t,u)\,dt$$

$$+\sum_{k=2}^{n} \frac{R_k^{n-1-\nu}}{n(1-R_k)} \int_0^s du \int_0^{s-u} \frac{(s-u-\xi)^\nu}{\nu!}$$

$$\times\left(q_\nu\left(\frac{\xi}{1-R_k}+x\right)Q\left(\frac{\xi}{1-R_k}+x,u\right)\right.$$

$$\left.-q_\nu\left(\frac{\xi}{1-R_k}\right)Q\left(\frac{\xi}{1-R_k},u\right)\right)d\xi\Bigg),$$

where

$$Q_1(x,s) = \sum_{\nu=0}^{n-2}\Bigg(-\frac{1}{n}\frac{s^\nu}{\nu!}\int_0^x q_\nu(t)\,dt + \sum_{k=2}^{n}\frac{R_k^{n-1-\nu}}{n(1-R_k)}$$

$$\times\int_0^s \frac{(s-\xi)^\nu}{\nu!}\left(q_\nu\left(\frac{\xi}{1-R_k}+x\right) - q_\nu\left(\frac{\xi}{1-R_k}\right)\right)d\xi\Bigg).$$

The function $u(x,t)$ defined by the equalities $u(x,t) = Q(x,t-x)$, $0 \le x \le t < \infty$; $u(x,t) = 0$, $t < x$ satisfies the following relations

$$\frac{\partial^n u(x,t)}{\partial x^n} + \sum_{\nu=0}^{n-2} p_\nu(x)\frac{\partial^\nu u(x,t)}{\partial x^\nu} = (-1)^n\,\frac{\partial^n u(x,t)}{\partial t^n}, \qquad u(0,t)=0,$$

$$\sum_{m=1}^{n-j} C_{n-m}^{j}\frac{d^{n-m-j}}{dx^{n-m-j}}u_{m-1}(x,x) + \sum_{\nu=j+1}^{n-2} p_\nu(x)\sum_{m=1}^{\nu-j} C_{\nu-m}^{j}\frac{d^{\nu-m-j}}{dx^{\nu-m-j}}u_{m-1}(x,x)$$

$$+(-1)^{n-j}\left.\frac{\partial^{n-j-1}u(x,t)}{\partial t^{n-j-1}}\right|_{t=x} = p_j(x), \qquad j=\overline{0,n-2},$$

where

$$u_\nu(x,t) = \frac{\partial^\nu}{\partial x^\nu}u(x,t), \qquad C_k^j = \frac{k!}{j!(k-j)!}.$$

The generalized Weyl solution $\Phi(x,\mu)$ is defined by the formula $\Phi(x,\mu) = \exp(i\mu x) + L_u(\mu)$, i.e.

$$\langle z(\mu), \Phi(x,\mu)\rangle = \int_{-\infty}^{\infty} z(\mu)\exp(i\mu x)\,d\mu + \int_{x}^{\infty} u(x,t)\,dt \int_{-\infty}^{\infty} z(\mu)\exp(i\mu t)\,d\mu, \qquad z(\mu)\in\mathcal{D}.$$

and $\ell\Phi(x,\mu) - (i\mu)^n\Phi(x,\mu) = 0$, $\Phi(0,\mu) = 1$. The functions $\mathfrak{M}_\nu(\mu) = \Phi^{(\nu)}(0,\mu)$, $\nu = \overline{1, n-1}$ are called the WGF of (11.40).

The inverse problem here is formulated as follows: find the coefficients $\{p_k(x)\}_{k=\overline{0,n-2}}$ of (11.40) via the given WGF $\{\mathfrak{M}_\nu(\mu)\}_{\nu=\overline{1,n-1}}$. The solution of this inverse problem can be obtained in exactly the same way as in the case of $n = 3$.

12. Discrete inverse problems. Applications to differential operators

Here we study an inverse problem of spectral analysis for operators of a triangular structure in an abstract space. The triangular structure, considered here, is an object of a very general form. As shown below, various inverse problems for difference, differential, integro-differential operators and pencils of operators can be reduced to the inverse problem for the triangular structure. In 12.1 the abstract inverse problem for the triangular structure is studied. We give the definition of operators of the triangular structure and consider their canonical form. We obtain the solution of the inverse problem for the triangular structure by its Weyl function. Futher, items 12.2 and 12.3 are devoted to applications of the result, obtained in 12.1, to the most important particular cases of the triangular structure — difference and differential operators.

12.1 Triangular structures

12.1.1. Preliminary information. Let LS denote a linear space, and LTS denote a complete separable linear topological space, which has a countable neighbourhood base of zero. If Γ_1 and Γ_2 are LS, then $\Gamma_1 \to \Gamma_2$ denotes the LS of linear operators mapping Γ_1 in Γ_2. For $A \in \Gamma_1 \to \Gamma_2$ let $\Delta^1(A)$ be the domain of definition of A, and $\Delta^2(A) = A(\Delta^1(A))$. An operator $A \in \Gamma_1 \to \Gamma_2$ is called non-singular (we shall write $A\#0$), if $\ker A = \{0\}$, i.e. the equation $Ax = 0$ only has the trivial solution. If Γ_1 and Γ_2 are LTS, then we denote by $[\Gamma_1 \to \Gamma_2]$ the set of continuous operators (possibly nonlinear) mapping Γ_1 in Γ_2, and by $\Gamma_1 \times \Gamma_2$ the LTS of vectors $x = [x_1, x_2]$, $x_j \in \Gamma_j$ with the coordinate convergence. If Γ is a LTS, then Γ^∞ denotes the LTS of sequences $x = [x_j]_{j\geq 0}$, $x_j \in \Gamma$ with the coordinate convergence.

Let Γ be a LS, $\widehat{\Gamma} = \Gamma \to \Gamma$, and let Λ be the set of polynomials of the form

$$F(\lambda) = \sum_{k=-j_1}^{j_2} F_k\lambda^k, \qquad F_k \in \widehat{\Gamma}$$

(with any natural j_1 and j_2). Set $\mathcal{F}(\Gamma) = \Lambda \to \Gamma$. Elements of the LS $\mathcal{F}(\Gamma)$ are called generalized functions (GF). If $P \in \mathcal{F}(\Gamma)$, then $P_{k+1} = (E\lambda^k, P) \in \Gamma$ are called the moments of P (E is the identity operator, and $(.,P)$ denotes the action of P). It is clear that a GF $P \in \mathcal{F}(\Gamma)$ is uniquely determined by its moments via $(F(\lambda), P) = \sum_k F_k P_{k+1}$, $F(\lambda) \in \Lambda$. It is convenient to write a GF $P \in \mathcal{F}(\Gamma)$ as the formal series

$$P(\lambda) = \sum_{k=-\infty}^{\infty} \frac{P_k}{\lambda^k}, \qquad P_k \in \Gamma. \tag{12.1.1}$$

We can multiply $P \in \mathcal{F}(\Gamma)$ by elements of Λ via $(F(\lambda), G(\lambda)P) = (F(\lambda)G(\lambda), P)$, $F(\lambda)$, $G(\lambda) \in \Lambda$. This means that the series (12.1.1) is formally multiplied by $G(\lambda)$. Let us denote by $\mathcal{F}^+(\Gamma)$ the LS of GF $P \in \mathcal{F}(\Gamma)$ such that $P_k = 0$ for $k < 0$. If Γ is LTS, then $\mathcal{F}[\Gamma]$ denotes the LTS of GF (12.1.1) with the convergence: $P^j \overset{\mathcal{F}[\Gamma]}{\longrightarrow} P$ if $P_k^j \overset{\Gamma}{\longrightarrow} P_k$ for all k. Let $\mathcal{F}^+[\Gamma]$ be the LTS of GF $P \in \mathcal{F}(\Gamma)$ such that $P_k = 0$ for $k < 0$.

Now let Γ, Γ_0, Γ_1, Γ_2 be LTS. For $r \in \Gamma_0$ we consider the following families of *GF*: $A = A(r) \in \mathcal{F}^+(\Gamma_1 \to \Gamma_2)$, $S = S(r) \in \mathcal{F}^+(\Gamma_1 \to \Gamma)$, $f = f(r) \in \mathcal{F}^+(\Gamma_2)$, with the moments A_k, S_k, f_k, respectively. Suppose the operator $A_0(r)$ is invertible for each fixed $r \in \Gamma_0$, and suppose the operators A_0^{-1}, A_k $(k \ge 1)$, S_k, f_k $(k \ge 0)$ are continuous with respect to all their arguments in the aggregate, i.e. $A_0^{-1} \in [\Gamma_0 \times \Gamma_2 \to \Gamma_1]$, $A_k \in [\Gamma_0 \times \Gamma_1 \to \Gamma_2]$ $(k \ge 1)$ $S_k \in [\Gamma_0 \times \Gamma_1 \to \Gamma]$, $f_k \in [\Gamma_0 \to \Gamma_2]$, $(k \ge 0)$. Without loss of generality we can take $\Delta^2(A_0) = \Gamma_2$, $\Delta^1(A_k) = \Gamma_1$ $(k \ge 1)$, $\Delta^1(S_k) = \Gamma_1$ $(k \ge 0)$ for each fixed $r \in \Gamma_0$. The triplet (A, S, f) is called the generalized spectral pencil (GSP).

For $r \in \Gamma_0$ we define the families of operators $\widehat{A} = \widehat{A}(r) \in \mathcal{F}^+(\Gamma_1) \to \mathcal{F}^+(\Gamma_2)$, $\widehat{S} = \widehat{S}(r) \in \mathcal{F}^+(\Gamma_1) \to \mathcal{F}^+(\Gamma)$ by

$$\widehat{A}y = \sum_{k=0}^{\infty} \frac{1}{\lambda^k} \sum_{\nu=0}^{k} A_\nu y_{k-\nu}, \qquad \widehat{S}y = \sum_{k=0}^{\infty} \frac{1}{\lambda^k} \sum_{\nu=0}^{k} S_\nu y_{k-\nu},$$

where

$$y = \sum_{k=0}^{\infty} \frac{1}{\lambda^k} y_k \in \mathcal{F}^+(\Gamma_1).$$

Lemma 12.1. *For each fixed $r \in \Gamma_0$ the operator $\widehat{A} = \widehat{A}(r)$ is invertible, and $\Delta^2(\widehat{A}(r)) = \mathcal{F}^+(\Gamma_2)$, $\widehat{A}^{-1} \in [\Gamma_0 \times \mathcal{F}^+[\Gamma_2] \to \mathcal{F}^+[\Gamma_1]]$, $\widehat{S} \in [\Gamma_0 \times \mathcal{F}^+[\Gamma_1] \to \mathcal{F}^+[\Gamma]]$, $f \in [\Gamma_0 \to \mathcal{F}^+[\Gamma_2]]$.*

Proof. The equation $\widehat{A}y = f$ is equivalent to the system

$$\sum_{\nu=0}^{k} A_\nu y_{k-\nu} = f_k, \qquad k \ge 0.$$

Hence we get

$$y_k = A_0^{-1}\left(f_k - \sum_{\nu=1}^{k} A_\nu y_{k-\nu}\right). \tag{12.1.2}$$

Consequently, for each $f \in \mathcal{F}^+(\Gamma_2)$ the equation $\widehat{A}y = f$ has a unique solution constructed via (12.1.2). Hence, and from the continuity of the operators A_0^{-1}, A_k, S_k, f_k, we obtain the assertion of Lemma 12.1.

Definition 12.1. The solution

$$\varphi = \sum_{k=0}^{\infty} \frac{1}{\lambda^k}\varphi_k \in \mathcal{F}^+[\Gamma_1]$$

of the equation $\widehat{A}\varphi = f$ is called the WS for the GSP (A, S, f). The GF

$$\Pi = \sum_{k=0}^{\infty} \frac{1}{\lambda^k}\Pi_k \in \mathcal{F}^+[\Gamma],$$

defined by $\Pi = \widehat{S}\varphi$, is called the WF for the GSP (A, S, f).

The IP is formulated as follows: given the WF $\Pi \in \mathcal{F}^+[\Gamma]$, construct $r \in \Gamma_0$. In this IP the operator A characterizes a mathemetical model of an object, $r \in \Gamma_0$ describes parameters of the object, f is an extertnal action, φ is the object reaction to the external action, Π is the result of measurement.

It is clear that solution of the IP is equivalent to solution of the equation $\Pi = \psi(r)$, where $\psi = \widehat{S}\widehat{A}^{-1}f \in [\Gamma_0 \to \mathcal{F}^+[\Gamma]]$. The operator ψ is called the GSP operator, and the operators $\psi_k = \psi_k(r) \in [\Gamma_0 \to \Gamma]$, defined by

$$\psi_k = \sum_{\nu=0}^{k} S_\nu \varphi_{k-\nu},$$

are called the moments of ψ.

Now we want to give a number of simple examples of GSP.

Example 1 (Linear pencil). Let $\Gamma_1 = \Gamma_2$ and let the families $B = B(r) \in \Gamma_1 \to \Gamma_1$, $\sigma = \sigma(r) \in \Gamma_1 \to \Gamma$, $g = g(r) \in \Gamma_1$ be given for $r \in \Gamma_0$. Consider the equation

$$\left(E - \frac{1}{\lambda}B\right)y = g. \tag{12.1.3}$$

Denote

$$\varphi = \sum_{k=0}^{\infty} \frac{1}{\lambda^k}\varphi_k \in \mathcal{F}^+[\Gamma_1], \qquad \Pi = \sum_{k=0}^{\infty} \frac{1}{\lambda^k}\Pi_k \in \mathcal{F}^+[\Gamma],$$

where $\varphi_k = B^k g$, $\Pi_k = \sigma B^k g$. In particular, if Γ, Γ_1 are Banach spaces, and B, σ are bounded operators, then $\varphi = \left(E - \frac{1}{\lambda}B\right)^{-1} g$ and $\Pi = \sigma\varphi$ are regular for $|\lambda| > ||B||$.

The IP of determining r from the given Π in the particular case of the IP for the GSP (A, S, f) with $A_0 = E$, $A_1 = -B$, $A_k = 0$ $(k \geq 2)$, $f_0 = g$, $S_0 = \sigma$, $S_k = f_k = 0$ $(k \geq 1)$. The triplet (B, σ, g) is called the linear GSP (LGSP).

Example 2. Consider the equation

$$b_\nu y_{\nu+1} + a_\nu y_\nu + b_{\nu-1} y_{\nu-1} = \lambda y_\nu, \qquad \nu \geq 1, \qquad (12.1.4)$$

where a_ν, b_ν are real numbers, and $b_\nu > 0$. It is known that for $\operatorname{Im}\lambda \neq 0$ there exists a solution $\Phi_\nu(\lambda) \in \ell_2$, $\Phi_0(\lambda) = 1$ of (12.1.4). Denote $M(\lambda) = \Phi_1(\lambda)$ The asymptotic formulas

$$\Phi_\nu(\lambda) = \sum_{k=\nu}^{\infty} \frac{1}{\lambda^k} \Phi_{k\nu}, \qquad M(\lambda) = \sum_{k=1}^{\infty} \frac{1}{\lambda^k} M_k, \qquad \lambda \to \infty$$

are valid, and M_k are the moments of a spectral function. The coefficients $\{a_\nu, b_\nu\}$ are uniquely determined by $\{M_k\}_{k \geq 1}$.

Consider the LGSP (B, σ, g), where $\Gamma_1 = \Gamma_2 = E_1^\infty$, $\Gamma = E_1$, $\Gamma_0 = E_2^\infty$, $y = [y_k]_{k \geq 0}$, $g = [\delta_{k0}]_{k \geq 0}$, $\sigma y = y_1$,

$$B = \begin{pmatrix} 0 & 0 & 0 & \dots & \dots \\ b_0 & a_1 & b_1 & \dots & \dots \\ 0 & b_1 & a_2 & b_2 & \dots \\ 0 & 0 & b_2 & a_3 & b_3 \\ \dots & \dots & \dots & \dots & \dots \end{pmatrix}$$

Here and in the sequel, E_p is the p-dimensional Euclidean space. Then $\varphi_k = [\Phi_{k\nu}]_{\nu \geq 0} \in \Gamma_1$, $\prod_k = M_k \in \Gamma$ are the moments of the WS and the WF, respectively. Thus, in this case the IP for the LGSP corresponds to the problem of recovery of $\{a_\nu, b_\nu\}$ from the given moments $\{M_k\}_{k \geq 1}$.

Example 3. Consider the Sturm–Liouville differential operator S:

$$\mathcal{L}_0 y \equiv -y'' + q(x) y, \qquad q(x) \in C[0, \pi],$$

$$U(y) \equiv y'(0) - hy(0) = 0; \qquad V(y) \equiv y'(\pi) + Hy(\pi) = 0.$$

Let $\Phi(x, \mu)$ be the solution of $\mathcal{L}_0 \Phi = \mu \Phi$, $x \in (0, \pi)$ under the conditions $U(\Phi) = 1$, $V(\Phi) = 0$, and let $M(\mu) = \Phi(0, \mu)$. Then $M(\mu)$ is the WF of S. If zero is not an eigenvalue, then $\Phi(x, \mu)$ satisfies $y - \mu G y = G(x, 0)$, where $G(x, t)$ is the kernel of the integral operator $G = S^{-1}$. In the other words, $\Pi = M(\mu)$ and $\varphi = \Phi(x, \mu)$ are the WF and WS of the LGSP (B, σ, g), respectively, where $\lambda = \frac{1}{\mu}$, $\Gamma_0 = \Gamma_1 = \Gamma_2 = C[0, \pi]$, $\Gamma = E_1$, $B = G$, $g = G(x, 0)$, $\sigma y(x) = y(0)$, $r = g(x)$. Thus, in this case, the IP for the GSP corresponds to the problem of determining the Sturm–Liouville operator from the WF $M(\mu)$.

Other more complicated examples of GSP are given in Section 12.2–12.3.

12.1.2. R-structures. Let γ_0, γ_1, γ_2, Γ be LTS, and let $\Gamma_\nu = \gamma_\nu^\infty$ $(\nu = 0, 1, 2)$ $r = [r_k]_{k \geq 0}$, $r_k \in \gamma_0$. Denote $x_p = [r_k]_{k=\overline{0,p}}$, $x'_p = [r_k]_{k > p}$. For $r \in \Gamma_0$ we consider the families $\alpha_{\nu j}^k = \alpha_{\nu j}^k(x_{\nu+k}) \in \gamma_1 \to \gamma_2$, $S_{kj} = S_{kj}(x_k) \in \gamma_1 \to \Gamma$, $f_{k\nu} = f_{k\nu}(x_{k+\nu}) \in \gamma_2$, $k, \nu, j \geq 0$, where $\alpha_{\nu j}^k = 0$ $(j > \nu + k)$, $S_{kj} = 0$ $(j > k)$, and for each fixed $r \in \Gamma_0$ the operators α_{jj}^0 are invertible. Suppose $(\alpha_{jj}^0)^{-1} \in [\gamma_0^{j+1} \times \gamma_2 \to \gamma_1]$, $\alpha_{\nu j}^k \in [\gamma_0^{\nu+k+1} \times \gamma_1 \to \gamma_2]$ $(k+|\nu-j| > 0)$, $f_{k\nu} \in [\gamma_0^{\nu+k+1} \to \gamma_2]$, $S_{kj} \in [\gamma_0^{k+1} \times \gamma_1 \to \Gamma]$. Without loss of generality we can take $f_{00} \neq 0$, $\Delta^2(\alpha_{jj}^0) = \gamma_2$, $\Delta^1(\alpha_{\nu j}^k) = \gamma_1$, $(k + |\nu - j| > 0)$, $\Delta^1(S_{kj}) = \gamma_1$ for each fixed $r \in \Gamma_0$.

For $r \in \Gamma_0$ we consider $f_k = [f_{k\nu}]_{\nu \geq 0} \in \Gamma_2$, $A_k \in \Gamma_1 \to \Gamma_2$, $S_k \in \Gamma_1 \to \Gamma$, where

$$A_k y = \left[\sum_{j=0}^{k+\nu} \alpha_{\nu j}^k y_j \right]_{\nu \geq 0}, \qquad S_k y = \sum_{j=0}^{k} S_{kj} y_j, \qquad y = [y_j]_{j \geq 0} \in \Gamma_1. \tag{12.1.5}$$

Lemma 12.2. *For each fixed $r \in \Gamma_0$ the operator A_0 is invertible, and*

$$\Delta^2(A_0) = \Gamma_2, \qquad A_0^{-1} \in [\Gamma_0 \times \Gamma_2 \to \Gamma_1], \qquad A_k \in [\Gamma_0 \times \Gamma_1 \to \Gamma_2], \qquad (k \geq 1),$$
$$S_k \in [\Gamma_0 \times \Gamma_1 \to \Gamma], \qquad f_k \in [\Gamma_0 \to \Gamma_2] \qquad (k \geq 0).$$

Proof. The equation $A_0 y = g$, $g = [g_\nu]_{\nu \geq 0} \in \Gamma_2$ is equivalent to the system

$$\sum_{j=0}^{\nu} \alpha_{\nu j}^0 y_j = g_\nu, \qquad \nu \geq 0.$$

Hence one has the recurrence relation

$$y_\nu = (\alpha_{\nu\nu}^0)^{-1} \left(g_\nu - \sum_{j=0}^{\nu-1} \alpha_{\nu j}^0 y_j \right). \tag{12.1.6}$$

Therefore, for each $g \in \Gamma_2$ the equation $A_0 y = g$ has a unique solution constructed via (12.1.6). Hence, and from the continuity of the operators $(\alpha_{jj}^0)^{-1}$, $\alpha_{\nu j}^k$, S_{kj}, $f_{k\nu}$, we obtain the assertion of Lemma 12.2.

The GSP (A, S, f), where

$$A = \sum_{k=0}^{\infty} \frac{A_k}{\lambda^k} \in \mathcal{F}^+(\Gamma_1 \to \Gamma_2), \qquad S = \sum_{k=0}^{\infty} \frac{S_k}{\lambda^k} \in \mathcal{F}^+(\Gamma_1 \to \Gamma),$$

$$f = \sum_{k=0}^{\infty} \frac{f_k}{\lambda^k} \in \mathcal{F}^+(\Gamma_2),$$

and A_k, S_k are defined via (12.1.5), is called the R-structure.

Let

$$\varphi = \sum_{k=0}^{\infty} \frac{\varphi_k}{\lambda^k} \in \mathcal{F}^+[\Gamma_1], \qquad \varphi_k = [\varphi_{k\nu}]_{\nu \geq 0} \in \Gamma_1, \qquad \Pi = \sum_{k=0}^{\infty} \frac{\Pi_k}{\lambda^k} \in \mathcal{F}^+[\Gamma],$$

be the WS and the WF of the R-structure (A, S, f), respectively. Then the moments of the WS and the WF satisfy

$$\sum_{s=0}^{k}\sum_{j=0}^{s+\nu} \alpha_{\nu j}^{s} \varphi_{k-s,j} = f_{k\nu}, \qquad \sum_{s=0}^{k}\sum_{j=0}^{s} S_{sj}\varphi_{k-s,j} = \Pi_k, \qquad k,\nu \geq 0. \tag{12.1.7}$$

For $r \in \gamma_0$ we consider

$$\widetilde{\varphi}^{\ell} = [\varphi_{\ell-\nu,\nu}]_{\nu=\overline{0,\ell}} \in \gamma_1^{\ell+1}, \qquad \widetilde{f}^{\ell} = [f_{\ell-\nu,\nu}]_{\nu=\overline{0,\ell}} \in \gamma_2^{\ell+1},$$

$$\widetilde{S}^{\ell\mu} \in \gamma_1^{\mu+1} \to \Gamma, \qquad \widetilde{A}^{\ell\mu} \in \gamma_1^{\mu+1} \to \gamma_2^{\ell+1}, \qquad \mu = \overline{0,\ell},$$

where

$$\widetilde{S}^{\ell\mu}\widetilde{y}^{\mu} = \sum_{j=0}^{\mu} S_{\ell-\mu+j,j}y_j, \qquad \widetilde{A}^{\ell\mu}\widetilde{y}^{\mu} = \left[\sum_{j=0}^{\mu} \alpha_{\nu j}^{\ell-\mu+j-\nu} y_j\right]_{\nu=\overline{0,\ell}},$$

$\widetilde{y}^{\mu} = [y_j]_{j=\overline{0,\mu}} \in \gamma_1^{\mu+1}$, $\alpha_{\nu j}^{k} = 0$ for $k < 0$. It is clear that for each fixed $r \in \Gamma_0$ the operators $\widetilde{A}^{\ell\ell}$ are invertible, and $\Delta^2(\widetilde{A}^{\ell\ell}) = \gamma_1^{\ell+1}$. The operators $\widetilde{f}^{\ell}$, $\widetilde{S}^{\ell\mu}$, $\widetilde{A}^{\ell\mu}$ do not depend on x'_{ℓ} and

$$\widetilde{f}^{\ell} = \widetilde{f}^{\ell}(x_\ell) \in [\gamma_0^{\ell+1} \to \gamma_2^{\ell+1}], \qquad \widetilde{S}^{\ell\mu} = \widetilde{S}^{\ell\mu}(x_\ell) \in [\gamma_0^{\ell+1} \times \gamma_1^{\mu+1} \to \Gamma],$$

$$(\widetilde{A}^{\ell\ell})^{-1} = (\widetilde{A}^{\ell\ell}(x_\ell))^{-1} \in [\gamma_0^{\ell+1} \times \gamma_2^{\ell+1} \to \gamma_1^{\ell+1}],$$

$$\widetilde{A}^{\ell\mu} = \widetilde{A}^{\ell\mu}(x_\ell) \in [\gamma_0^{\ell+1} \times \gamma_1^{\mu+1} \to \gamma_2^{\ell+1}], \qquad (\mu < \ell).$$

We can transform (12.1.7) to the form

$$\sum_{\mu=0}^{\ell} \widetilde{A}^{\ell\mu}\widetilde{\varphi}^{\mu} = \widetilde{f}^{\ell}, \qquad \sum_{\mu=0}^{\ell} \widetilde{S}^{\ell\mu}\widetilde{\varphi}^{\mu} = \Pi_\ell, \qquad \ell \geq 0. \tag{12.1.8}$$

Since $\widetilde{A}^{\ell\ell}$ are invertible, we find

$$\widetilde{\varphi}^{\ell} = (\widetilde{A}^{\ell\ell})^{-1}\left(\widetilde{f}^{\ell} - \sum_{\mu=0}^{\ell-1} \widetilde{A}^{\ell\mu}\widetilde{\varphi}^{\mu}\right), \qquad \ell \geq 0. \tag{12.1.9}$$

Define the operators ψ_ℓ by

$$\psi_\ell = \sum_{\mu=0}^{\ell} \widetilde{S}^{\ell\mu}\widetilde{\varphi}^{\mu}, \quad \ell \geq 0. \tag{12.1.10}$$

With the help of (12.1.9) and (12.1.10) it is easily proved by the method of induction that $\widetilde{\varphi}^\ell$ and ψ_ℓ do not depend on x'_ℓ, and $\widetilde{\varphi}^\ell = \widetilde{\varphi}^\ell(x_\ell) \in [\gamma_0^{\ell+1} \to \gamma_1^{\ell+1}]$, $\psi_\ell = \psi_\ell(x_\ell) \in [\gamma_0^{\ell+1} \to \Gamma]$. Thus, the following theorem holds

Theorem 12.1. *Let $\psi = \psi(r)$ with the moments $\psi_\ell(r)$, $\ell \geq 0$ be the GSP operator for a R-structure (A, S, f). Then*

(1) *The operators $\psi_\ell = \psi_\ell(x_\ell)$ do not depend on x'_ℓ. They are constructed via (12.1.9), (12.1.10), and $\psi_\ell(x_\ell) \in [\gamma_0^{\ell+1} \to \Gamma]$. The equation of the inverse problem $\Pi = \psi(r)$ has the form*

$$\Pi_\ell = \psi_\ell(r_0, r_1, \dots, r_1), \qquad \ell \geq 0. \tag{12.1.11}$$

(2) *For the GF $\Pi \in \mathcal{F}^+[\Gamma]$ to be the WF for a R-structure (A, S, f) it is necessary and sufficient that $\Pi \in \Delta^2(\psi)$.*

(3) *If the operator $\psi(r)$ is invertible, then the solution of the IP is unique. In particular, if for any fixed r_0, r_1, ..., $r_{\ell-1}$, $\ell \geq 0$ the operators $\psi_\ell(x_{\ell-1}, r_\ell)$ are invertible with respect to r_ℓ, then the solution of the IP is unique.*

We observe that the GSP operator for the R-structure has the triangular form $\psi_\ell = \psi_\ell(r_0, r_1, \dots, r_\ell)$, i.e. the moments ψ_ℓ do not depend on x'_ℓ (we shall call for this property the V-property). The construction of $r = [r_k]_{k\geq 0}$ can be realized by solution of (12.1.11) with respect to r_ℓ successively for $\ell = 0, 1, 2, \dots$. We note that the R-structure is a certain canonical form of discrete GSP with the V-property. For simplicity, we shall explain this remark in the particular case of the LGSP (12.1.3) in a sequence space.

Let us consider the LGSP (B, σ, g), where $\Gamma_0 = \Gamma_1 = \Gamma_2 = E_1^\infty$, $g = [g_\nu]_{\nu\geq 0}$, $\sigma = [\sigma_\nu]^T_{\nu\geq 0}$, $\sigma_0 \neq 0$, $B = [\alpha_{\nu j}]_{\nu,j\geq 0}$ (T denotes transposition, i.e. g is the column, σ is the row). We denote the rows of B by $b_\nu = [\alpha_{\nu j}]^T_{j\geq 0}$. The moments ψ_k, $k \geq 0$ of the GSP operator are defined by $\psi_k = \sigma\varphi_k$, $\varphi_k = B^k g$.

The LGSP is called non-degenerate, if for any $p \geq 0$ and x_p the operator $\psi = \psi(x_p, x'_p)$ depends on x'_p, i.e. there exist x'_p and $\widetilde{x}'_p$ such that $\psi(x_p, x'_p) \neq \psi(x_p, \widetilde{x}'_p)$. The family of operators $B(r)$ is called non-degenerate (for a given $\sigma(r)$), if there exists a $g \in \Gamma_1$ such that the LGSP (B, σ, g) is non-degenerate. Denote the set of elements $g = g(r) \in [\Gamma_0 \to \Gamma_1]$ such that g_k do not depend on x'_k by Ω.

Theorem 12.2. *Let $B(r)$ be non-degenerate. For the operators ψ_ℓ, $\ell \geq 0$ not to depend on x'_ℓ for all $g(r) \in \Omega$ (i.e. for the LGSP (B, σ, g) to have the V-property) it is necessary and sufficient that $\alpha_{\nu j} = 0$ $(j > \nu + 1)$, $\sigma_k = 0$ $(k \geq 1)$, and $\alpha_{\nu j}$ do not depend on $x'_{\nu+1}$, and σ_0 does not depend on x'_0 (i.e. the LGSP (B, σ, g) is the R-structure).*

Proof. The sufficiency has been proved above. We shall prove the necessity. We assume that the operators ψ_ℓ, $\ell \geq 0$ do not depend on x'_ℓ for all $g(r) \in \Omega$. Since $\psi_0 = \sigma g$, we obtain $\sigma_k = 0$ $(k \geq 1)$, $\sigma_0 \neq 0$, and σ_0 does not depend on x'_0.

Let us show by induction that b_ν, $b_j\varphi_{\nu-j}$ $(j = \overline{0, \nu})$ do not depend on $x'_{\nu+1}$, and $\alpha_{\nu j} = 0$ $(j > \nu + 1)$, $\alpha_{\nu,\nu+1}(x_{\nu+1}) \neq 0$ $(*)$. Since $\psi_1 = \sigma_0 b_0 g$, we get that $b_0 g$,

b_0 do not depend on x_1', and $\alpha_{0j} = 0$ $(j > 1)$. It can be shown that $\alpha_{01}(x_1) \neq 0$. Indeed, if we assume that $\alpha_{01}(x_1^*) = 0$ for a certain x_1^*, then $\psi_k = \sigma_0(\alpha_{00})^k g_0$, $k \geq 0$. Consequently the LGSP operator $\psi = \psi(x_1^*)$ does not depend on x_1' for all g.This contradicts the non-degeneracy of $B(r)$. Thus, $(*)$ is proved for $\nu = 0$.

Now we assume that $(*)$ is proved for $\nu = 0, \dots, \mu - 1$. Since $\psi_{\mu+1} = \sigma_0 b_0 \varphi_\mu$ one sees that $b_0 \varphi_\mu$ does not depend on $x'_{\mu+1}$. Using the assumption of induction we have

$$b_j \varphi_{\mu-j} = \sum_{k=0}^{j+1} \alpha_{jk} b_k \varphi_{\mu-j-1}, \qquad j = \overline{0, \mu - 1}.$$

Hence we find successively for $j = 1, 2, \dots$ that $b_j \varphi_{\mu-j}$ do not depend on $x'_{\mu+1}$. Consequently, by virtue of the arbitrariness of $g \in \Omega$, we obtain that b_μ does not depend on $x'_{\mu+1}$, and $\alpha_{\mu j} = 0$, $j > \mu + 1$. It can be shown that $\alpha_{\mu,\mu+1}(x_{\mu+1}) \neq 0$. Indeed, if we assume that $\alpha_{\mu,\mu+1}(x^*_{\mu+1}) = 0$ for a certain $x^*_{\mu+1}$, then the LGSP operator $\psi = \psi(x^*_{\mu+1})$ does not depend on $x'_{\mu+1}$ for all g. This contradicts the non-degeneracy of $B(r)$. Thus, $(*)$ is proved for $\nu = \mu$. Theorem 12.2 is proved.

12.1.3. Now we will study the R_1-structure which is the important particular case of the R-structure. For $r = [r_k]_{k \geq 0} \in \Gamma_0$ we consider the families $\beta_\nu^k = \beta_\nu^k(x_{\nu+k}) \in \gamma_1 \to \gamma_2$, $z_{k\nu} = z_{k\nu}(x_{\nu+k}) \in \gamma_2$ $(k + \nu > 0)$, $c_k = c_k(x_k) \in \gamma_1 \to \Gamma$ $(k \geq 0)$, and assume that for fixed $x_{\nu+k-1} \in \gamma_0^{\nu+k}$ the operators $\beta_\nu^k \in \gamma_0 \to (\gamma_1 \to \gamma_2)$, $z_{k\nu} \in \gamma_0 \to \gamma_2$ are linear with respect to $r_{\nu+k}$, and for fixed $x_{k-1} \in \gamma_0^k$ the operators $c_k \in \gamma_0 \to (\gamma_1 \to \Gamma)$ are linear with respect to r_k. We also assume that

$$\beta_\nu^k \in [\gamma_0^{k+\nu+1} \times \gamma_1 \to \gamma_2], \qquad c_k \in [\gamma_0^{k+1} \times \gamma_1 \to \Gamma], \qquad z_{k\nu} \in [\gamma_0^{k+\nu+1} \to \gamma_2].$$

Let $\alpha_{\nu 0}^k = \beta_\nu^k + \alpha_{\nu 0}^{*,k}$, $f_{k\nu} = z_{k\nu} + f_{k\nu}^*$ $(k + \nu > 0)$, $S_{k0} = c_k + S_{k0}^*$ $(k \geq 0)$ and let $\alpha_{\nu 0}^{*,k}$, $f_{k\nu}^{*,k}$, $\alpha_{\nu j}^k$ $(j \geq 1)$ not depend on $r_{\nu+k}$ and let S_{k0}^*, S_{kj} $(j \geq 1)$ not depend on r_k. The R-structure with these properties is called the R_1-structure.

For fixed $x_{\ell-1} \in \gamma_0^\ell$ we define the operators

$$c_\ell^* r_\ell = c_\ell(\alpha_{00}^0)^{-1} f_{00}, \qquad \widetilde{\theta}^\ell r_\ell = [\beta_\nu^{\ell-\nu}(\alpha_{00}^0)^{-1} f_{00} - z_{\ell-\nu,\nu}]_{\nu=\overline{0,\ell}}\ .$$

The operators c_ℓ^*, $\widetilde{\theta}^\ell$ are linear with respect to r_ℓ. For R_1-structures we can transform (12.1.8) to the form

$$\left.\begin{aligned} c_1^* r_\ell + \widetilde{S}^{\ell\ell}\widetilde{\varphi}^\ell &= \Pi_\ell - \widetilde{S}_*^{\ell 0}\widetilde{\varphi}^0 - \sum_{\mu=1}^{\ell-1} \widetilde{S}^{\ell\mu}\widetilde{\varphi}^\mu, \qquad \ell \geq 1, \\ \widetilde{\theta}^\ell r_\ell + \widetilde{A}^{\ell\ell}\widetilde{\varphi}^\ell &= \widetilde{f}_*^\ell - \widetilde{A}_*^{\ell 0}\widetilde{\varphi}^0 - \sum_{\mu=1}^{\ell-1} \widetilde{A}^{\ell\mu}\widetilde{\varphi}^\mu, \qquad \ell \geq 1, \end{aligned}\right\} \tag{12.1.12}$$

where $\widetilde{A}_*^{\ell 0}$, $\widetilde{S}_*^{\ell 0}$, $\widetilde{f}_*^\ell$ are constructed by $\alpha_{\nu 0}^{*,k}$, S_{k0}^*, $f_{k\nu}^*$ in the same way as $\widetilde{A}^{\ell 0}$, $\widetilde{S}^{\ell 0}$, $\widetilde{f}^\ell$ are constructed by $\alpha_{\nu 0}^k$, S_{k0}, $f_{k\nu}$. From (12.1.9) and (12.1.10) we obtain for R_1-structures

$$\psi_\ell = d_\ell r_\ell + \psi_\ell^*, \qquad \ell \geq 1, \tag{12.1.13}$$

where

$$d_\ell = c_\ell^* - \widetilde{S}^{\ell\ell}(\widetilde{A}^{\ell\ell})^{-1}\widetilde{\theta}^\ell, \tag{12.1.14}$$

$$\psi_\ell^* = \widetilde{S}^{\ell\ell}\widetilde{h}^\ell + \widetilde{S}_*^{\ell 0}\widetilde{\varphi}^0 + \sum_{\mu=1}^{\ell-1}\widetilde{S}^{\ell\mu}(\widetilde{h}^\mu - (\widetilde{A}^{\mu\mu})^{-1}\widetilde{\theta}^\mu r_\mu), \tag{12.1.15}$$

$$\widetilde{h}^\ell = (\widetilde{A}^{\ell\ell})^{-1}\{\widetilde{f}_*^\ell - \widetilde{A}_*^{\ell 0}\widetilde{\varphi}^0 - \sum_{\mu=1}^{\ell-1}\widetilde{A}^{\ell\mu}(\widetilde{h}^\mu - (\widetilde{A}^{\mu\mu})^{-1}\widetilde{\theta}^\mu r_\mu)\}. \tag{12.1.16}$$

Let us denote

$$D_\ell = \begin{bmatrix} c_\ell^* & \widetilde{S}^{\ell\ell} \\ \widetilde{\theta}^\ell & \widetilde{A}^{\ell\ell} \end{bmatrix}.$$

For each fixed $x_{\ell-1} \in \gamma_0^\ell$ the operators $d_\ell = d_\ell(x_{\ell-1}) \in \gamma_0 \to \Gamma$ and $D_\ell = D_\ell(x_{\ell-1}) \in \gamma_0 \times \gamma_1^{\ell+1} \to \Gamma \times \gamma_1^{\ell+1}$ are linear with respect to r_ℓ and $[r_\ell, \widetilde{\varphi}^\ell]$, respectively. Furthermore, $d_\ell(x_{\ell-1})r_\ell \in [\gamma_0^{\ell+1} \to \Gamma]$, $\psi_\ell^* = \psi_\ell^*(x_{\ell-1}) \in [\gamma_0^\ell \to \Gamma]$. By virtue of the equivalence of following systems

$$\left.\begin{aligned} &c_\ell^* y_1 + \widetilde{S}^{\ell\ell} y_2 = w_1, \qquad y_1 \in \gamma_0, \qquad y_2 \in \gamma_1^{\ell+1}, \\ &\widetilde{\theta}^\ell y_1 + \widetilde{A}^{\ell\ell} y_2 = w_2, \qquad w_1 \in \Gamma_1, \qquad w_2 \in \gamma_2^{\ell+1}, \end{aligned}\right\}$$

$$\left.\begin{aligned} &y_2 + (\widetilde{A}^{\ell\ell})^{-1}\widetilde{\theta}^\ell y_1 = (\widetilde{A}^{\ell\ell})^{-1} w_2, \\ &d_\ell y_1 = w_1 - \widetilde{S}^{\ell\ell}(\widetilde{A}^{\ell\ell})^{-1} w_2, \end{aligned}\right\}$$

we conclude that for fixed $\ell \geq 1$, $x_{\ell-1} \in \gamma_0^\ell$ the linear operator $d_\ell \in \gamma_0 \to \Gamma$ is invertible (non-singular) if and only if the linear operator $D_\ell \in \gamma_0 \times \gamma_1^{\ell+1} \to \Gamma \times \gamma_1^{\ell+1}$ is invertible (non-singular). For convenience, we suppose in the sequel that r_0 is known (in applications it is almost always true), i.e. (12.1.11) will be considered for $\ell \geq 1$. From the previous reasonings we obtain the following theorem, which gives the solution of the IP for R_1-structures. Let $\psi = \psi(r)$ with the moments ψ_ℓ be the GSP operator for the R_1-structure (A, S, f).

Theorem 12.3. (1) *The operators* ψ_ℓ, $\ell \geq 1$ *are constructed via* (12.1.13)–(12.1.16). *They do not depend on* x'_ℓ, *and* $\psi_\ell(x_\ell) \in [\gamma_0^{\ell+1} \to \Gamma]$.

(2) *Let* $D_\ell \# 0$, $\ell \geq 1$ *for each fixed* $x_{\ell-1} \in \gamma_0^\ell$. *Then for any* $\Pi \in \Delta^2(\psi)$ *the IP has a unique solution, which can be found by the following algorithm.*

Algorithm 12.1. *For* $\ell = 1, 2, \ldots$: (i) *construct* $\widetilde{h}^\ell$, ψ_ℓ^*, d_ℓ *via* (12.1.14)–(12.1.16); (ii) *find* r_ℓ *from the relation* $d_\ell r_\ell + \psi_\ell^* = \Pi_\ell$. *In particular, if* D_ℓ *are invertible, then* $r_\ell = d_\ell^{-1}\Pi_\ell - d_\ell^{-1}\psi_\ell^*$.

(3) *Moreover, if the operators* d_ℓ^{-1} *are continuous with respect to all their arguments in the aggregate, then* $\psi^{-1} \in [\mathcal{F}^+[\Gamma] \to \Gamma_0]$ *(stability of the solution of the IP).*

Now we consider the particular case of R_1-structure, when γ_0, γ_1, γ_2, Γ are finite-dimentional spaces. Let $\gamma_1 = \gamma_2 = E_m$, $\gamma_0 = \Gamma = E_N$, and let

$$\alpha_{\nu j}^k = [\alpha_{k\nu j}^{wz}]_{w,z=\overline{1,m}}, \qquad f_{k\nu} = [f_{k\nu}^z]_{z=\overline{1,m}}, \qquad S_{kj} = [S_{kj}^{\Delta z}]_{\Delta=\overline{1,N};z=\overline{1,m}},$$

$$r_\ell = [r_{\ell j}]_{j=\overline{1,N}}, \qquad \Pi_k = [\Pi_{k\Delta}]_{\Delta=\overline{1,N}}, \qquad \beta_{\nu j}^k = [\beta_{k\nu j}^{wz}]_{w,z=\overline{1,m}},$$

$$c_{kj} = [c_{kj}^{\Delta z}]_{\Delta=\overline{1,N};z=\overline{1,m}}, \qquad z_{k\nu} = [z_{k\nu}^j]_{j=\overline{1,N}} = [z_{k\nu}^{wj}]_{w=\overline{1,m};j=\overline{1,N}},$$

$$\alpha_{\nu 0}^k = \sum_{j=1}^N \beta_{\nu j}^k r_{\nu+k,j} + \alpha_{\nu 0}^{*,k}, \quad f_{k\nu} = \sum_{j=1}^N z_{k\nu}^j r_{\nu+k,j} + f_{k\nu}^*, \quad S_{k0} = \sum_{j=1}^N c_{kj} r_{kj} + S_{k0}^*,$$

$$\beta_{\nu j}^{*,k} = \beta_{\nu j}^k (\alpha_{00}^0)^{-1} f_{00}, \qquad c_{\ell j}^* = c_{\ell j}\varphi_{00}, \qquad \beta_\nu^{*,k} = [\beta_{\nu j}^{*,k}]_{j=\overline{1,N}} = [\beta_{\nu wj}^{*,k}]_{w=\overline{1,m};j=\overline{1,N}},$$

$$c_\ell^* = [c_{\ell j}^*]_{j=\overline{1,N}} = [c_{\ell\Delta j}^*]_{\Delta=\overline{1,N};j=\overline{1,N}}, \qquad \widetilde{\theta}^\ell = [\beta_\nu^{*,\ell-\nu} - z_{\ell-\nu,\nu}]_{\nu=\overline{0,\ell}}.$$

Then the matrix D_ℓ has the form $D_\ell = [D_{\ell\nu k}]_{\nu,k=\overline{-1,\ell}}$, where $D_{\ell\nu k} = \alpha_{\nu k}^{k-\nu}$ $(k,\nu \geq 0)$, $D_{\ell,-1,k} = S_{kk}$, $D_{\ell,\nu,-1} = \beta_\nu^{*,\ell-\nu} - z_{\ell-\nu,\nu}$, $D_{\ell,-1,-1} = c_\ell^*$. The matrices d_ℓ and D_ℓ are square matrices of order N and $N+m(\ell+1)$, respectively. The condition $D_\ell \# 0$, $\ell \geq 1$ of Theorem 12.3 goes over into the condition $\det D_\ell \neq 0$, $\ell \geq 1$.

12.2 Difference operators.

In this section we study one of the important particular cases of the triangular structures — difference operators of an arbitrary order.

12.2.1. For fixed p, $q \geq 1$ we consider the equation

$$(Ly)_\nu \equiv \sum_{j=-q}^p a_{\nu j} y_{\nu+j} = \lambda y_\nu, \qquad \nu \geq q, \qquad a_{\nu,-q} = 1, \tag{12.2.1}$$

where $y = [y_\nu]_{\nu \geq 0}$, and $a_{\nu j} \in \mathbf{C}$ are complex numbers. The operator L is called non-degenerate if $a_{\nu p} \neq 0$, $\nu \geq q$. Denote by $\mathcal{F}^+ = \mathcal{F}^+(\mathbf{C})$ the set of GF

$$P(\lambda) = \sum_{k=0}^\infty \frac{1}{\lambda^k} P_k, \qquad P_k \in \mathbf{C},$$

and by $\mathcal{F}_0^+$ the set of GF $P(\lambda) \in \mathcal{F}^+$ such that $P_0 = 0$.

Let $\Phi_\nu(\lambda) = [\Phi_\nu^i(\lambda)]^T_{i=\overline{1,q}}$, $\nu \geq 0$ be a solution of (12.2.1) under the conditions

$$\Phi_\nu^i(\lambda) = \sum_{k=0}^{\infty} \frac{1}{\lambda^k} \Phi_{k\nu}^i \in \mathcal{F}^+, \qquad \Phi_\nu^i(\lambda) = \delta_{i,\nu+1} \qquad (\nu = \overline{0, q-1}).$$

The $\{\Phi_\nu(\lambda)\}_{\nu\geq 0}$ is called the WS, and the matrix $M(\lambda) = [M_j^i(\lambda)]_{j=\overline{1,p};i=\overline{1,q}}$, $M_j^i(\lambda) = \Phi_{j+q-1}^i(\lambda)$ is called the WM of L.The IP is formulated as follows.

Problem 12.1. Construct L from the given WM $M(\lambda)$.

Denote $\lambda = \rho^q$,

$$H_\nu(\rho) = \sum_{i=1}^{q} \rho^{1-i} \Phi_\nu^i(\lambda), \qquad \nu \geq 0;$$

$$\Pi_j(\rho) = H_{j+q-1}(\rho), \qquad j = \overline{1,p}; \qquad \Pi(\rho) = [\Pi_j(\rho)]_{j=\overline{1,p}} .$$

Then

$$\left.\begin{aligned} H_\nu(\rho) &= \sum_{k=0}^{\infty} \frac{1}{\rho^k} H_{k\nu}, \quad H_{qk+i-1,\nu} = \Phi_{k\nu}^i, \\ \Pi_j(\rho) &= \sum_{k=0}^{\infty} \frac{1}{\rho^k} \Pi_{kj}, \quad \Pi_j(\rho) = \sum_{i=1}^{q} \rho^{1-i} M_j^i(\lambda). \end{aligned}\right\} \qquad (12.2.2)$$

Substituting $H_\nu(\rho)$ in (12.2.1) we obtain the equalities

$$H_{k\nu} = \sum_{j=-q}^{p} a_{\nu j} H_{k-q,\nu+j}, \qquad (k > \nu \geq q), \qquad (12.2.3)$$

$$H_{\xi\nu} = \delta_{\xi\nu}, \qquad (\xi \leq \nu), \qquad H_{\xi\nu} = 0, \qquad (\nu = \overline{0, q-1}; \xi > \nu), \qquad (12.2.4)$$

from which the coefficients $H_{k\nu}$ (and consequently $\Phi_{k\nu}^i$) are uniquely determined. Thus, for L there exists exactly one WS which can be found via (12.2.3), (12.2.4).

Let us denote by M^* the set of matrices $M(\lambda) = [M_j^i(\lambda)]_{j=\overline{1,p};i=\overline{1,q}}$, $M_j^i(\lambda) \in \mathcal{F}^+$ such that $\Pi_{kj} = \delta_{k,j+q-1}$ $(j = \overline{1,p}; k = \overline{1, j+q-1})$. It is evident that if $M(\lambda)$ is the WM of L, then $M(\lambda) \in M^*$.

We want to solve Problem 12.1. This problem is a particular case of the IP for the R-structure, studied in Section 12.1. We apply the method, given in Section 12.1, to solve Problem 12.1. For this we set

$$\chi_{ki} = \left[\frac{k-q-i}{p+q}\right], \qquad \mu_{ki} = p\chi_{ki} + q + i - 1,$$

$$\gamma_{ki} = k - 2q - i + 1(\operatorname{mod}(p+q)), \qquad r_{ki} = a_{\mu_{ki},\gamma_{ki}},$$

$$\pi_{ji} = a_{jp+q+i-1,p}, \qquad W_s^{k,i} = H_{k-sq,sp+q+i-1}, \qquad (s = \overline{0,\chi_{ki}}), \qquad W_{\chi_{ki}+1}^{k,i} = r_{ki},$$

where [.] denotes the greatest integer in number. Regroup the relations (12.2.3) and add the equalities $H_{k,i+q-1} = \Pi_{ki}$ we obtain

$$\left.\begin{aligned} &H_{k,i+q-1} = \Pi_{ki}, \\ &H_{k-qs,sp+q+i-1} = \sum_{j=-q}^{p} a_{sp+q+i-1,j} H_{k-q(s-1),sp+q+i+j-1}, \\ &\qquad s = \overline{0,\chi_{ki}}. \end{aligned}\right\} \tag{12.2.5}$$

For $k \geq q+1$ we consider linear algebraic systems $X_k = (X_{ki})$, $i = \overline{1, \min(p, k-q)}$ of the form (12.2.4), (12.2.5) with respect to $W_s^{k,i}$, $s = \overline{0, \chi_{ki}+1}$. The determinant of the system X_{ki}, is equal to

$$\prod_{j=0}^{\chi_{ki}-1} \pi_{ji}.$$

Solving the systems X_k successively for $k = q+1, q+2, \ldots$ we find the coefficients $H_{k\nu}$, $a_{\nu j}$. Thus, the algorithm of the solution of Problem 12.1 is obtained, and the following theorem holds.

Theorem 12.4. *For the matrix $M(\lambda) \in M^*$ to be the WM for L it is necessary and sufficient that the systems X_k, $k \geq q+1$ are solvable. For non-degenerate operators the solution of the IP is unique, i.e. if L and L^0 are non-degenerate, $M(\lambda)$ and $M^0(\lambda)$ are their WM, and $M(\lambda) = M^0(\lambda)$, then $L = L^0$.*

12.2.2. For non-degenerate operators L one can obtain more convenient necessary and sufficient conditions of solvability of the IP. In the sequel we assume that L is non-degenerate. Let us denote

$$R_k(\lambda) = [R_k^j(\lambda)]_{j=\overline{1,p}}^T, \qquad R_{pk+i-1}^j(\lambda) = \delta_{ij}\lambda^k, \qquad i = \overline{1,p},$$

$$R_k^*(\lambda) = [R_k^{*,j}(\lambda)]_{j=\overline{1,q}}, \qquad R_{qk+i-1}^{*,j}(\lambda) = \delta_{ij}\lambda^k, \qquad i = \overline{1,q}.$$

Let

$$P_k(\lambda) = [P_k^j(\lambda)]_{j=\overline{1,p}}^T, \qquad Q_k(\lambda) = [Q_k^i(\lambda)]_{i=\overline{1,q}}^T, \qquad k \geq 0$$

be solutions of (12.2.1) under the initial conditions

$$P_k(\lambda) = 0, \qquad (k = \overline{0, q-1}), \qquad P_{k+q}(\lambda) = R_k(\lambda), \qquad (k = \overline{0, p-1});$$

$$Q_k^i(\lambda) = \delta_{k+1,i}, \qquad (k = \overline{0, q-1}), \qquad Q_{k+q}(\lambda) = 0, \qquad (k = \overline{0, p-1}).$$

Then

$$P_{k+q}(\lambda) = \sum_{i=0}^{k} c_{ik} R_i(\lambda), \qquad k \geq 0, \qquad c_{kk} \neq 0. \tag{12.2.6}$$

Substituting (12.2.6) in (12.2.1) and comparing the corresponding coeffitients, we arrive at

$$\sum_{j=-q}^{p} a_{kj} c_{i,k+j-q} = c_{i-p,k-q}, \qquad i = \overline{0, k+p-q}.$$

Hence we obtain the recurrence formulae for determining $a_{k\nu}$ from c_{ik}:

$$a_{k\nu} = (c_{k+\nu-q,k+\nu-q})^{-1} (c_{k+\nu-q-p,k-q} - \sum_{j=\nu+1}^{p} a_{kj} c_{k+\nu-q,k+j-q}), \tag{12.2.7}$$

$$\nu = \overline{p, -q+1}.$$

Here $k \geq q$ for $\nu \geq 0$ and $k \geq q - \nu$ for $\nu < 0$. Furthermore, $c_{ik} = 0$ here for $i < 0$. In particular, one has $a_{kp} = (c_{k+p-q,k+p-q})^{-1} c_{k-q,k-q}$.

Lemma 12.3. *The following relations hold*

$$(1, P_{k+q}(\lambda) M(\lambda) R_\nu^*(\lambda)) = \theta_{k\nu}, \qquad \theta_{k\nu} = a_{q+k,\nu-q-k}, \tag{12.2.8}$$

$$0 \leq k < \nu \leq q-1,$$

$$(1, P_{k+q}(\lambda) M(\lambda) R_\nu^*(\lambda)) = \delta_{\nu k}, \qquad 0 \leq \nu \leq k. \tag{12.2.9}$$

Proof. Let us denote $\psi_k(\lambda) = \Phi_k(\lambda) - Q_k(\lambda) - P_k(\lambda) M(\lambda)$, $k \geq 0$. It is obvious that $\{\psi_k(\lambda)\}_{k \geq 0}$ is a solution of (12.2.1), and $\psi_0(\lambda) = \ldots = \psi_{p+q-1}(\lambda) = 0$. Hence we get $\psi_k(\lambda) = 0$ for all $k \geq 0$ and consequently

$$\Phi_k(\lambda) = Q_k(\lambda) + P_k(\lambda) M(\lambda), \qquad k \geq 0. \tag{12.2.10}$$

Furthermore, using (12.2.10) we have for $\nu = \overline{0, q-1}$, $k \geq 0$

$$(1, P_{k+q}(\lambda)M(\lambda)R_\nu^*(\lambda)) = (1, \Phi_{k+q}(\lambda)R_\nu^*(\lambda)) = (1, \Phi_{k+q}^{\nu+1}(\lambda)) = \Phi_{1,k+q}^{\nu+1}.$$

Consequently, by virtue of (12.2.2)–(12.2.4), we obtain (12.2.8) and (12.2.9) for $\nu = \overline{0, q-1}$. Now we use the method of induction. Let us assume that (12.2.9) is proved for $\nu = \overline{0, \mu - 1}$; $\mu \geq q$. Then for $k \geq \mu$ one has

$$(1, P_{k+q}(\lambda)M(\lambda)R_\mu^*(\lambda)) = (1, \lambda P_{k+q}(\lambda)M(\lambda)R_{\mu-q}^*(\lambda))$$

$$= \sum_{j=-q}^{p} a_{k+q,j}(1, P_{k+q+j}(\lambda)M(\lambda)R_{\mu-q}^*(\lambda)) = (1, P_k(\lambda)M(\lambda)R_{\mu-q}^*(\lambda)) = \delta_{k\mu}.$$

Lemma 12.3 is proved.

Lemma 12.4. *Let* $T(\lambda) = [T_j^i(\lambda)]_{j=\overline{1,p};i=\overline{1,q}}$ $T_j^i(\lambda) \in \mathcal{F}_0^+$, *and let* $(1, P_{k+q}(\lambda)T(\lambda)R_\nu^*(\lambda)) = 0$ *for* $k \geq 0$, $\nu = \overline{0, q-1}$. *Then* $T(\lambda) = 0$.

Indeed, by induction, as in the proof of Lemma 12.3, we get

$$(1, P_{k+q}(\lambda)T(\lambda)R_\nu^*(\lambda)) = 0, \qquad k, \nu \geq 0.$$

Hence and from (12.2.6) it follows that

$$(1, R_i(\lambda)T(\lambda)R_\nu^*(\lambda)) = 0, \qquad i, \nu \geq 0,$$

and consequently $T(\lambda) = 0$.

Let us denote

$$\mu_{i\nu} = (1, R_i(\lambda)M(\lambda)R_\nu^*(\lambda)), \qquad i, \nu \geq 0; \qquad \Delta_k = \det\,[\mu_{i\nu}]_{i,\nu=\overline{0,k}}, \qquad k \geq 0.$$

Substituting (12.2.6) in (12.2.9) we calculate

$$\sum_{i=0}^{k} c_{ik}\mu_{i\nu} = \delta_{\nu k}, \qquad 0 \leq \nu \leq k. \tag{12.2.11}$$

For each fixed $k \geq 0$, system (12.2.11) is a linear algebraic system with respect to c_{ik}. The determinant of (12.2.11) is equal to Δ_k. Let us show that $\Delta_k \neq 0$ for all $k \geq 0$. We assume that $\Delta_j \neq 0$ ($j = \overline{0, k_0 - 1}$), $\Delta_{k_0} = 0$ for a certain $k_0 \geq 0$. Since $\Delta_{k_0 - 1} \neq 0$, the rank of the extended matrix of (12.2.1) is equal to $k_0 + 1$. But this contradicts the solvability of (12.2.11). Since $\Delta_0 = 1$, we conclude that $\Delta_k \neq 0$ for all $k \geq 0$. Solving (12.2.11) we find

$$c_{ik} = (-1)^{k-i}\Delta_k^{-1}\det\,[\mu_{j\nu}]_{j=\overline{0,k}\setminus i;\nu=\overline{0,k-1}}\ (i = \overline{0,k}), \quad c_{kk} = \Delta_{k-1}\Delta_k^{-1}. \tag{12.2.12}$$

Since $P_{k+q}(\lambda) = R_k(\lambda)$, $k = \overline{0, p-1}$ we get, by virtue of (12.2.9), that $\mu_{i\nu} = \delta_{i\nu}$, $0 \le \nu \le i \le p-1$. Note that the above-mentioned set M^* can be defined in a different way: M^* is the set of matrices $M(\lambda) = [M_j^i(\lambda)]_{j=\overline{1,p};i=\overline{1,q}}$, $M_j^i(\lambda) \in \mathcal{F}_0^+$ such that $\mu_{i\nu} = \delta_{i\nu}$ for $0 \le \nu \le i \le p-1$. The following theorem gives necessary and sufficient conditions of solvability of the IP for non-degenerate operators.

Theorem 12.5. *For the matrix $M(\lambda) \in M^*$ to be the WM for a non-degenerate operator L it is necessary and sufficient that $\Delta_k \neq 0$ for all $k \ge 0$. The operator L can be found by the algorithm.*

Algorithm 12.2. (i) Construct c_{ik} $(0 \le i \le k)$ and $P_{k+q}(\lambda)$ for $k \ge 0$ via (12.2.12), (12.2.6); and set $P_k(\lambda) \equiv 0$ for $k = \overline{0, q-1}$.

(ii) Compute $a_{k\nu}$ by (12.2.7), (12.2.8).

The necessity of Theorem 12.5 has been proved above. We shall prove the sufficiency. Let $M(\lambda) \in M^*$ be given. We construct c_{ik}, $P_k(\lambda)$, $a_{k\nu}$ via Algorithm 12.2. In the other words we construct the non-degenerate operator L. Furthermore, the equalities $P_{k+q}(\lambda) = R_k(\lambda)$, $k = \overline{0, p-1}$ are valid.

Let us show that the $P_k(\lambda)$ satisfy (12.2.1). Indeed, since $c_{kk} \neq 0$, one gets

$$\lambda P_k(\lambda) = \sum_{j=q}^{k+p} \gamma_{jk} P_j(\lambda), \qquad \gamma_{k+p,k} \neq 0. \tag{12.2.13}$$

By construction, the numbers c_{ik} satisfy (12.2.11). Hence and from (12.2.6) we obtain (12.2.9). Using (12.2.9) and (12.2.13), we find successively for $\mu = q, q+1, \ldots, k-q$

$$\begin{aligned}
\delta_{\mu,k-q} &= (1, P_k(\lambda)M(\lambda)R_\mu^*(\lambda)) = (1, \lambda P_k(\lambda)M(\lambda)R_{\mu-q}^*(\lambda)) \\
&= \sum_{j=\mu}^{k+p} \gamma_{jk}(1, P_j(\lambda)M(\lambda)R_{\mu-q}^*(\lambda)) = \gamma_{\mu k}.
\end{aligned}$$

Thus,

$$\lambda P_k(\lambda) = \sum_{j=-q}^{p} a_{kj}^0 P_{k+j}(\lambda),$$

i.e. $P_k(\lambda)$ satisfy $(L^0 y)_\nu = \lambda y_\nu$ with certain coeffitients a_{kj}^0. Consequently

$$P_{k+q}(\lambda) = \sum_{i=0}^{k} c_{ik}^0 R_i(\lambda), \qquad k \ge 0.$$

Comparing this relation with (12.2.6) we obtain $c_{ik}^0 = c_{ik}$ $(0 \le i \le k)$ and, by virtue of (12.2.7), $a_{k\nu}^0 = a_{k\nu}$. Thus, $P_k(\lambda)$ satisfy (12.2.1).

Let $M^0(\lambda)$ be the WM for the above-constructed operator L. Then, by Lemma 12.3, we conclude $M^0(\lambda) = M(\lambda)$. Theorem 12.5 is proved.

Corollary 12.1. *For the matrix $M(\lambda) \in M^*$ to be the WM for a non-degenerate operator L with real coefficients $a_{\nu j}$, $a_{\nu p} > 0$ it is necessary and sufficient that the $\mu_{i\nu}$ are real and $\Delta_k > 0$ for all $k \geq 0$.*

Remark. Let us consider the well-known case $p = q = 1$.Then

$$M(\lambda) = \sum_{k=1}^{\infty} \frac{1}{\lambda^k} M_k \in \mathcal{F}^+, \qquad M_1 = 1, \qquad \mu_{i\nu} = M_{i+\nu+1},$$

$$\Delta_k = \det\left[M_{i+\nu+1}\right]_{i,\nu=\overline{0,k}}.$$

In this case the WF $M(\lambda)$ coincide with the generalized spectral function considered in [109]. It is known that if $a_{\nu 1} > 0$ and $a_{\nu 0}$ are real, then there exists at least one spectral function $\sigma(\lambda)$ of L such that

$$M_{k+1} = (\lambda^k, M(\lambda)) = \int_{-\infty}^{\infty} \lambda^k \, d\sigma(\lambda), \qquad k \geq 0.$$

Thus, the numbers $\{M_k\}_{k\geq 1}$ are the moments of $\sigma(\lambda)$. Therefore, Corollary 12.1 for $p = q = 1$ coincides with the theorem of solvability of the classical problem of moments. We note that for $p = q = 1$ IP's for difference operators were studied in [107]–[110], [116] and other works.

12.3 Applications to differential operators

12.3.1. Let us consider the DE and LF (1.1)–(1.2). Let S be a fixed sector with the property (1.3). Denote

$$S^\delta = \{\rho : \rho \in S, \operatorname{Re}(\rho(R_{j+1} - R_j)) > \delta|\rho|, j = \overline{1, n-1}\}, \qquad \delta > 0,$$

$$T_{j\nu}(x, \rho) = C^\nu_{n-j} R^{n-j-\nu} p_{n-j}(x), \qquad F_\nu(R) = C^\nu_n R^{n-\nu},$$

$$V^0_{j\nu}(\mu, R) = C^\nu_{\sigma_{\mu 0}-j} R^{\sigma_{\mu 0}-j-\nu} u_{\mu,\sigma_{\mu 0}-j,0}, \qquad E_\nu(\mu, R) = C^\nu_{\sigma_{\mu 0}} R^{\sigma_{\mu 0}-\nu},$$

$$E^{1,\tau}_\nu(\mu, R) = \sum_{j=0}^{n-\nu-1} C^\nu_{j+\nu} q^\tau_{\mu,j+\nu} R^j, \qquad a^0_{mj} = (-1)^{m+j}(\Omega(\overline{1,m}))^{-1}\Omega_j(\overline{1, m-1}),$$

where $C^\nu_j = j!(\nu!(j-\nu)!)^{-1}$ for $0 \leq \nu \leq j$, and $C^\nu_j = 0$ otherwise. Let $\psi_\tau(x, \lambda) = \Phi_{m_\tau}(x, \lambda)$ (see §9).

Lemma 12.5. *Let $p_j(x) \in PA[0, a]$, $A[a, b)$, $0 \leq a < b < T$, $j = \overline{0, n-2}$. Then for $|\rho| \to \infty$, $\rho \in \bar{S}^\delta$, the following asymptotic formulae hold uniformly in $x \in [a, b-\varepsilon]$, $\varepsilon > 0$*

$$\begin{aligned} \psi^{(\nu)}_\tau(x, \lambda) &= \sum_{\xi=1}^{m_\tau} \exp\left(\rho(R_{m_\tau} a + R_\xi(x-a))\right) \\ &\quad \times \sum_{k=0}^{\infty} \rho^{\nu-\sigma_{m_\tau,0}-k} \sum_{j=0}^{\nu} C^j_\nu R^{\nu-j}_\xi g^{(j)}_{k-j,\xi,m_\tau}(x, a), \end{aligned} \qquad (12.3.1)$$

$$\mathfrak{M}_{m_\tau,\mu}(\lambda) = \rho^{\sigma_{\mu 0}-\sigma_{m_\tau,0}} \sum_{k=0}^{\infty} \frac{1}{\rho^k} \mathfrak{M}^k_{m_\tau,\mu},$$

$$\mathfrak{M}^0_{m_\tau \mu} = (\Omega(\overline{1,m_\tau}))^{-1}\Omega(\overline{1,m_\tau - 1},\mu), \tag{12.3.2}$$

$$Q^\tau_{\xi\nu}(\lambda,a) = \rho^{\xi-1-\nu} \sum_{k=0}^{\infty} \frac{1}{\rho^k} q^{\tau k}_{\xi\nu}(a), \qquad q^{\tau 0}_{\xi\nu} = q^\tau_{\xi\nu}, \tag{12.3.3}$$

$$\mathfrak{N}^\tau_\xi(\lambda,a) = \rho^{\xi-1-\sigma_{m_\tau},0}\exp(\rho R_{m_\tau} a) \sum_{k=0}^{\infty} \frac{1}{\rho^k} \mathfrak{N}^{\tau,k}_\xi(a), \tag{12.3.4}$$

where $\mathfrak{N}^{\tau,0}_\xi(a) = a^0_{m_\tau,m_\tau}\omega^1_{\xi\tau}(R_{m_\tau})$, $g_{k,\xi,m_\tau}(x,a) \in A[a,b)$, $g_{k,\xi,m_\tau}(x,a) \equiv 0$ $(k<0)$, $g_{0,\xi,m_\tau}(x,0) \equiv a^0_{m_\tau,\xi}$, *and* $g_{0,\xi,m_\tau}(x,a) = \delta_{\xi,m_\tau} a^0_{m_\tau,\xi}$ *for* $a>0$. *Moreover*

$$\sum_{s=0}^{k} \sum_{j=0}^{s+1} T_{s+1-j,j}(x,R_\xi) g^{(j)}_{k-s,\xi,m_\tau}(x,a) = 0, \tag{12.3.5}$$

$$x \in [a,b), \qquad \xi = \overline{1,m_\tau}, \qquad k \ge 0, \qquad a \ge 0,$$

$$\sum_{\xi=1}^{m_\tau} \sum_{s=0}^{k} \sum_{j=0}^{s} V^{a,\tau}_{s-j,j}(\mu,R_\xi) g^{(j)}_{k-s,\xi,m_\tau}(a+0,a) = b^a_{\tau k\mu}, \qquad k \ge 0, \qquad a \ge 0, \tag{12.3.6}$$

$$V^{0,\tau}_{j\nu}(\mu,R) = V^0_{j\nu}(\mu,R), \qquad V^{a,\tau}_{j\nu}(\mu,R) = \sum_{k=0}^{n-\nu-1} q^{\tau,j}_{\mu,k+\nu}(a) R^k C^\nu_{k+\nu}, \qquad (a>0),$$

$$b^0_{\tau k\mu} = \mathfrak{M}^k_{m_\tau,\mu}, \qquad (\mu = \overline{1,n}), \qquad b^a_{\tau k\mu} = \mathfrak{N}^{\tau,k}_\mu(a), \qquad (a>0,\ \mu = \overline{1,\beta_\tau}).$$

Proof. Let us take the FSS $B_0 = \{y_k(x,\rho)\}_{k=\overline{1,n}}$. Using the integral equation for $y_k(x,\rho)$ we obtain the asymptotic formula, for $|\rho| \to \infty$, $\rho \in \bar{S}^\delta$ uniformly in $x \in [a, b-\varepsilon]$, $\varepsilon > 0$

$$y^{(\nu)}_k(x,\rho) = (\rho R_k)^\nu \sum_{j=1}^{k} \exp(\rho(R_k a - R_j(x-a))) \sum_{s=0}^{\infty} \frac{1}{\rho^s} f_{\nu k s j}(x), \tag{12.3.7}$$

where $f_{\nu k s j}(x) \in A[a,b)$, $f_{\nu k 0 j}(x) \equiv \delta_{jk}$.

Let $a=0$. First we will obtain (12.3.1). For definiteness suppose $T=\infty$. For $T<\infty$ the proof is analogous. Representing $\psi_\tau(x,\lambda)$ with the help of $\{y_k(x,\rho)\}_{k=\overline{1,n}}$ and using the boundary conditions on $\psi_\tau(x,\lambda)$ we calculate

$$\psi_\tau(x,\lambda) = \sum_{k=1}^{m_\tau} a_{m_\tau,k}(\rho) y_k(x,\rho),$$

$$a_{m_\tau,k}(\rho) = (-1)^{m_k+k}(\det[U_{\xi 0}(y_\nu)]_{\xi,\nu=\overline{1,m_\tau}})^{-1} \det[U_{\xi 0}(y_\nu)]_{\xi=\overline{1,m_\tau-1};\nu=\overline{1,m_\tau}\setminus k}.$$

From this and from (12.3.7) it follows that

$$a_{m_\tau,\xi}(\rho) = \sum_{k=0}^{\infty} \rho^{-k-\sigma_{m_\tau,0}} a^k_{m_\tau,\xi}, \qquad \xi = \overline{1, m_\tau}.$$

Hence we get (12.3.1). The relations (12.3.2), (12.3.5) and (12.3.6) can be obtained by substituting (12.3.1) into (1.1) and (1.2). Indeed, since

$$p_\nu(x)\psi_\tau^{(\nu)}(x,\lambda) = \sum_{\xi=1}^{m_\tau} \exp(\rho R_\xi x) \sum_{k=0}^{\infty} \rho^{n-k-\sigma_{m_\tau,0}} p_\nu(x) \sum_{j=0}^{\nu} C_\nu^j R_\xi^{\nu-j} g^{(j)}_{k-n+\nu-j,\xi,m_\tau}(x,0),$$

we have

$$\sum_{s=0}^{n}\sum_{j=0}^{s} C^j_{n-j+s} R_\xi^{n-s} g^{(j)}_{k-s,\xi,m_\tau}(x,0) p_{n+j-s}(x) = g_{k,\xi,m_\tau}(x,0), \qquad k \ge 0,$$

where $p_n(x) \equiv 1$, $p_{n-1}(x) \equiv 0$. After cancellation of $g_{k,\xi,m_\tau}(x,0)$ we get

$$\sum_{s=0}^{n-1}\sum_{j=0}^{s+1} C^j_{n+j-s-1} R_\xi^{n-s-1} p_{n+j-s-1}(x) g^{(j)}_{k-s,\xi,m_\tau}(x,0) = 0, \qquad k \ge 0.$$

This yelds (12.3.5). Substituting (12.3.1) in (1.2) we calculate

$$\mathfrak{M}_{m_\tau,\mu}(\lambda) = \sum_{k=0}^{\infty} \rho^{\sigma_{\mu 0}-\sigma_{m_\tau,0}-k} \sum_{\xi=1}^{m_\tau}\sum_{\nu=0}^{\sigma_{\mu 0}}\sum_{j=0}^{\nu} u_{\mu\nu 0} C_\nu^j R_\xi^{\nu-j} g^{(j)}_{k-\sigma_{\mu 0}+\nu-j,\xi,m_\tau}(0,0).$$

Since

$$\sum_{\nu=0}^{\sigma_{\mu 0}}\sum_{j=0}^{\nu} u_{\mu\nu 0} C_\nu^j R_\xi^{\nu-j} g^{(j)}_{k-\sigma_{\mu 0}+\nu-j,\xi,m_\tau}(0,0)$$

$$= \sum_{s=0}^{\sigma_{\mu 0}}\sum_{j=0}^{s} u_{\mu,\sigma_{\mu 0}-s+j,0} C^j_{\sigma_{\mu 0}-s+j} R_\xi^{\sigma_{\mu 0}-s} g^{(j)}_{k-s,\xi,m_\tau}(0,0),$$

we obtain

$$\mathfrak{M}_{m_\tau,\mu}(\lambda) = \rho^{\sigma_{\mu 0}-\sigma_{m_\tau,0}} \sum_{k=0}^{\infty} \frac{1}{\rho^k} \sum_{\xi=1}^{m_\tau}\sum_{s=0}^{k}\sum_{j=0}^{s} V^0_{s-j,j}(\mu, R_\xi) g^{(j)}_{k-s,\xi,m_\tau}(0,0), \qquad \mu = \overline{1,n}.$$

This yields (12.3.2) and (12.3.6).

For $a > 0$ the relations (12.3.1) and (12.3.5) are proved analogously. Setting $z_k(x,\lambda) = y_k(x,\rho)$ and using (12.3.7) we obtain (12.3.3). Furthermore, by virtue of Lemma 9.2, we have

$$\mathfrak{N}^\tau_\mu(\lambda, a) = U^\tau_{\mu a}(\psi_\tau) = \sum_{\nu=0}^{n-1} Q^\tau_{\mu\nu}(\lambda, a)\psi_\tau^{(\nu)}(a,\lambda).$$

It follows from (12.3.1) and (12.3.3) that

$$\mathfrak{N}_\mu^\tau(\lambda, a) = \rho^{\mu-1-\sigma_{m_\tau},0} \exp(\rho R_{m_\tau} a) \sum_{\nu=0}^{n-1} \sum_{i=0}^{\infty} q_{\mu\nu}^{\tau,i}(a) \sum_{k=i}^{\infty} \frac{1}{\rho^k}$$

$$\times \sum_{\xi=1}^{m_\tau} \sum_{j=0}^{\nu} C_\nu^j R_\xi^{\nu-j} g_{k-i-j,\xi,m_\tau}^{(j)}(a+0,a),$$

and consequently

$$\mathfrak{N}_\mu^\tau(\lambda, a) = \rho^{\mu-1-\sigma_{m_\tau},0} \exp(\rho R_{m_\tau} a) \sum_{k=0}^{\infty} \frac{1}{\rho^k} \sum_{i=0}^{k} \sum_{\nu=0}^{n-1} \sum_{\xi=1}^{m_\tau} \sum_{j=0}^{\nu} C_\nu^j R_\xi^{\nu-j}$$

$$\times q_{\mu\nu}^{\tau,i}(a) g_{k-i-j,\xi,m_\tau}^{(j)}(a+0,a).$$

Hence we get (12.3.4), where

$$\mathfrak{N}_\mu^{\tau,k}(a) = \sum_{i=0}^{k} \sum_{\nu=0}^{n-1} \sum_{\xi=1}^{m_\tau} \sum_{j=0}^{\nu} C_\nu^j R_\xi^{\nu-j} q_{\mu\nu}^{\tau,i}(a) g_{k-i-j,\xi,m_\tau}^{(j)}(a+0,a).$$

From this equality we obtain (12.3.6) for $a > 0$. Lemma 12.5 is proved.

Let us denote $m = m_1 + \ldots + m_\tau$, $N = N_1 + \ldots + N_\tau$. In the following natural numbers z, w, Δ $(1 \le z, w \le m, 1 \le \Delta \le N)$ are represented in the form

$$z = \sum_{k=1}^{\tau-1} m_k + \xi \quad (1 \le \xi \le m_\tau); \qquad w = \sum_{k=1}^{\mu-1} m_k + \eta \qquad (1 \le \eta \le m_\mu),$$

$$\Delta = \sum_{k=1}^{\mu-1} N_k + i, \qquad (1 \le i \le N_\mu).$$

Let $p_\nu(x) \in PA[0,T]$, $\nu = \overline{0, n-2}$. We denote

$$y_{jz}(a) = \sum_{k=0}^{\infty} \frac{1}{\lambda^k} g_{k,\xi,m_\tau}^{(j)}(a+0,a) \in \mathcal{F}^+, \qquad y_j(a) = [y_{jz}(a)]_{z=\overline{1,m}}, \qquad a \ge 0.$$

Let us show that

$$\sum_{k=0}^{\infty} \frac{1}{\lambda^k} \sum_{j=0}^{\nu+k} \left(\sum_{w=0}^{k+1} C_{\nu-1}^{j-w} T_{k+1-w,w}^{(\nu-j+w-1)}(a+0, R_\xi) \right) y_{jz}(a) = 0, \qquad \nu \ge 1, \quad (12.3.8)$$

$$\sum_{k=0}^{\infty} \frac{1}{\lambda^k} \sum_{j=0}^{k} \left(\sum_{\xi=1}^{m_\tau} V_{k-j,j}^{a,\tau}(\mu, R_\xi) y_{jz}(a) \right) = \sum_{k=0}^{\infty} \frac{1}{\lambda^k} b_{\tau k\mu}^a. \qquad (12.3.9)$$

Indeed, differentiating (12.3.5) ν times with respect to x and setting $x = a + 0$, we obtain

$$\sum_{s=0}^{k} \sum_{w=0}^{s+1} \sum_{j=0}^{\nu} C_\nu^j T_{s+1-w,w}^{(\nu-j)}(a+0, R_\xi) g_{k-s,\xi,m_\tau}^{(w+j)}(a+0,a) = 0, \qquad k, \nu \ge 0.$$

Hence one gets

$$\sum_{k=0}^{\infty}\frac{1}{\lambda^k}\sum_{s=0}^{k}\sum_{w=0}^{s+1}\sum_{j=0}^{\nu}C_\nu^j T_{s+1-w,w}^{(\nu-j)}(a+0,R_\xi)g_{k-s,\xi,m_\tau}^{(w+j)}(a+0,a)=0,\qquad \nu\ge 0,$$

or

$$\sum_{s=0}^{\infty}\frac{1}{\lambda^s}\sum_{w=0}^{s+1}\sum_{j=0}^{\nu}C_\nu^j T_{s+1-w,w}^{(\nu-j)}(a+0,R_\xi)\sum_{k=s}^{\infty}\frac{1}{\lambda^{k-s}}g_{k-s,\xi,m_\tau}^{(w+j)}(a+0,a)=0,\qquad \nu\ge 0.$$

Consequently

$$\sum_{k=0}^{\infty}\frac{1}{\lambda^k}\sum_{w=0}^{k+1}\sum_{j=w}^{\nu+w-1}C_{\nu-1}^{j-w}T_{k+1-w,w}^{(\nu-j+w-1)}(a+0,R_\xi)y_{jz}(a)=0,\qquad \nu\ge 1.$$

This yields (12.3.8). Further, it follows from (12.3.6) that

$$\sum_{k=0}^{\infty}\frac{1}{\lambda^k}\sum_{\xi=1}^{m_\tau}\sum_{s=0}^{k}\sum_{j=0}^{s}V_{s-j,j}^{a,\tau}(\mu,R_\xi)g_{k-s,\xi,m_\tau}^{(j)}(a+0,a)=\sum_{k=0}^{\infty}\frac{1}{\lambda^k}b_{\tau k\mu}^a.$$

Consequently

$$\sum_{s=0}^{\infty}\frac{1}{\lambda^s}\sum_{j=0}^{s}\sum_{\xi=1}^{m_\tau}V_{s-j,j}^{a,\tau}(\mu,R_\xi)\sum_{k=s}^{\infty}\frac{1}{\lambda^{k-s}}g_{k-s,\xi,m_\tau}^{(j)}(a+0,a)=\sum_{k=0}^{\infty}\frac{1}{\lambda^k}b_{\tau k\mu}^a.$$

This yields (12.3.9).

For a, ν, j, k, $\ell\ge 0$ we consider the matrices $\alpha_{\nu j}^k(a)=[\alpha_{k\nu j}^{wz}(a)]_{w,z=\overline{1,m}}$,

$$S_{kj}(a)=[S_{kj}^{\Delta z}(a)]_{\Delta=\overline{1,N};z=\overline{1,m}},\qquad f_{k\nu}(a)=[f_{k\nu}^z(a)]_{z=\overline{1,m}},\qquad \mathfrak{N}_k(a)=[\mathfrak{N}_{k\Delta}(a)]_{\Delta=\overline{1,N}},$$

$$r_\ell^a=[r_{\ell j}^a]_{j=\overline{1,N}},\qquad \varphi_k(a)=[\varphi_{k\nu}(a)]_\nu\ge 0,\qquad \varphi_{k\nu}(a)=[g_{k,\xi,m_\tau}^{(\nu)}(a+0,a)]_{z=\overline{1,m}},$$

$$\beta_{\nu j}^k=[\beta_{k\nu j}^{wz}]_{w,z=\overline{1,m}};\qquad \beta_{\nu j}^{*,k}=\beta_{\nu j}^k\varphi_{00}(a),$$

$$\beta_\nu^{*,k}(a)=[\beta_{\nu j}^{*,k}(a)]_{j=\overline{1,N}}=[\beta_{\nu wj}^{*,k}(a)]_{w=\overline{1,m};j=\overline{1,N}},$$

where

$$\alpha_{k\nu j}^{wz}(a)=\delta_{wz}\sum_{w=0}^{k+1}C_{\nu-1}^{j-w}T_{k+1-w,w}^{(\nu-j+w-1)}(a+0,R_\xi),\qquad \nu\ge 1,\qquad a\ge 0,$$

$$\alpha_{k0j}^{wz}(0)=\delta_{\tau\mu}V_{k-j,j}^{0,\tau}(\eta,R_\xi),\qquad \alpha_{k0j}^{wz}(a)=\delta_{\tau\mu}V_{k-j,j}^{a,\tau}(\varepsilon_{\eta\mu},R_\xi),\qquad a>0,$$

$$S_{kj}^{\Delta z}(0)=\delta_{\tau\mu}V_{k-j,j}^{0,\tau}(\gamma_{\mu i},R_\xi),\qquad S_{kj}^{\Delta z}(a)=\delta_{\tau\mu}V_{k-j,j}^{a,\tau}(\varepsilon_{m_\tau+i,\mu},R_\xi),\qquad a>0,$$

$$f_{00}^{z}(0)=\delta_{\xi,m_\tau},\qquad f_{k\nu}(0)=0,\qquad k+\nu>0,$$

$$f_{k0}^{z}(a)=\mathfrak{N}_{\varepsilon_{\xi\tau}}^{\tau,k}(a),\qquad f_{k\nu}(a)=0,\qquad \nu\geq 1,\qquad a>0,$$

$$\mathfrak{N}_{k\Delta}(0)=\mathfrak{M}_{m_\mu,\gamma_{\mu i}}^{k},\ \mathfrak{N}_{k\Delta}(a)=\mathfrak{N}_{\varepsilon_{m_\mu+i},\mu}^{\mu k}(a),\ a>0;\ r_{\ell j}^{a}=p_{n-k_j}^{(\ell-k_j)}(a+0),\ \ell\geq\kappa_j,$$

$$r_{\ell j}^{a}=0,\ \ell<\kappa_j;\ \beta_{k,\ell-k,j}^{wz}=\delta_{wz}\delta_{ij}R_\xi^{n-\kappa_i},\ (k+1=\kappa_i),\ \beta_{k,\ell-k,j}^{wz}=0,\quad (k+1\notin\kappa).$$

It is obvious that the matrices $\alpha_{jj}^{0}(a)$ are invertible and $f_{00}(a)\neq 0$. The equalities (12.3.8) and (12.3.9) become

$$\sum_{k=0}^{\infty}\frac{1}{\lambda^k}\sum_{j=0}^{k+\nu}\alpha_{\nu j}^{k}(a)y_j(a)=\sum_{k=0}^{\infty}\frac{1}{\lambda^k}f_{k\nu}(a),\qquad \nu\geq 0,\tag{12.3.10}$$

$$\sum_{k=0}^{\infty}\frac{1}{\lambda_k}\sum_{j=0}^{k}S_{kj}(a)y_j(a)=\sum_{k=0}^{\infty}\frac{1}{\lambda^k}\mathfrak{N}_k(a).\tag{12.3.11}$$

For each fixed $a\geq 0$ the relations (12.3.10) and (12.3.11) define the R_1-structure (A,S,f) (in the same notations as in Section 12.1), where $\gamma_1=\gamma_2=E_m$, $\gamma_0=\Gamma=E_N$, $z_{k\nu}=c_k=0$, and $\varphi_k(a)$, $\mathfrak{N}_k(a)$, $k\geq 0$ are the moments of the WS and the WF, respectively, for this R_1-structure.

We shall use the results of Section 12.1 for R_1-structures. For $\ell\geq 1$, $\varepsilon=0,1$ we define the matrices

$$D_\ell^\varepsilon=[D_{\ell\nu k}^\varepsilon]_{\nu,k=\overline{-1,\ell}},\qquad d_\ell^\varepsilon=[D_{\ell,-1,k}^\varepsilon]_{k=\overline{0,\ell}}[D_{\ell\nu k}^\varepsilon]_{\nu,k=\overline{0,\ell}}^{-1}[D_{\ell,\nu,-1}^\varepsilon]_{\nu=\overline{0,\ell}},$$

$$D_{\ell\nu,k}^\varepsilon=[D_{\ell\nu k}^{\varepsilon wz}]_{w,z=\overline{1,m}},\quad (\nu,k\geq 0),\quad D_{\ell,-1,k}^\varepsilon=[D_{\ell,-1,k}^{\varepsilon\Delta z}]_{\Delta=\overline{1,N};z=\overline{1,m}},\quad (k\geq 0),$$

$$D_{\ell,\nu,-1}^\varepsilon=[D_{\ell,\nu,-1}^{\varepsilon wj}]_{w=\overline{1,m};j=\overline{1,\rho}},\quad (\nu\geq 0),\quad D_{\ell,-1,-1}^\varepsilon=[D_{\ell,-1,-1}^{\varepsilon\Delta j}]_{\Delta=\overline{1,N};j=\overline{1,\rho}},$$

(p is chosen from the condition $\ell\in[\kappa_p,\kappa_{p+1})$, $\kappa_0=1$, $\kappa_{N+1}=\infty$), where

$$D_{\ell,-1,-1}^\varepsilon=D_{\ell,0,-1}^\varepsilon=D_{\ell\nu k}^{0}=0,\ (0\leq k\leq\nu),\ D_{\ell\nu k}^{0wz}=\delta_{wz}F_{k-\nu+1}(R_\xi),\ (k\geq\nu\geq 1),$$

$$D^{0wz}_{\ell 0 k} = \delta_{\tau\mu} E_k(\eta, R_\xi), \qquad D^{0\Delta z}_{\ell,-1,k} = \delta_{\tau\mu} E_k(\gamma_{\mu i}, R_\xi), \qquad (k \geq 0),$$

$$D^{0wj}_{\ell,\nu,-1} = \beta^{*,\ell-\nu}_{\nu w j}(0), \qquad (\nu \geq 1),$$

$$D^1_{\ell\nu k} = D^0_{\ell\nu k}, \qquad (\nu \geq 1, \, k \geq 0), \qquad D^{1wz}_{\ell 0 k} = \delta_{\tau\mu} E^{1,\tau}_k(\varepsilon_{\eta\mu}, R_\xi),$$

$$D^{1\Delta z}_{\ell,-1,k} = \delta_{\tau\mu} E^{1,\tau}_k(\varepsilon_{m_\mu+i,\mu}, R_\xi), \quad (k \geq 0), \quad D^{1wj}_{\ell,\nu,-1} = \beta^{*,\ell-\nu}_{\nu w j}(a), \quad (a > 0, \nu \geq 1),$$

and D^ε_ℓ do not depend on a. It is clear that $D^\varepsilon_\ell \# 0$ if and only if $d^\varepsilon_\ell \# 0$. Denote

$$\Lambda^{\mu 0}_{ij} = (\Omega^1_\mu(\varepsilon_{1\mu}, \ldots, \varepsilon_{m_\mu,\mu}))^{-1} \Omega^1_\mu(\varepsilon_{1\mu}, \ldots, \varepsilon_{j-1,\mu}, \varepsilon_{m_\mu+i,\mu}, \varepsilon_{j+1,\mu}, \ldots, \varepsilon_{m_\tau,\mu}).$$

Since for $a > 0$

$$\alpha^0_{00}(a) = [\delta_{\tau\mu}\omega^1_{\varepsilon_{\eta\mu},\tau}(R_\xi)]_{w,z=\overline{1,m}}, \qquad S_{00}(a) = [\delta_{\tau\mu}\omega^1_{\varepsilon_\mu+i,\mu,\tau}(R_\xi)]_{\Delta=\overline{1,N}; z=\overline{1,m}},$$

we have

$$S_{00}(a)(\alpha^0_{00}(a))^{-1} = [\delta_{\tau\mu}\Lambda^{\mu 0}_{i\xi}]_{\Delta=\overline{1,N}; z=\overline{1,m}},$$

and consequently

$$S_{00}(a)(\alpha^0_{00})^{-1} f_{\ell 0}(a) = \left[\sum_{j=1}^{m_\mu} \Lambda^{\mu 0}_{ij} \mathfrak{N}^{\mu\ell}_{\varepsilon_{j\mu}}(a) \right]_{\Delta=\overline{1,N}}, \qquad a > 0. \qquad (12.3.12)$$

Let us denote

$$\widetilde{S}^{\ell\mu}(a) = [S_{\ell-\mu+j,j}(a)]^T_{j=\overline{0,\mu}}, \qquad \widetilde{A}^{\ell\mu}(a) = [\alpha^{\ell-\mu+j-\nu}_{\nu j}(a)]_{\nu=\overline{0,\ell}; j=\overline{0,\mu}},$$

$$\widetilde{A}^{\ell 0}_*(a) = [\alpha^{*,\ell-\nu}_{\nu 0}(a)]_{\nu=\overline{0,\ell}}, \qquad \widetilde{\beta}^{*,\ell}(a) = [\beta^{*,\ell-\nu}_\nu(a)]_{\nu=\overline{0,\ell}},$$

$$\widetilde{h}^\ell_0(a) = [\delta_{\nu 0}(\alpha^0_{00}(a))^{-1} f_{\ell 0}(a)]_{\nu=\overline{0,\ell}},$$

$$\psi^a_{\ell 1} = \widetilde{S}^{\ell\ell}\widetilde{h}^\ell_1 + \widetilde{S}^{\ell 0}\varphi_{00} + \sum_{\mu=1}^{\ell-1} \widetilde{S}^{\ell\mu}(\widetilde{h}^\mu_0 + \widetilde{h}^\mu_1 - (\widetilde{A}^{\mu\mu})^{-1}\widetilde{\beta}^{*,\mu} r^a_\mu),$$

$$\widetilde{h}^\ell_1 = -(\widetilde{A}^{\ell\ell})^{-1} + (\widetilde{A}^{\ell 0}_*)\varphi_{00} + \sum_{\mu=1}^{\ell-1} \widetilde{A}^{\ell\mu}(\widetilde{h}^\mu_0 + \widetilde{h}^\mu_1 - (\widetilde{A}^{\mu\mu})^{-1}\widetilde{\beta}^{*,\mu} r^a_\mu),$$

$$H_\ell^0 = \left[\mathfrak{M}^\ell_{m_\mu,\gamma_{\mu i}}\right]_{\Delta=\overline{1,N}}, \quad H_\ell^a = \left[\mathfrak{N}^{\mu\ell}_{\varepsilon_{m_\tau+i},\mu}(a) - \sum_{j=1}^{m_\tau} \Lambda_{ij}^{\mu 0}\mathfrak{N}^{\mu\ell}_{\varepsilon_{j\mu}}(a)\right]_{\Delta=\overline{1,N}}, \quad (a>0).$$

In view of (12.3.12), the equation of the IP for the R_1-structure become

$$\left.\begin{aligned} &d_\ell^0 r_\ell^0 + \psi_{\ell 1}^0(r_1^0,\dots,r_{\ell-1}^0) = H_\ell^0, && \ell \ge 1,\\ &d_\ell^1 r_\ell^a = \psi_{\ell 1}^a(r_1^a,\dots,r_{\ell-1}^a) = H_\ell^a, && \ell\ge 1, \quad a>0. \end{aligned}\right\} \tag{12.3.13}$$

Thus, the following theorem is proved.

Theorem 12.6. *Let $D_\ell^0\#0$, $\ell\ge 1$. Then Problem 9.2 has a unique solution in the class $p_\nu(x)\in A[0,T)$. If, moreover, $D_\ell^1\#0$, $\ell\ge 1$, then Problem 9.2 has a unique solution in the class $p_\nu(x)\in PA[0,T)$. The solution of Problem 9.2 can be obtained by the following algorithm.*

Algorithm 12.3. (1) Set $a=0$.

(2) Compute $\{r_\ell^a\}_{\ell\ge 1}$ from the recurrence relations (12.3.13).

(3) Construct L for $x\in(a,\alpha)$ via

$$p_{n-\kappa_j}(x) = \sum_{\ell=0}^{\infty} r^a_{\ell+\kappa_j,j}\frac{(x-a)^\ell}{\ell!}, \qquad j=\overline{1,N}.$$

(4) If $\alpha<T$ then we set $a:=\alpha$ and go to step 2.

Lemma 12.6. *The following relations hold*

$$A_\ell = d_\ell^0, \qquad A_\ell^1 = d_\ell^1, \qquad \ell\ge 1.$$

Proof. The matrices A_ℓ, A_ℓ^1, d_ℓ^0, d_ℓ^1 do not depend on the coefficients of (1.1) and (1.2). Let $p_\nu(x)\equiv 0$ for $n-\nu\notin\kappa$ and let $r_j^0=0$ for $j=\overline{1,\ell-1}$. Then $\psi_{\ell 1}^0(r_1^0,\dots,r_{\ell-1}^0)=0$, and from (12.3.13) we obtain $d_\ell^0 r_\ell^0 = H_\ell^0$. On the other hand, by virtue of Lemma 9.3, $A_\ell r_\ell^0 = X_\ell^0$. In our case we have $X_\ell^0 = [M^\ell_{m_\mu,\gamma_{\mu i}}]_{\Delta=\overline{1,N}}$. Thus, $X_\ell^0 = H_\ell^0$, and consequently $(A_\ell - d_\ell^0)r_\ell^0 = 0$. Since r_ℓ^0 is arbitrary, we conclude that $A_\ell = d_\ell^0$. The equality $A_\ell^1 = d_\ell^1$ is proved analogously.

Corollary 12.2. *For $A_\ell\#0$ ($A_\ell^1\#0$) it is necesary and sufficient that $D_\ell^0\#0$ ($D_\ell^1\#0$).*

We note that Algorithm 12.3 is simpler than Algorithm 9.1, because in Algorithm 9.1 we must construct auxiliary differential operators in each step.

12.3.2. Let us study some particular cases of Problem 9.2.

Case 1. We consider the IP of determining all the coefficients $p_\nu(x)$, $\nu = \overline{0, n-2}$ of (1.1), i.e. Problem 9.4. For Problem 9.4 we formulated Theorem 9.3, but its proof was put off, because it was difficult to check the information condition $A_\ell \# 0$, $\ell \geq 1$. The results obtained in §12 allow us to prove Theorem 9.3. Indeed, it is obvious that for Problem 9.4 the conditions $D_\ell^0 \# 0$, $D_\ell^1 \# 0$, $\ell \geq 1$ of Theorem 12.6 are fulfilled. In particular, for $\ell \geq n$

$$\det D_\ell^0 = \det D_\ell^1 = \pm n R_1^{(n^2-2n)/2} \neq 0.$$

By virtue of Corollary 12.2, we obtain $A_\ell \# 0$, $A_\ell^1 \# 0$, $\ell \geq 1$ for Problem 9.4. Thus, Theorem 9.3 is proved.

Case 2. We now consider Problem 9.3. As in §9, let $n = 2q$, $\sigma_\xi = \xi - 1$. Denote

$$B_\alpha(R_1) = \det\,[F_{j-i+2}(R_1)]_{i,j=\overline{1,2}}, \qquad (\alpha \geq 1), \qquad B_0(R_1) = 1.$$

For Problem 9.3 we calculate

$$d_\ell^0 = (F_1(R_1))^{-1} \det D_\ell^0 = (-1)^{\ell-\kappa+1} R_1^{n-\kappa} (F_1(R_1))^{-\ell-1+\kappa} B_{\ell-\kappa}(R_1), \;\; \ell \geq \kappa. \tag{12.3.14}$$

On the other hand, using (9.12) and Lemma 12.6,we have

$$d_{\alpha+\kappa}^0 = \frac{(-R_1)^{1-\alpha}}{n 2^\alpha R_1^\kappa} a_\alpha, \qquad \alpha \geq 0, \tag{12.3.15}$$

where a_α is defined by (9.17). Comparing (12.3.14) and (12.3.15), we obtain the following theorem.

Theorem 12.7.

$$\left(\frac{2}{n}\right)^\alpha \det\,[C_n^{j-i+2}]_{i,j=\overline{1,\alpha}}$$
$$= 1 + 2\sum_{\nu=1}^{q-1} \left(\cos\frac{\nu\pi}{n}\right)^{-\alpha} \cos\frac{\nu\pi(\alpha-2)}{n}, \qquad \alpha \geq 1. \tag{12.3.16}$$

Case 3. We consider the DE and the LF

$$-y'' + q(x)y = \lambda y = \rho^2 y, \quad U_{\xi 0}(y) = y^{(\xi-1)}(0), \quad \xi = 1,2; \quad \operatorname{Im}\rho > 0. \tag{12.3.17}$$

Let $\Phi(x,\lambda)$ be the WS for (12.3.17) such that $\Phi(0,\lambda) = 1$ and let $M(\lambda) = \Phi'(0,\lambda)$. Then $M(\lambda)$ is the WF for (12.3.17). Let $q(x)$ be analytic at the point $x = 0$ (in this case we shall write: $q(x) \in A$). It is easy to see that the asymptotic formulas

$$\Phi(x,\lambda) = \exp(i\rho x)\left(1 + \sum_{k=1}^{\infty} \frac{b_k(x)}{(2i\rho)^k}\right), \qquad |\rho| \to \infty,$$

$$M(\lambda) = (i\rho)\left(1 + \sum_{k=1}^{\infty} \frac{M_k}{(i\rho)^k}\right), \qquad |\rho| \to \infty, \qquad M_1 = 0.$$

are valid. The sequence $\{M_k\}_{k\ge 1}$ is called the Weyl sequence. The following theorem gives necessary and sufficient conditions on the Weyl sequence.

Theorem 12.8. (1) *If $q(x) \in A$, then there exists a $\delta > 0$ such that*

$$M_k = O\left(\left(\frac{k}{\delta}\right)^k\right). \tag{12.3.18}$$

(2) *Let an arbitrary sequence $\{M_k\}_{k\ge 1}$, $M_1 = 0$, satisfying* (12.3.18) *for a certain $\delta > 0$, be given. Then there exists a unique function $q(x) \in A$ for which $\{M_k\}_{k\ge 1}$ is the Weyl sequence.*

Proof. (1) Let $q(x) \in A$. We denote $\alpha_k = 2^k M_{k+1}$, $q_k = q^{(k)}(0)$, $b_{k\nu} = b_k^{(\nu)}(0)$. Then (12.1.12) for this problem becomes

$$\left.\begin{aligned} &b_{\nu+1,k-\nu+1} = -b_{\nu,k-\nu+2} + \sum_{j=1}^{k-\nu} C_{k-\nu}^j\, q_{k-\nu-j}\, b_{\nu j}, \quad 0 \le \nu \le k \\ &b_{k+1,0} = 0, \quad b_{k+1,1} = \alpha_{k+1}, \quad b_{0k} = \delta_{0k}, \quad k \ge 0. \end{aligned}\right\} \tag{12.3.19}$$

Since $q(x) \in A$, there exist $\delta_0 > 0$ and $C > 0$ such that

$$|q_k| \le C\left(\frac{k}{\delta_0}\right)^k.$$

We denote $a = 1 + C\delta_0^2$. Let us show by induction that

$$|b_{\nu+1,k-\nu+1}| \le Ca^\nu\left(\frac{k}{\delta_0}\right)^k, \qquad 0 \le \nu \le k. \tag{12.3.20}$$

For $\nu = 0$ (12.3.20) is obvious by virtue of $b_{1,k+1} = q_k$. We assume that (12.3.20) holds for $\nu = 1, \ldots, s-1$. Then, using (12.3.19) and the estimate $C_k^\nu = k!(\nu!(k-\nu)!)^{-1} \le k^k(\nu^\nu(k-\nu)^{k-\nu})^{-1}$ we obtain

$$|b_{s+1,k-s+1}| \le Ca^{s-1}\left(\frac{k}{\delta_0}\right)^k$$
$$\times\left(1 + C\delta_0^2\,\frac{(k-s)^{k-s}}{k^k}\sum_{j=1}^{k-s}\frac{(j+s-2)^{j+s-2}}{j^j}\right) \le Ca^s\left(\frac{k}{\delta_0}\right)^k,$$

i.e. (12.3.20) holds for $\nu = s$.

It follows from (12.3.20) for $s = k$, that $|\alpha_{k+1}| \le Ca^k \left(\frac{k}{\delta_0}\right)^k$ and consequently $M_k = O\left(\left(\frac{k}{\delta}\right)^k\right)$, $\delta = \frac{2\delta_0}{a}$.

(2) Let $\{M_k\}_{k\ge1}$ such that $M_k = O\left(\left(\frac{k}{\delta}\right)^k\right)$, $\delta > 0$, $M_1 = 0$ be given. Then $|\alpha_{k+1}| \le C\left(\frac{k}{\delta_1}\right)^k$. Using (12.3.19) successively for $k = 1, 2, \ldots$ we construct the numbers q_k, $b_{k\nu}$, k, $\nu \ge 0$. We obtain the recurrent relations

$$\left.\begin{aligned} q_k &= (-1)^{k+1}\alpha_{k+1} + \sum_{\nu=1}^{k-1}\sum_{\xi=1}^{\nu}(-1)^{k-\nu-1}C_\nu^\xi\, q_{\nu-\xi} b_{k-\nu,\xi}\,, \\ b_{k-j+1,j+1} &= (-1)^j\alpha_{k+1} + \sum_{\nu=1}^{j-1}\sum_{\xi=1}^{\nu}(-1)^{j-\nu-1}C_\nu^\xi\, q_{\nu-\xi}\, b_{k-\nu,\xi}. \end{aligned}\right\} \tag{12.3.21}$$

By induction from (12.3.21) we derive the estimates

$$|q_k| \le C\, a_0^k \left(\frac{k}{\delta_1}\right)^k, \qquad |b_{k-j+1,j+1}| \le Ca_0^k \left(\frac{k}{\delta_1}\right)^k, \qquad 0 \le j \le k,$$

where $a_0 = \max(2, \sqrt{6C}\,\delta_1)$. Consequently, $q_k = O\left(\left(\frac{k}{\delta_2}\right)^k\right)$. We construct the function $q(x) \in A$ by

$$q(x) = \sum_{k=0}^{\infty} q_k \frac{x^k}{k!}.$$

It is clear that $\{M_k\}_{k\ge1}$ is the Weyl sequence for this function. The uniqueness of the solution of the IP is obvious. Theorem 12.8 is proved.

13. Inverse problems for integro-differential operators

In this section, a perturbation of the Sturm–Liouville operator by a Volterra integral operator is considered. The presence of an "aftereffect" in a mathematical model produced qualitative changes in the study of the IP. The main results of the section are expressed by Theorems 13.1–13.3. Note that the IP for integro-differential operators in various formulations has been studied in [119]–[122]. Among other things, a connection is pointed out in [120] between the IP under consideration here and the completeness of the eigen- and associated functions of a bundle of fourth-order integrodifferential operators.

13.1. Let $\{\lambda_n\}_{n\ge1}$ be the eigenvalues of a boundary value problem $L = L(q, M)$ of the form

$$\ell y(x) \equiv -y''(x) + q(x)y(x) + \int_0^x M(x-t)y(t)\,dt = \lambda y(x) = \rho^2 y(x), \tag{13.1}$$

$$y(0) = y(\pi) = 0. \tag{13.2}$$

Considered the following problem:

Problem 13.1. Given the function $q(x)$ and the spectrum $\{\lambda_n\}_{n\geq 1}$, find the function $M(x)$.

Put

$$M_0(x) = (\pi - x)M(x), \qquad M_1(x) = \int_0^x M(t)\,dt, \qquad Q(x) = M_0(x) - M_1(x).$$

We shall assume that $q(x)$, $Q(x) \in \mathcal{L}_2(0,\pi)$, $M_k(x) \in \mathcal{L}(0,\pi)$, $k = 0,1$.

Let $S(x,\lambda)$ be the solution of (13.1) under the initial conditions $S(0,\lambda) = 0$, $S'(0,\lambda) = 1$.

Lemma 13.1. *The representation*

$$S(x,\lambda) = \frac{\sin \rho x}{\rho} + \int_0^x K(x,t)\,\frac{\sin \rho t}{\rho}\,dt, \tag{13.3}$$

holds, where $K(x,t)$ is a continuous function, and $K(x,0) = 0$.

Proof. The function $S(x,\lambda)$ is the solution of the integral equation

$$S(x,\lambda) = \frac{\sin\,\rho x}{\rho} + \int_0^x \frac{\sin\,\rho(x-\tau)}{\rho}\left(q(\tau)S(\tau,\lambda) + \int_0^\tau M(\tau - s)S(s,\lambda)\,ds\right)d\tau. \tag{13.4}$$

Since

$$\int_0^x \frac{\sin\,\rho(x-\tau)}{\rho}\,f(\tau)\,d\tau = \int_0^x dt \int_0^t f(\tau)\cos\,\rho(t-\tau)\,d\tau,$$

then (13.4) is transformable to the form

$$S(x,\lambda) = \frac{\sin\,\rho x}{\rho} + \int_0^x dt \int_0^t \left(q(\tau)S(\tau,\lambda) + \int_0^\tau M(\tau - s)S(s,\lambda)\,ds\right)\cos\,\rho(t-\tau)\,d\tau. \tag{13.5}$$

Apply the method of successive approximations to solving the equation (13.5):

$$S_0(x,\lambda)=\frac{\sin\rho x}{\rho},$$

$$S_{n+1}(x,\lambda)=\int_0^x dt\int_0^t\left(q(\tau)S_n(\tau,\lambda)+\int_0^\tau M(\tau-s)S_n(s,\lambda)\,ds\right)\cos\rho(t-\tau)\,d\tau.$$

Transform $S_1(x,\lambda)$:

$$\begin{aligned}S_1(x,\lambda)&=\frac{1}{2\rho}\int_0^x\sin\rho t\,dt\int_0^t q(\tau)\,d\tau+\frac{1}{2\rho}\int_0^x dt\int_0^t q(\tau)\sin\rho(2\tau-t)\,d\tau\\&+\frac{1}{2\rho}\int_0^x dt\int_0^t d\tau\int_0^\tau M(\tau-s)\sin\rho(s+t-\tau)\,ds\\&+\frac{1}{2\rho}\int_0^x dt\int_0^t d\tau\int_0^\tau M(\tau-s)\sin\rho(s-t+\tau)\,ds.\end{aligned}$$

Carrying out the change of variables $\xi=2\tau-t$, $\xi=s+t-\tau$, $\xi=s-t+\tau$, respectively, in the last three integrals and reversing the order of integration, we obtain

$$S_1(x,\lambda)=\int_0^x K_1(x,\xi)\frac{\sin\rho\xi}{\rho}\,d\xi,$$

$$\begin{aligned}K_1(x,\xi)&=\frac{1}{2}\int_0^\xi q(\tau)\,d\tau+\frac{1}{4}\int_\xi^x\left(q\left(\frac{\xi+t}{2}\right)-q\left(\frac{t-\xi}{2}\right)\right)dt\\&+\frac{1}{2}\int_\xi^x dt\left(\xi M(t-\xi)+\int_{(\xi+t)/2}^{t} M(2\tau-\xi-t)\,d\tau-\int_{(t-\xi)/2}^{t-\xi} M(2\tau+\xi-t)\,d\tau\right).\end{aligned}\tag{13.6}$$

Clearly $K_1(x,0)=0$. In an analogous manner we calculate

$$S_n(x,\lambda)=\int_0^x K_n(x,\xi)\frac{\sin\rho\xi}{\rho}\,d\xi,$$

where the functions $K_n(x,\xi)$ are determined by the recurrence formula

$$\begin{aligned}K_{n+1}(x,\xi) &= \frac{1}{2}\int_{\xi}^{x}\Big(\int_{t-\xi}^{t} q(\tau)K_n(\tau,\xi+\tau-t)\,d\tau + \int_{(\xi+t)/2}^{t} q(\tau)K_n(\tau,\xi+t-\tau)\,d\tau \\ &\quad - \int_{(t-\xi)/2}^{t-\xi} q(\tau)K_n(\tau,-\xi+t-\tau)\,d\tau \\ &\quad + \int_{t-\xi}^{t} d\tau \int_{\xi-t+\tau}^{\tau} M(\tau-s)K_n(s,\xi-t+\tau)\,ds \\ &\quad + \int_{(\xi+t)/2}^{\xi} d\tau \int_{\xi+t-\tau}^{\tau} M(\tau-s)K_n(s,\xi+t-\tau)\,ds \\ &\quad - \int_{(t-\xi)/2}^{t-\xi} d\tau \int_{-\xi+t-\tau}^{\tau} M(\tau-s)K_n(s,-\xi+t-\tau)\,ds\Big)\,dt.\end{aligned} \tag{13.7}$$

One has, $K_n(x,0) = 0$. From (13.6) and (13.7) we use induction to obtain the estimate

$$|K_n(x,\xi)| \le \frac{1}{n!}(Cx)^n, \qquad 0 \le \xi \le x \le \pi.$$

Thus,

$$S(x,\lambda) = \sum_{n=0}^{\infty} S_n(x,\lambda) = \frac{\sin\rho x}{\rho} + \int_0^x K(x,\xi)\frac{\sin\rho\xi}{\rho}\,d\xi,$$

$$K(x,\xi) = \sum_{n=1}^{\infty} K_n(x,\xi), \tag{13.8}$$

and the series (13.8) converges absolutely and uniformly for $0 \le \xi \le x \le \pi$. Lemma 13.1 is proved.

Denote $\Delta(\lambda) = S(\pi,\lambda)$. The eigenvalues $\{\lambda_n\}_{n\ge 1}$ of the boundary value problem L coincide with the zeros of the function $\Delta(\lambda)$, and as $n\to\infty$,

$$\rho_n = \sqrt{\lambda_n} = n + \frac{A_1}{n} + \frac{\kappa_n}{n}, \qquad \kappa_n \in \ell_2, \qquad A_1 = \frac{1}{2\pi}\int_0^{\pi} q(t)\,dt. \tag{13.9}$$

The following assertion is obvious.

Lemma 13.2. *The function $\Delta(\lambda)$ is uniquely determined by its zeros. Moreover,*

$$\Delta(\lambda) = \pi \prod_{n=1}^{\infty} \frac{\lambda_n - \lambda}{n^2}. \tag{13.10}$$

We will now prove the uniqueness theorem for the solution of Problem 13.1. Let $\{\widetilde{\lambda}_n\}_{n\geq 1}$ be the eigenvalues of the boundary value problem $\widetilde{L} = L(q, \widetilde{M})$.

Theorem 13.1. *If $\lambda_n = \widetilde{\lambda}_n$, $n \geq 1$, then $M(x) \overset{a.e.}{=} \widetilde{M}(x)$, $x \in (0, \pi)$.*

Proof. Let the function $S^*(x,\lambda)$ be the solution of the equation

$$\ell^* z(x) = -z''(x) + q(x)z(x) + \int_x^\pi M(t-x)z(t)\,dt = \lambda z(x) \tag{13.11}$$

under the conditions $S^*(\pi,\lambda) = 0$, $S'^*(\pi,\lambda) = -1$. Put $\Delta^*(\lambda) = S^*(0,\lambda)$. Then

$$\int_0^\pi S^*(x,\lambda)\,dx \int_0^x \widehat{M}(x-t)\widetilde{S}(t,\lambda)\,dt$$

$$= \int_0^\pi S^*(x,\lambda)\ell\widetilde{S}(x,\lambda)\,dx - \int_0^\pi S^*(x,\lambda)\widetilde{\ell}\widetilde{S}(x,\lambda)\,dx$$

$$= \int_0^\pi \ell^* S^*(x,\lambda)\widetilde{S}(x,\lambda)\,dx - \int_0^\pi S^*(x,\lambda)\widetilde{\ell}\widetilde{S}(x,\lambda)\,dx$$

$$+\Big|_0^\pi (\widetilde{S}(x,\lambda)S'^*(x,\lambda) - \widetilde{S}'(x,\lambda)S^*(x,\lambda)) = \Delta^*(\lambda) - \widetilde{\Delta}(\lambda).$$

For $\widetilde{\ell} = \ell$ we have $\Delta^*(\lambda) \equiv \Delta(\lambda)$, and consequently

$$\int_0^\pi S^*(x,\lambda)\,dx \int_0^x \widehat{M}(x-t)\widetilde{S}(t,\lambda)\,dt = \widehat{\Delta}(\lambda). \tag{13.12}$$

Transform (13.12) into

$$\int_0^\pi \widehat{M}(x)\,dx \int_x^\pi S^*(t,\lambda)\widetilde{S}(t-x,\lambda)\,dt = \widehat{\Delta}(\lambda). \tag{13.13}$$

Denote $w(x,\lambda) = S^*(\pi - x,\lambda)$, $N(x) = M(\pi - x)$,

$$\varphi(x,\lambda) = \int_0^x w(t,\lambda)\widetilde{S}(x-t,\lambda)\,dt. \tag{13.14}$$

Then (13.13) takes the form

$$\int_0^\pi \widehat{N}(x)\varphi(x,\lambda)\,dx = \widehat{\Delta}(\lambda). \tag{13.15}$$

Lemma 13.3. *The representation*

$$\varphi(x,\lambda) = \frac{1}{2\rho^2}\left(-x\cos\rho x + \int_0^x V(x,t)\cos\rho t\,dt\right), \tag{13.16}$$

holds, where $V(x,t)$ is a continuous function.

Proof. Since $w(x,\lambda) = S^*(\pi - x,\lambda)$, the function $w(x,\lambda)$ is the solution of the Cauchy problem

$$-w''(x,\lambda) + q(\pi - x)w(x,\lambda) + \int_0^x M(x-t)w(t,\lambda)\,dt = \lambda w(x,\lambda),$$

$$w(0,\lambda) = 0, \qquad w'(0,\lambda) = 1.$$

Therefore, by Lemma 13.1, the representation

$$w(x,\lambda) = \frac{\sin\rho x}{\rho} + \int_0^x K^0(x,t)\frac{\sin\rho t}{\rho}\,dt, \tag{13.17}$$

holds, where $K^0(x,t)$ is a continuous function. Substituting (13.3) and (13.17) into (13.14), we obtain

$$\varphi(x,\lambda) = \varphi_1(x,\lambda) + \varphi_2(x,\lambda) + \varphi_3(x,\lambda) + \varphi_4(x,\lambda),$$

where

$$\varphi_1(x,\lambda) = \frac{1}{\rho^2}\int_0^x \sin\rho t\,\sin\rho(x-t)\,dt,$$

$$\varphi_2(x,\lambda) = \frac{1}{\rho^2}\int_0^x \sin\rho(x-t)\,dt\int_0^t K^0(t,\xi)\sin\rho\xi\,d\xi,$$

$$\varphi_3(x,\lambda) = \frac{1}{\rho^2}\int_0^x \sin\rho t\,dt\int_0^{x-t}\widetilde{K}(x-t,\eta)\sin\rho\eta\,d\eta,$$

$$\varphi_4(x,\lambda) = \frac{1}{\rho^2}\int_0^x dt\int_0^t K^0(t,\xi)\sin\rho\xi\,d\xi\int_0^{x-t}\widetilde{K}(x-t,\eta)\sin\rho\eta\,d\eta.$$

For $\varphi_1(x,\lambda)$ we have

$$\varphi_1(x,\lambda) = \frac{1}{2\rho^2}\int_0^x (\cos\,\rho(x-2t) - \cos\,\rho x)\,dt = \frac{1}{2\rho^2}\left(-x\cos\,\rho x + \int_0^x \cos\,\rho t\,dt\right).$$

Transform $\varphi_2(x,\lambda)$:

$$\begin{aligned}\varphi_2(x,\lambda) &= \frac{1}{2\rho^2}\int_0^x dt \int_0^t K^0(t,\xi)(\cos\,\rho(x-t-\xi) - \cos\,\rho(x-t+\xi))\,d\xi \\ &= \frac{1}{2\rho^2}\int_0^x dt\Bigg(\int_{x-2t}^{x-t} K^0(t,x-t-s)\cos\,\rho s\,ds \\ &\quad - \int_{x-t}^{x} K^0(t,s+t-x)\cos\,\rho s\,ds\Bigg).\end{aligned}$$

Reversing the integration order, we obtain

$$\varphi_2(x,\lambda) = \frac{1}{2\rho^2}\int_0^x V_2(x,t)\cos\,\rho t\,dt,$$

$$V_2(x,t) = \int_{(x-t)/2}^{x-t} K^0(s,x-t-s)\,ds + \int_{(x+t)/2}^{x} K^0(s,x+t-s)\,ds - \int_{x-t}^{x} K^0(s,t+s-x)\,ds.$$

The integrals $\varphi_3(x,\lambda)$ and $\varphi_4(x,\lambda)$ are transformable in an analogous manner. Lemma 13.3 is proved.

Let us return to proving Theorem 13.1. Since $\lambda_n = \tilde{\lambda}_n$, $n \geq 1$, we have by Lemma 13.2 $\Delta(\lambda) \equiv \tilde{\Delta}(\lambda)$. Then, substituting (13.16) into (13.15), we obtain

$$\int_0^\pi \cos\,\rho x\left(-x\widehat{N}(x) + \int_x^\pi V(t,x)\widehat{N}(t)\,dt\right)dx \equiv 0,$$

and consequently,

$$-x\widehat{N}(x) + \int_x^\pi V(t,x)\widehat{N}(t)\,dt = 0. \tag{13.18}$$

For each fixed $\varepsilon > 0$ (13.18) is a Volterra homogeneous integral equation of the second kind in the interval (ε,π). Consequently $\widehat{N}(x) = 0$ a.e. in (ε,π) and, since ε is arbitrary, in the whole interval $(0,\pi)$. Thus, $M(x) = \widetilde{M}(x)$ a.e. in $(0,\pi)$, and the theorem is proved.

Also, (13.15) makes it possible to obtain an algorithm for solving Problem 13.1 in the case when $M(x) \in PA[0,\pi]$. Consider $L(q,M)$ and $L(q,\widetilde{M})$, and assume that $q(x) \in \mathcal{L}_2(0,\pi)$; $M(x),\widetilde{M}(x) \in PA$. Let for some fixed $a > 0$

$$\widehat{N}(x) = 0, \qquad x \in (a,\pi); \qquad \widehat{N}(x) \sim \widehat{N}_\alpha^a (\alpha!)^{-1}(a-x)^\alpha, \qquad x \to a-0. \quad (13.19)$$

It follows from (13.16) that as $|\rho| \to \infty$, $\arg\rho \in [\delta, \pi-\delta]$, $x \in [\varepsilon,\pi]$, $\delta > 0$, $\varepsilon > 0$, the asymptotic formula

$$\varphi(x,\lambda) = -x(4\rho^2)^{-1}\exp(-i\rho x)(1 + O(\rho^{-1})) \quad (13.20)$$

holds. Furthermore, it follows from (13.16) that

$$|\varphi(x,\lambda)| < C|\rho^{-2}\exp(-i\rho x)|, \qquad x \in [0,\pi], \qquad \operatorname{Im}\rho \geq 0. \quad (13.21)$$

Using (13.21) we obtain the estimate

$$\left| \int_0^\varepsilon \widehat{N}(x)\varphi(x,\lambda)\,dx \right| < C|\rho^{-2}\exp(-i\rho\varepsilon)|, \qquad \operatorname{Im}\rho \geq 0. \quad (13.22)$$

Using (13.19), (13.20) and Lemma 9.1, we obtain for $|\rho| \to \infty$,

$$\int_\varepsilon^a \widehat{N}(x)\varphi(x,\lambda)\,dx = \frac{a}{4(-i\rho)^{\alpha+3}}\exp(-i\rho a)(\widehat{N}_\alpha^a + o(1)). \quad (13.23)$$

Since $\widehat{N}(x) = 0$ for $x \in (a,\pi)$, it follows from (13.15), (13.22) and (13.23) that as $|\rho| \to \infty$, $\arg\rho \in [\delta, \pi-\delta]$

$$\widehat{\Delta}(\lambda) = \frac{a}{4}(-i\rho)^{-\alpha-3}\exp(-i\rho a)(\widehat{N}_\alpha^a + o(1)),$$

and consequently

$$\widehat{N}_\alpha^a = \frac{4}{a}\lim \widehat{\Delta}(\lambda)(-i\rho)^{\alpha+3}\exp(i\rho a), \qquad |\rho| \to \infty, \qquad \arg\rho \in [\delta,\pi-\delta]. \quad (13.24)$$

Thus we have proved the following theorem.

Theorem 13.2. *Let $\{\lambda_n\}_{n\geq 1}$ be the eigenvalues of $L(q,M)$, where $q(x) \in \mathcal{L}_2(0,\pi)$, $M(x) \in PA$. Then the solution of Problem 13.1 can be found by the following algorithm:*

(1) *Construct the function $\Delta(\lambda)$ from $\{\lambda_n\}_{n\geq 1}$ by formula* (13.10).

(2) *Take $a = \pi$.*

(3) *Carry out the operations successively for $\alpha = 0,1,2,\ldots$: construct a function $\widetilde{M}(x) \in KA$ so that $\widehat{N}(x) = 0$, $x \in (a,\pi)$; $\widehat{N}^{(k)}(a-0) = 0$, $k = \overline{0,\alpha-1}$ and find $N_\alpha^a = (-1)^\alpha N^{(\alpha)}(a-0)$ from* (13.24).

(4) *Construct* $N(x)$ *for* $x \in (a^+, a)$ *by the formula*

$$N(x) = \sum_{\alpha=0}^{\infty} N_\alpha^a \frac{(a-x)^\alpha}{\alpha!}.$$

(5) *If* $a^+ > 0$ *set* $a := a^+$ *and pass to step* (3).

13.2. We will now investigate the question of solving Problem 13.1 "in the small" and of stability. First let us prove an auxiliary assertion.

Lemma 13.4. *In a Banach space* B *consider the nonlinear equation*

$$r = f + \sum_{j=2}^{\infty} \psi_j(r), \tag{13.25}$$

$$||\psi_j(r)|| \le (C||r||)^j, \qquad ||\psi_j(r) - \psi_j(r^*)|| \le ||r - r^*||(C\max(||r||, ||r^*||))^{j-1}.$$

There exists $\delta > 0$ *such that if* $||f|| < \delta$, *then in the ball* $||r|| < 2\delta$. *Eq.*(13.25) *has a unique solution* $r \in B$, *for which* $||r|| \le 2||f||$.

Proof. Assume that $C \ge 1$. Put

$$\psi(r) = \sum_{j=2}^{\infty} \psi_j(r), \qquad C_0 = 2C^2, \qquad \delta = \frac{1}{4C_0}.$$

If $||r||, ||r^*|| \le (2C_0)^{-1}$, then

$$\begin{gathered} ||\psi(r)|| \le \sum_{j=2}^{\infty} (C||r||)^j \le C_0||r||^2 \le \frac{1}{2}||r||, \\ ||\psi(r) - \psi(r^*)|| \le ||r - r^*|| \sum_{j=2}^{\infty} (C(2C_0)^{-1})^{j-1} \le \tfrac{1}{2}||r - r^*||. \end{gathered} \tag{13.26}$$

Let $||f|| \le \delta$; construct $r_0 = f$, $r_{k+1} = f + \psi(r_k)$, $k \ge 0$. By induction, using (13.26), we obtain the estimates

$$||r_k|| \le 2||f||, \qquad ||r_{k+1} - r_k|| \le \frac{1}{2^{k+1}}||f||, \qquad k \ge 0.$$

Consequently, the series

$$r = r_0 + \sum_{k=0}^{\infty} (r_{k+1} - r_k)$$

converges to the solution of (13.25), and $||r|| \le 2||f||$. Lemma 13.4 is proved.

Theorem 13.3. *For the boundary value problem $L = L(q, M)$ with the spectrum $\{\lambda_n\}_{n\geq 1}$ there exists $\delta > 0$ (which depends on L) such that if the numbers $\{\tilde\lambda_n\}_{n\geq 1}$ satisfy the condition*

$$\Lambda \overset{df}{=} \left(\sum_{n=1}^{\infty} |\lambda_n - \tilde\lambda_n|^2\right)^{1/2} < \delta,$$

then there exists a unique $\tilde L = L(q, \tilde M)$ for which the numbers $\{\tilde\lambda_n\}_{n\geq 1}$ are the eigenvalues, and

$$\|Q(x) - \tilde Q(x)\|_{\mathcal{L}_2(0,\pi)} \leq C\Lambda,$$

$$\|M_k(x) - \tilde M_k(x)\|_{\mathcal{L}(0,\pi)} \leq C\Lambda, \qquad k = 0, 1.$$

Here and below, C denotes various constants dependent on L.

Proof. For brevity, we confine ourselves to the case when all the eigenvalues are simple. The Cauchy problem $\ell y(x) - \lambda y(x) + f(x) = 0$, $y(0) = y'(0) = 0$ has a unique solution

$$y(x) = \int_0^x g(x, t, \lambda) f(t)\, dt,$$

where $g(x, t, \lambda)$ is the Green's function satisfying the relations

$$-g_{xx}(x, t, \lambda) + q(x) g(x, t, \lambda) - \lambda g(x, t, \lambda) + \int_t^x M(x - \tau) g(\tau, t, \lambda)\, d\tau = 0, \qquad x > t$$

$$g(t, t, \lambda) = 0, \qquad g_x(x, t, \lambda)\Big|_{x=t} = 1.$$

Denote

$$\mathcal{G}(x, t, \lambda) = g_t(x, t, \lambda), \qquad y_n(x) = S(x, \tilde\lambda_n), \qquad \varepsilon_n = n^2 \Delta(\tilde\lambda_n),$$

$$v_n(x, t) = \begin{cases} w'(\pi - x - t, \tilde\lambda_n), & 0 < t < \pi - x, \\ 0, & \pi - x < t < \pi, \end{cases}$$

$$\mathcal{G}_n(x, t, s) = \begin{cases} \mathcal{G}(x, s + t, \tilde\lambda_n), & s + t \leq x, \\ 0, & s + t > x \end{cases}$$

$$\varphi_n(x) = \int_0^x w(t, \tilde\lambda_n) S(x - t, \tilde\lambda_n)\, dt, \qquad \xi_n(x) = \int_0^\pi v_n(x, t) y_n(t)\, dt,$$

$$\psi_n(x) = \frac{n}{x}\varphi_n'(x), \qquad \psi_{n0}(x) = \frac{n}{2\widetilde{\rho}_n}\sin\widetilde{\rho}_n x, \qquad \eta_n(x) = \frac{n}{\pi - x}\overline{\xi_n(x)}.$$

Let W_2^1 be the space of functions $f(x)$ be absolutely continuous on $[0,\pi]$ and such that $f'(x) \in \mathcal{L}_2(0,\pi)$, with the norm $\|f\|_{W_2^1} = \|f\|_{\mathcal{L}_2(0,\pi)} + \|f'\|_{\mathcal{L}_2(0,\pi)}$, and let $W_{20}^1 = \{f(x) : f(x) \in W_2^1, f(0) = f(\pi) = 0\}$.

Lemma 13.5. *The functions $\{\psi_n(x)\}_{n\geq 1}$ constitute a Riesz basis in $\mathcal{L}_2(0,\pi)$, and the biorthogonal basis $\{\psi_n^*(x)\}_{n\geq 1}$ possesses the following properties:*
(1) $\psi_n^*(x) \in W_{20}^1$;
(2) $|\psi_n(x)| \leq C, \quad n \geq 1, \quad x \in [0,\pi]$;
(3) *for any* $\{\theta_n\} \in \ell_2$

$$\theta(x) \overset{df}{=} \sum_{n=1}^{\infty} \frac{\theta_n}{n}\psi_n^*(x) \in W_{20}^1, \qquad \|\theta(x)\|_{W_2^1} \leq C\left(\sum_{n=1}^{\infty}|\theta_n|^2\right)^{1/2}.$$

To prove this, we shall use the well-known results for the Sturm–Liouville IP. Since $\Lambda < \infty$, it follows from (13.9) that

$$\widetilde{\rho}_n = \sqrt{\widetilde{\lambda}_n} = n + \frac{A_1}{n} + \frac{\widetilde{\kappa}_n}{n}, \qquad \widetilde{\kappa}_n \in \ell_2. \tag{13.27}$$

Consequently, there exists a function $\widetilde{q}(x)$ (not unique) such that the numbers $\{\widetilde{\lambda}_n\}_{n\geq 1}$ are the eigenvalues of the Sturm–Liouville boundary value problem

$$-y'' + \widetilde{q}(x)y = \lambda y, \qquad y(0) = y(\pi) = 0. \tag{13.28}$$

Let $\widetilde{s}_n(x)$ be the eigenfunctions of (13.28) normalized by the condition $\widetilde{s}'(0) = \frac{n}{2}$. The functions $\{\widetilde{s}_n(x)\}_{n\geq 1}$ constitute a Riesz basis in $\mathcal{L}_2(0,\pi)$, and

$$\int_0^{\pi} \widetilde{s}_n(x)\widetilde{s}_m(x)\,dx = \delta_{nm}\widetilde{\alpha}_n. \tag{13.29}$$

Using Lemma 13.1, we obtain

$$\widetilde{s}_n(x) = \psi_{n0}(x) + \int_0^x \widetilde{K}(x,t)\psi_{n0}(t)\,dt, \qquad \widetilde{K}(x,0) = 0. \tag{13.30}$$

In particular, it follows from (13.30), (13.29) and (13.27) that

$$\widetilde{s}_n(x) = \frac{1}{2}\sin nx + O\left(\frac{1}{n}\right), \qquad \widetilde{\alpha}_n = \int_0^{\pi} \widetilde{s}_n^2(x)\,dx = \frac{\pi}{8} + O\left(\frac{1}{n}\right), \qquad n \to \infty.$$

Due to (13.30), the functions $\{\psi_{n0}(x)\}_{n\geq 1}$ constitute a Riesz basis in $\mathcal{L}_2(0,\pi)$. Denote

$$\psi_{n0}^{**}(x) = \widetilde{s}_n(x) + \int_x^\pi \widetilde{K}(t,x)\widetilde{s}_n(t)\,dt. \tag{13.31}$$

It follows from (13.29)–(13.31) that

$$\begin{aligned}
\int_0^\pi \psi_{n0}(x)\psi_{m0}^{**}(x)\,dx &= \int_0^\pi \psi_{n0}\Big(\widetilde{s}_m(x) + \int_x^\pi \widetilde{K}(t,x)\widetilde{s}_m(t)\,dt\Big)\,dx \\
&= \int_0^\pi \widetilde{s}_m(x)\Big(\psi_{n0}(x) + \int_0^x \widetilde{K}(x,t)\psi_{n0}(t)\,dt\Big)dx \\
&= \int_0^\pi \widetilde{s}_n(x)\widetilde{s}_m(x)\,dx = \delta_{nm}\widetilde{\alpha}_n.
\end{aligned} \tag{13.32}$$

Also, we compute

$$\psi_n(x) = \frac{n}{x}\int_0^x w(t,\widetilde{\lambda}_n)S'(x-t,\widetilde{\lambda}_n)\,dt. \tag{13.33}$$

Since

$$S'(x,\lambda) = \cos\rho x + \int_0^x K^1(x,t)\cos\rho t\,dt, \tag{13.34}$$

we obtain, substituting (13.34) and (13.17) into (13.33) as in the proof of Lemma 13.3,

$$\psi_n(x) = \psi_{n0}(x) + \int_0^x V_0(x,t)\psi_{n0}(t)\,dt, \tag{13.35}$$

where $V_0(x,t)$ is a continuous function, $V_0(x,0)=0$. Solving the integral equation (13.35), we find

$$\psi_{n0}(x) = \psi_n(x) + \int_0^x V_1(x,t)\psi_n(t)\,dt, \qquad V_1(x,0)=0. \tag{13.36}$$

Consider the functions

$$\psi_n^{**}(x) = \psi_{n0}^{**}(x) + \int_x^\pi V_1(t,x)\psi_{n0}^{**}(t)\,dt. \tag{13.37}$$

It follows from (13.32), (13.36) and (13.37) that

$$\int_0^\pi \psi_n(x)\psi_m^{**}(x)\,dx = \delta_{nm}\widetilde{\alpha}_n. \tag{13.38}$$

By virtue of (13.35) and (13.38), the functions $\{\psi_n(x)\}_{n\geq 1}$ constitute a Riesz basis in $\mathcal{L}_2(0,\pi)$, and the biorthogonal basis $\{\psi_n^*(x)\}_{n\geq 1}$ has the form $\psi_n^*(x) = \overline{\widetilde{\alpha}_n^{-1}\psi_n^{**}(x)}$. Substituting (13.31) into (13.37), we have

$$\psi_n^{**}(x) = \widetilde{s}_n(x) + \int_x^\pi V_1^0(t,x)\widetilde{s}_n(t)\,dt, \qquad V_1^0(t,0) = 0.$$

Hence we obtain the required properties of the biorthogonal basis. Lemma 13.5 is proved.

Since $\eta_n(x) = \psi_n(\pi - x)$, Lemma 13.5 implies

Corollary 13.1. *The functions $\{\eta_n(x)\}_{n\geq 1}$ constitute a Riesz basis in $\mathcal{L}_2(0,\pi)$, and the biorthogonal basis $\{\chi_n(x)\}_{n\geq 1}$ possesses the properties:*
(1) $\chi_n(x) \in W_{20}^1$;
(2) $|\chi_n(x)| \leq C, \ n \geq 1, \ x \in [0,\pi]$;
(3) *for any* $\{\theta_n\} \in \ell_2$

$$\theta(x) \stackrel{df}{=} \sum_{n=1}^\infty \frac{\theta_n}{n}\chi_n(x) \in W_{20}^1, \qquad \|\theta(x)\|_{W_2^1} \leq C\left(\sum_{n=1}^\infty |\theta_n|^2\right)^{1/2}$$

Let us return to proving Theorem 13.3. Put

$$\varepsilon(x) = \sum_{n=1}^\infty \frac{\varepsilon_n}{n}\chi_n(x). \tag{13.39}$$

Using Lemma 13.1, the relations $\Delta(\lambda) = S(\pi,\lambda)$, $\Delta(\lambda_n) = 0$, and the formulae (13.9), (13.27), we obtain the estimate $|\varepsilon_n| = n^2|\Delta(\widetilde{\lambda}_n) - \Delta(\lambda_n)| \leq C|\lambda_n - \widetilde{\lambda}_n|$. Now by Corollary 13.1 we have

$$\varepsilon(x) \in W_{20}^1, \qquad \|\varepsilon(x)\|_{W_2^1} \leq C\Lambda.$$

Consider the nonlinear equation in W_{20}^1

$$r = \varepsilon + \sum_{j=2}^\infty \psi_j(r), \tag{13.40}$$

where $\varepsilon(x)$ is defined by (13.39), and the operators $z_j = \psi_j(r)$ act from W_{20}^1 to W_{20}^1 according to the formula

$$z_j(x) = -\sum_{n=1}^\infty \Bigl(\underbrace{\int_0^\pi \dots \int_0^\pi}_{j} r(t_1)\dots r(t_j)B_{nj}(t_1,\dots,t_j)\,dt_1\dots dt_j\Bigr)\chi_n(x),$$

$$B_{nj}(t_1,\dots,t_j)=\frac{n}{(\pi-t_1)\dots(\pi-t_j)}$$

$$\times\underbrace{\int_0^\pi\dots\int_0^\pi}_{j} v_n(t_1,s_1)\mathcal{G}_n(s_1,t_2,s_2)\dots\mathcal{G}_n(s_{j-1},t_j,s_j)y_n(s_j)\,ds_1\dots ds_j,$$

$$r(x)\in W_{20}^1,$$

and

$$\|\psi_j(r)\|_{W_2^1}\le (C\|r\|_{W_2^1})^j,$$

$$\|\psi_j(r)-\psi_j(r^*)\|_{W_2^1}\le\|r-r^*\|_{W_2^1}\left(C\max\left(\|r\|_{W_2^1},\|r^*\|_{W_2^1}\right)\right)^{-1}.$$

By Lemma 13.4, there exists $\delta>0$ such that for $\Lambda<\delta$ equation (13.40) has a solution $r(x)\in W_{20}^1$, $\|r(x)\|_{W_2^1}\le C\Lambda$. Put $\widetilde{M}(x)=M(x)-((\pi-x)^{-1}r(x))'$, and consider the boundary value problem $\widetilde{L}=L(q,\widetilde{M})$. Clearly

$$\widetilde{Q}(x)=Q(x)-r'(x)\in\mathcal{L}_2(0,\pi),\qquad \|Q(x)-\widetilde{Q}(x)\|_{\mathcal{L}_2(0,\pi)}\le C\Lambda.$$

Since

$$\widehat{M}_1(x)=-\frac{1}{\pi-x}\int_x^\pi\widehat{Q}(t)\,dt,\qquad \widehat{M}_0(x)=\widehat{Q}(x)+\widehat{M}_1(x),$$

we have

$$\|M_k(x)-\widetilde{M}_k(x)\|_{\mathcal{L}(0,\pi)}\le C\Lambda,\qquad k=0,1.$$

It remains to be shown that the numbers $\{\widetilde{\lambda}_n\}_{n\ge1}$ are the eigenvalues of the problem $\widetilde{L}$. To do this, consider the functions $\widetilde{y}_n(x)$ which are solutions of the integral equations

$$\widetilde{y}_n(x)=y_n(x)+\int_0^\pi\widehat{M}_1(t)\,dt\int_0^\pi\mathcal{G}_n(x,t,s)\widetilde{y}_n(s)\,ds \tag{13.41}$$

or, which is the same,

$$\widetilde{y}_n(x)=y_n(x)+\int_0^x\widehat{M}_1(t)\,dt\int_0^{x-t}\mathcal{G}(x,s+t,\widetilde{\lambda}_n)\widetilde{y}_n(s)\,ds. \tag{13.42}$$

After integration by parts, (13.42) takes the form

$$\widetilde{y}_n(x) = y_n(x) - \int_0^x \widehat{M}_1(t)\,dt \int_t^x g(x,s,\widetilde{\lambda}_n)\widetilde{y}_n'(s-t)\,ds.$$

Reverse the integration order:

$$\widetilde{y}_n(x) = y_n(x) - \int_0^x g(x,t,\widetilde{\lambda}_n)\,dt \int_0^t \widehat{M}_1(s)\widetilde{y}_n'(t-s)\,ds.$$

Integrate by parts:

$$\widetilde{y}_n(x) = y_n(x) - \int_0^x g(x,t,\widetilde{\lambda}_n)\,dt \int_0^t \widehat{M}(t-s)\widetilde{y}_n(s)\,ds. \tag{13.43}$$

It follows from (13.43) that

$$\ell(\widetilde{y}_n(x) - y_n(x)) = \int_0^x \widehat{M}(t-s)\widetilde{y}_n(s)\,ds = (\ell - \widetilde{\ell})\widetilde{y}_n(x),$$

and consequently,

$$\widetilde{\ell}\widetilde{y}_n(x) = \widetilde{\lambda}_n\widetilde{y}_n(x), \qquad \widetilde{y}_n(0) = 0, \qquad \widetilde{y}_n'(0) = 1.$$

Since the solution of the Cauchy problem is unique, we have $\widetilde{y}_n(x) = \widetilde{S}(x,\widetilde{\lambda}_n)$.

Write (13.13) in the form

$$\int_0^\pi \widehat{M}(x)\,dx \int_0^{\pi-x} w(\pi - x - t,\lambda)\widetilde{S}(t,\lambda)\,dt = \widehat{\Delta}(\lambda).$$

Integrating by parts, we obtain for $\lambda = \widetilde{\lambda}_n$

$$\int_0^\pi \widehat{M}_1(x)\,dx \int_0^\pi v_n(x,t)\widetilde{y}_n(t)\,dt = \widehat{\Delta}(\widetilde{\lambda}_n). \tag{13.44}$$

Solving (13.41) by the method of successive approximations, we have

$$\widetilde{y}_n(x) = y_n(x) + \mathcal{Y}_n(x), \tag{13.45}$$

$$\mathcal{Y}_n(x) = \sum_{j=1}^{\infty} \underbrace{\int_0^\pi \cdots \int_0^\pi}_{j} \widehat{M}_1(t_1)\ldots\widehat{M}_1(t_j)\Big(\underbrace{\int_0^\pi \cdots \int_0^\pi}_{j} \mathcal{G}_n(x,t_1,s_1)$$

$$\times \mathcal{G}_n(s_1,t_2,s_2)\ldots\mathcal{G}_n(s_{j-1},t_j,s_j)y_n(s_j)\,ds_1\ldots ds_j\Big)dt_1\ldots dt_j.$$

Further, multiplying (13.40) by $\overline{\eta_n(x)}$ and integrating from 0 to π, we obtain

$$\int_0^\pi r(x)\overline{\eta_n(x)}\,dx + \sum_{j=2}^\infty \underbrace{\int_0^\pi \dots \int_0^\pi}_{j} r(t_1)\dots r(t_j)B_{nj}(t_1,\dots,t_j)\,dt_1\dots dt_j = \frac{\varepsilon_n}{n}. \quad (13.46)$$

Since $r(x) = (\pi - x)\widehat{M_1}(x)$, $\overline{\eta_n(x)} = n(\pi - x)^{-1}\xi_n(x)$, we can transform (13.46) to the form

$$\int_0^\pi \widehat{M_1}(x)\xi_n(x)\,dx + \sum_{j=2}^\infty \underbrace{\int_0^\pi \dots \int_0^\pi}_{j} \widehat{M_1}(t_1)\dots\widehat{M_1}(t_j)\Big(\underbrace{\int_0^\pi \dots \int_0^\pi}_{j} v_n(t_1,s_1)$$

$$\times \mathcal{G}_n(s_1,t_2,s_2)\dots\mathcal{G}_n(s_{j-1},t_j,s_j)y_n(s_j)\,ds_1\dots ds_j\Big)dt_1\dots dt_j = \frac{\varepsilon_n}{n^2}.$$

Hence, taking (13.45) into account, we obtain

$$\int_0^\pi \widehat{M_1}(x)\,dx \int_0^\pi v_n(x,t)\widetilde{y}_n(t)\,dt = \Delta(\widetilde{\lambda}_n). \quad (13.47)$$

Comparing (13.44) with (13.47), we find that $\widetilde{\Delta}(\widetilde{\lambda}_n) = 0$. Hence the numbers $\{\widetilde{\lambda}_n\}_{n\ge1}$ are the eigenvalues of the boundary value problem $\widetilde{L}$. Theorem 13.3. is proved.

APPENDIX I.

SOLUTION OF THE BOUSSINESQ EQUATION ON THE HALF-LINE BY THE INVERSE PROBLEM METHOD

We study a mixed problem for the nonlinear Boussinesq equation on the half-line. An algorithm for the solution, and the necessary and sufficient conditions for the solvability of this problem are obtained, and uniqueness is proved.

Let us consider the following problem

$$u_t = i(2v_x - u_{xx}), \qquad v_t = i\left(v_{xx} - \frac{2}{3}u_{xxx} - \frac{2}{3}uu_x\right), \qquad x > 0, \qquad t > 0 \quad \text{(I.1)}$$

$$u\big|_{t=0} = u_0(x), \qquad v\big|_{t=0} = v_0(x), \qquad u_0(x),\ v_0(x) \in \mathcal{L}(0,\infty) \qquad \text{(I.2)}$$

$$u\big|_{x=0} = u_1(t), \qquad u_x\big|_{x=0} = u_2(t), \qquad v\big|_{x=0} = v_1(t), \qquad v_x\big|_{x=0} = v_2(t). \qquad \text{(I.3)}$$

System (I.1) after elimination of $v(x,t)$ reduced to the Boussinesq equation

$$3u_{tt} = u_{xxxx} + 2(u^2)_{xx}.$$

In this section the mixed problem (I.1)–(I.3) is solved by the inverse problem method. For this we use the results on the IP for third-order differential operators on the half-line by its WM, obtained in §2. We note that in [111]–[114], using the IP for second-order equations, the evolution of spectral data for difference and differential nonlinear equations on the half-line is obtained, and in [65], [115] the Boussinesq equation on the line is studied by the inverse scattering method.

Let $\mathcal{D} = \{(x,t) : x \geq 0, t \geq 0\}$, and let J_n be the set of functions $z(x,t)$ such that functions

$$\left(\frac{\partial^{j+k}}{\partial x^j \partial t^k}\right) z(x,t), \qquad 0 \leq j + 2k \leq n$$

are continuous in $\mathcal{D}$, integrable on the half-line $x \in (0,\infty)$ for any fixed $t > 0$, and $x\left(\frac{\partial^n z}{\partial x^n}\right) \in \mathcal{L}(0,\infty)$. We shall write $\{u(x,t), v(x,t)\} \in M$, if $u(x,t) \in J_3$, $v(x,t) \in J_2$. We denote elements of matrices $A, B\ldots$ by $A_{ij}, B_{ij}, \ldots$, where i is the number of the row, and j is the number of the column.

1. Auxiliary statements. Let $\{u(x,t), v(x,t)\} \in M$. For a fixed $t \geq 0$ we consider the DE with respect to x

$$\ell y \equiv y''' + uy' + vy = \lambda y = \rho^3 y. \tag{I.4}$$

Let $\Phi(x,t,\lambda) = [\Phi_k^{(j-1)}(x,t,\lambda)]_{j,k=\overline{1,3}}$, where $\Phi_k(x,t,\lambda)$ is the solution of (I.4) under the conditions $\Phi_k^{(j-1)}(0,t,\lambda) = \delta_{jk}$, $j = \overline{1,k}$; $\Phi_k(x,t,\lambda) = O(\exp(\rho r_k x))$, $x \to \infty$. Here r_k are roots of the equation $r^3 - 1 = 0$ such that

$$\operatorname{Re}(\rho r_1) < \operatorname{Re}(\rho r_2) < \operatorname{Re}(\rho r_3). \tag{I.5}$$

We set $\mathfrak{M}(t,\lambda) = \Phi(0,t,\lambda)$, i.e.

$$\mathfrak{M}^T(t,\lambda) = [\mathfrak{M}_{kj}(t,\lambda)]_{k,j=\overline{1,3}}, \qquad \mathfrak{M}_{kj}(t,\lambda) = \delta_{kj}, \qquad (j \leq k), \tag{I.6}$$

where $\mathfrak{M}_{kj}(t,\lambda) = \Phi_k^{(j-1)}(0,t,\lambda)$, $k < j$. The functions $\mathfrak{M}_{kj}(t,\lambda)$ are the WF's, and the matrix $\mathfrak{M}(t,\lambda)$ is the WM for ℓ.

Let $\mathfrak{M}^* = \mathfrak{M}^{-1}$, $\Phi_j^* = \Phi_1\Phi_{j+1}' - \Phi_{j+1}\Phi_j'$, $j = 1,2$. It is clear that the functions Φ_j^* are solutions of the equation

$$\ell^* z = -z''' - (uz)' + vz = \lambda z.$$

For fixed $t > 0$, $k = \overline{1,3}$, $j = \left[\frac{k+1}{2}\right]$, $p = \left[\frac{5-k}{2}\right]$ we consider the functions

$$\begin{aligned} \psi_k(x,\lambda) &= (\psi_{k1}(x,\lambda), \psi_{k2}(x,\lambda)) \\ &= \left(\Phi_j(x,t,\lambda)\Phi_p^*(x,t,\lambda), -\int_x^\infty \Phi_j'(s,t,\lambda)\Phi_p^*(s,t,\lambda)\,ds\right), \end{aligned}$$

where $[\cdot]$ denotes the greatest integer in the number. At the end of Appendix I the following theorem of completeness is proved.

Theorem I.1. *If*

$$\int_0^\infty \psi_k(x,\lambda) f(x)\,dx = 0, \qquad k = \overline{1,3}, \qquad f(x) = (f_1(x), f_2(x))^T \in \mathcal{L}(0,\infty), \tag{I.7}$$

then $f(x) = 0$ *a.e.*

Let us now denote

$$\mathcal{G}(x,t,\lambda) = \begin{bmatrix} 0 & 1 & 0 \\ 0 & 0 & 1 \\ \lambda - v & -u & 0 \end{bmatrix},$$

$$F(x,t,\lambda) = i \begin{bmatrix} \frac{2}{3}u & 0 & 1 \\ \lambda - v + \frac{2}{3}u_x & -\frac{1}{3}u & 0 \\ \frac{2}{3}u_{xx} - v_x & \lambda - v + \frac{1}{3}u_x & -\frac{1}{3}u \end{bmatrix}$$

$$Q = \mathcal{G}_t - F_x + \mathcal{G}F - F\mathcal{G}, \qquad q = [Q_1, Q_2]^T,$$

$$Q_1 = -v_t + i\left(v_{xx} - \frac{2}{3}u_{xxx} - \frac{2}{3}uu_x\right), \qquad Q_2 = -u_t + i(2v_x - u_{xx}).$$

Then

$$Q(x,t) = \begin{bmatrix} 0 & 0 & 0 \\ 0 & 0 & 0 \\ Q_1 & Q_2 & 0 \end{bmatrix},$$

i.e. the system (I.1) is equivalent to the equality $Q = 0$. We define the matrices $W(x,t,\lambda)$ and $S(x,t,\lambda)$ from the relations

$$W_x = \mathcal{G}(x,t,\lambda)W, \qquad W\big|_{x=0} = E \tag{I.8}$$

$$S_t = F(x,t,\lambda)S, \qquad S\big|_{t=0} = E \tag{I.9}$$

where $E = [\delta_{jk}]_{j,k=\overline{1,3}}$ is the identity matrix. Then

$$\Phi(x,t,\lambda) = W(x,t,\lambda)\mathfrak{M}(t,\lambda). \tag{I.10}$$

Consider the matrices

$$C^0(t,\lambda) = (\mathfrak{M}_t^*(t,\lambda) + \mathfrak{M}^*(t,\lambda)F^0(t,\lambda))\mathfrak{M}(t,\lambda), \quad F^0(t,\lambda) = F(0,t,\lambda) \tag{I.11}$$

$$d(x,t,\lambda) = F^0(t,\lambda) - \int_0^x W^{-1}(s,t,\lambda)Q(s,t)W(s,t,\lambda)\,ds \tag{I.12}$$

$$C(x,t,\lambda) = C^0(t,\lambda) - \int_0^x \Phi^{-1}(s,t,\lambda)Q(s,t)\Phi(s,t,\lambda)\,ds. \tag{I.13}$$

Lemma I.1. *The following equality holds:*

$$\Phi_t(x,t,\lambda) = F(x,t,\lambda)\Phi(x,t,\lambda) - \Phi(x,t,\lambda)C(x,t,\lambda). \tag{I.14}$$

Proof. By virtue of (I.8), we have

$$(W_t - FW)_x - \mathcal{G}(W_t - FW) = QW, \qquad (W_t - FW)\big|_{x=0} = -F^0(t,\lambda).$$

Consequently

$$W_t(x,t,\lambda) = F(x,t,\lambda)W(x,t,\lambda) - W(x,t,\lambda)d(x,t,\lambda). \tag{I.15}$$

Hence, according to (I.10), we obtain (I.14).

Lemma I.2. *The following relations are valid*

$$C^0_{kj}(t,\lambda) = (-1)^{k-1}\int\limits_0^\infty \varphi_{k+j-2}(x,t,\lambda)q(x,t)\,dx, \qquad 1 \le j < k \le 3, \tag{I.16}$$

where $\varphi_k = (\psi_{k1}, \psi'_{k2})$, $k = \overline{1,3}$.

Proof. Rewriting (I.13) in coordinates and using properties of the functions Φ_k we obtain in particular that with fixed t, λ

$$C_{kj}(x,t,\lambda) = O(1), \quad k \ge j; \quad C_{12}(x,t,\lambda) = O(\exp(\rho(r_2 - r_1)x)), \quad x \to \infty, \tag{I.17}$$

$$C_{kj}(x,t,\lambda) = C^0_{kj}(t,\lambda) + (-1)^k \int\limits_0^x \varphi_{k+j-2}(s,t,\lambda)q(s,t)\,ds, \qquad 1 \le j < k \le 3 \tag{I.18}$$

and

$$\Phi_k^{(j-1)}(x,t,\lambda) = \rho^{1-k}\sum_{m=1}^{k}(\rho r_m)^{j-1}\exp(\rho r_m x)(a_{km} + o(1)),$$

$$|\rho|x \to \infty, \qquad a_{mm} \ne 0.$$

Now by virtue of (I.14)

$$\sum_{m=1}^{3}\Phi_m^{(j-1)}(x,t,\lambda)C_{mk}(x,t,\lambda) = -\frac{\partial}{\partial t}\Phi_k^{(j-1)}(x,t,\lambda) + \sum_{m=1}^{3}F_{jm}(x,t,\lambda)\Phi_k^{(m-1)}(x,t,\lambda).$$

Hence, using (I.17) we calculate with fixed t, λ ($\operatorname{Im}\lambda \ne 0$)

$$\lim_{x\to\infty} C_{kj}(x,t,\lambda) = 0, \qquad j < k.$$

With (I.18) it gives (I.16). Lemma I.2 is proved

Rewriting (I.11) in coordinates, substituting into (I.16) and solving with respect to $\frac{\partial}{\partial t}\mathfrak{M}_{jk}(t,\lambda)$, we obtain

$$\begin{aligned}\frac{\partial}{\partial t}\mathfrak{M}_{jk} \;&=\; \sum_{m=0}^{3-j}\mathfrak{M}_{j,j+m}(F^0_{k,j+m} - F^0_{j,j+m}\mathfrak{M}_{jk} + \delta_{j2}F^0_{1,2+m}(-\mathfrak{M}_{13} + \mathfrak{M}_{12}\mathfrak{M}_{23}))\\ &\quad + (-1)^k\int\limits_0^\infty(\varphi_{k+j-2} - \delta_{k-j,2}\mathfrak{M}_{23}\varphi_1)q\,dx, \qquad 0 \le j < k \le 3.\end{aligned} \tag{I.19}$$

2. Solution of the problem (I.1)–(I.3). In the following theorem evolution of the WM with respect to t is obtained.

Theorem I.2. *Let $\{u(x,t), v(x,t)\}$ be the solution of the problem (I.1)–(I.3). We denote $u_3(t) = u_{xx}\big|_{x=0}$ and $\mathfrak{M}^0_{jk}(\lambda) = \mathfrak{M}_{jk}(0,\lambda)$ are the WF for $\{u_0(x), v_0(x)\}$, and*

$$\widetilde{F}(t,\lambda) = i\begin{bmatrix} \frac{2}{3}u_1(t) & 0 & 1 \\ \lambda - v_1(t) + \frac{2}{3}u_2(t) & -\frac{1}{3}u_1(t) & 0 \\ \frac{2}{3}u_3(t) - v_2(t) & \lambda - v_1(t) + \frac{1}{3}u_2(t) & -\frac{1}{3}u_1(t) \end{bmatrix}. \tag{I.20}$$

Let the matrix $R(t,\lambda)$ be the solution of the Cauchy problem

$$R_t(t,\lambda) = -R(t,\lambda)\widetilde{F}(t,\lambda), \qquad R\Big|_{t=0} = E. \tag{I.21}$$

We define

$$\left.\begin{aligned} \Delta_k(t,\lambda) &= R_{3k}(t,\lambda) - \mathfrak{M}^0_{23}(\lambda)R_{2k}(t,\lambda) + (\mathfrak{M}^0_{12}(\lambda)\mathfrak{M}^0_{23}(\lambda) - \mathfrak{M}^0_{13}(\lambda))R_{1k}(t,\lambda), \\ \Delta_{mk}(t,\lambda) &= \det\left[R_{jp}(t,\lambda) - \mathfrak{M}^0_{1j}(\lambda)R_{1p}(t,\lambda)\right]_{j=2,3;p=m,k}. \end{aligned}\right\} \tag{I.22}$$

Then

$$\mathfrak{M}_{12}(t,\lambda) = -\frac{\Delta_{13}(t,\lambda)}{\Delta_{23}(t,\lambda)}, \qquad \mathfrak{M}_{13}(t,\lambda) = -\frac{\Delta_{21}(t,\lambda)}{\Delta_{23}(t,\lambda)},$$
$$\mathfrak{M}_{23}(t,\lambda) = -\frac{\Delta_2(t,\lambda)}{\Delta_3(t,\lambda)}, \tag{I.23}$$

$$\frac{\partial}{\partial t}\mathfrak{M}_{jk} = \sum_{m=0}^{3-j}\mathfrak{M}_{j,j+m}(\widetilde{F}_{k,j+m} - \widetilde{F}_{j,j+m}\mathfrak{M}_{jk} + \delta_{j2}\widetilde{F}_{1,2+m}(-\mathfrak{M}_{13} + \mathfrak{M}_{12}\mathfrak{M}_{23})),$$
$$0 \le j < k \le 3. \tag{I.24}$$

Proof. Since $\{u(x,t), v(x,t)\}$ is the solution of (I.1)–(I.3), then $Q(x,t) = 0$, $q(x,t) = 0$, $\widetilde{F}(t,\lambda) = F^0(t,\lambda)$. Consequently, by virtue of (I.12)–(I.16), we have

$$C^0_{kj}(t,\lambda) = 0, \qquad 1 \le j < k \le 3, \tag{I.25}$$

$$W_t(x,t,\lambda) = F(x,t,\lambda)W(x,t,\lambda) - W(x,t,\lambda)F^0(t,\lambda), \tag{I.26}$$

$$\Phi_t(x,t,\lambda) = F(x,t,\lambda)\Phi(x,t,\lambda) - \Phi(x,t,\lambda)C^0(t,\lambda). \tag{I.27}$$

It follows from (I.26) that, in accordance with (I.9), (I.10) and (I.21), $W(x,t,\lambda) = S(x,t,\lambda)W(x,0,\lambda)R(t,\lambda)$ and

$$\Phi(x,t,\lambda) = S(x,t,\lambda)\Phi(x,0,\lambda)B(t,\lambda), \tag{I.28}$$

where

$$B(t,\lambda) = \mathfrak{M}^*(0,\lambda)R(t,\lambda)\mathfrak{M}(t,\lambda). \tag{I.29}$$

Differentiating (I.28) with respect to t and comparing with (I.27), we obtain $B_t(t,\lambda) = -B(t,\lambda)C^0(t,\lambda)$, $B(0,\lambda) = E$. Hence, from (I.25) we find $B_{kj}(t,\lambda) = 0$, $j < k$. Now rewriting (I.29) in coordinates for $j < k$ and solving with respect to $\mathfrak{M}_{jk}(t,\lambda)$, we obtain (I.23). Equalities (I.24) follow from (I.19), since $F^0(t,\lambda) = \widetilde{F}(t,\lambda)$, $q(x,t) = 0$. We note that (I.24) can be obtained directly from (I.23) by differentiating with respect to t. Theorem I.2 is proved.

Using evolution relations (I.23) and the solution of the IP for equation (I.4), we obtain the following algorithm for the solution of the mixed problem (I.1)–(I.3).

Algorithm I.1. For $x \geq 0$, $t \geq 0$ continuous functions $u_0(x)$, $v_0(x)$, $u_1(t)$, $v_1(t)$, $u_2(t)$, $v_2(t)$ are given. Let $u_0(x), v_0(x) \in \mathcal{L}(0,\infty)$, $u_0(0) = u_1(0)$, $v_0(0) = v_1(0)$, and $\frac{\partial}{\partial t}u_1(t)$ be continuous. We then:

(1) compute the function $u_3(t) = 2v_2(t) + i\frac{\partial}{\partial t}u_1(t)$;

(2) find the WF's $\mathfrak{M}^0_{kj}(\lambda)$, $1 \leq k < j \leq 3$ for $\{u_0, v_0\}$;

(3) find the matrix $R(t,\lambda)$ from (I.20) and (I.21);

(4) compute the matrix $\mathfrak{M}(t,\lambda)$ using formulae (I.6), (I.22) and (I.23);

(5) find the functions $\{u(x,t), v(x,t)\}$ by solving the IP by the method described in §2.

Let us now find the conditions of existence of the solution of (I.1)–(I.3). The following theorem shows that the existence of the solution of (I.1)–(I.3) is equivalent to solvability of the corresponding IP.

Theorem I.3. *Let the matrix $\mathfrak{M}(t,\lambda)$ be constructed from the given function u_j, v_j, $j = \overline{0,2}$ according to the steps (1)–(4) of Algorithm I.1. We assume that there exist functions $\{u(x,t), v(x,t)\} \in M$ for which $\mathfrak{M}(t,\lambda)$ is the WM. Then $\{u(x,t), v(x,t)\}$ is the solution of (I.1)–(I.3).*

Proof. From (I.23) with $t = 0$ we find $\mathfrak{M}_{jk}(0,\lambda) = \mathfrak{M}^0_{jk}(\lambda)$, $j < k$. The coefficients of the DE (I.4) are uniquely determined from its WM. Then we have $u(x,0) = u_0(x)$, $v(x,0) = v_0(x)$, i.e. $u(x,t)$, $v(x,t)$ satisfy the initial conditions (I.2).

Differentiating (I.23) with respect to t, we obtain (I.24). Comparing (I.24) with (I.19), we get

$$\left.\begin{aligned} &h_{21}(t) + (h_{22}(t) - h_{11}(t))\mathfrak{M}_{12}(t,\lambda) + T_1(t,\lambda), \qquad h_{32}(t) - T_3(t,\lambda) = 0, \\ &h_{31}(t) + h_{32}(t)\mathfrak{M}_{12}(t,\lambda) + (h_{33}(t) - h_{11}(t))\mathfrak{M}_{13}(t,\lambda) - T_2(t,\lambda) = 0, \end{aligned}\right\} \tag{I.30}$$

where

$$h(t) = F^0(t,\lambda) - \widetilde{F}(t,\lambda),$$

$$T_k(t,\lambda) = \int_0^\infty (\varphi_k(x,t,\lambda) - \delta_{k2}\mathfrak{M}_{23}(t,\lambda)\varphi_1(x,t,\lambda))q(x,t)\,dx.$$

Using the above-mentioned asymptotic properties of the functions Φ_k, we compute for the fixed $t > 0$

$$\mathfrak{M}_{jk}(t,\lambda) = (-1)^{j-1}(\rho r_{2j-1})^{k-j}(1 + o(1)), \qquad j < k,$$

$$T_k(t,\lambda) = o(1), \qquad |\lambda| \to \infty.$$

Then (I.30) yields

$$h_{32}(t) = h_{22}(t) - h_{11}(t) = h_{33}(t) - h_{11}(t) = h_{21}(t) = h_{31}(t) = 0$$

and consequently

$$\left.\begin{aligned} &u(0,t) = u_1(t), \quad u_x(0,t) = u_2(t), \quad v(0,t) = v_1(t), \\ &\frac{2}{3}(u_{xx}(0,t) - u_3(t)) - (v_x(0,t) - v_2(t)) = 0, \end{aligned}\right\} \tag{I.31}$$

$$\int_0^\infty \varphi_k(x,t,\lambda) q(x,t)\, dx = 0, \qquad k = \overline{1,3}. \tag{I.32}$$

From (I.32) we have

$$\int_0^\infty \psi_k(x,\lambda) f(x)\, dx + f^0 \psi_{k2}(0,\lambda) = 0, \qquad k = \overline{1,3}, \tag{I.33}$$

with fixed $t > 0$, where $f(x) = [Q_1(x,t), \frac{\partial}{\partial x} Q_2(x,t)]^T$, $f^0 = Q_2(0,t)$. If $|\lambda| \to \infty$, then from (I.33) one gets $f^0 = 0$, i.e.

$$\int_0^\infty \psi_k(x,\lambda) f(x)\, dx = 0, \qquad k = \overline{1,3}.$$

Hence, by virtue of Theorem I.1, we conclude that $f(x) = 0$ or $Q_1(x,t) = Q_2(x,t) = 0$. Thus, the functions $u(x,t)$, $v(x,t)$ are the solutions of system (I.1). Finally, since

$$u_{xx}(0,t) = 2v_x(0,t) + iu_t(0,t), \qquad u_3(t) = 2v_2(t) + i\frac{\partial}{\partial t} u_1(t),$$

then in accordance with (I.31) we obtain $v_x(0,t) = v_2(t)$. Theorem I.3 is proved.

It follows that problem (I.1)–(I.3) has a solution if and only if the solution of the corresponding IP exists. Necessary and sufficient conditions of solvability of the IP from the WM and algorithm for the solution are given in §2.

3. Proof of Theorem I.1. For a fixed t, we consider the DE's with respect to x:

$$L_\lambda \mathcal{Y}(x) = \mathcal{Y}^{(5)}(x) + \sum_{j=0}^{3} P_j(x)\mathcal{Y}^{(j)}(x) - \lambda\Omega\mathcal{Y}''(x) = 0,$$

$$L_\lambda^* Z(x) = -Z^{(5)}(x) - \sum_{j=0}^{3} Z^{(j)}(x)P_j^*(x) - \lambda Z''(x)\Omega = 0.$$

where

$$\mathcal{Y} = [\mathcal{Y}_1, \mathcal{Y}_2]^T, \qquad Z = [Z_1, Z_2], \qquad \Omega = \begin{bmatrix} -9 & -6 \\ 18 & 9 \end{bmatrix}$$

$$P_k^* = [P_{kjm}^*]_{j,m=1,2}, \qquad P_{3jm}^* = \delta_{jm}u, \qquad P_{211}^* = 3u_x - 9v_x,$$

$$P_{212}^* = 6v, \qquad P_{221}^* = 9u_x - 18v, \qquad P_{222}^* = 6u_x - 9v, \qquad P_{112}^* = 4v_x,$$

$$P_{1j1}^* = (-1)^j(3v_x - 3u_{xx}), \qquad P_{122}^* = 2u_{xx} - 3v_x, \qquad P_{0j1}^* = \delta_{j1}u_{xxx},$$

$$P_{0j2}^* = \delta_{j1}v_{xx}, \qquad P_3 = P_3^*, \qquad P_2 = -P_2^* + 3P_3^{*\prime},$$

$$P_1 = P_1^* - 2P_2^{*\prime} + 3P_3^{*\prime\prime}, \qquad P_0 = -P_0^* + P_1{*}' - P_2^{*\prime\prime} + P_3^{*\prime\prime\prime}.$$

Lemma I.3. $L_\lambda^*\psi_k(x,\lambda) = 0,\ k = \overline{1,3}.$

Proof. Since

$$\varphi_k = [\varphi_{k1}, \varphi_{k2}] = [\Phi_j\Phi_p^*, \Phi_j'\Phi_p^*], \qquad k = \overline{1,3}, \qquad j = \left[\frac{k+1}{2}\right], \qquad p = \left[\frac{5-k}{2}\right],$$

and the functions Φ_j, Φ_p^* are solutions of the equations $\ell\Phi_j = \lambda\Phi_j$, $\ell^*\Phi_p^* = \lambda\Phi_p^*$, we have

$$\Phi_j\Phi_p^{*\prime} = \varphi_{k1}' - \varphi_{k2} \tag{I.34}$$

$$2\Phi_j''\Phi_p^{*\prime} + \Phi_j'\Phi_p^{*\prime\prime} = \varphi_{k2}'' - (\lambda - v)\varphi_{k1} + u\varphi_{k2} \tag{I.35}$$

$$\Phi_j''\Phi_p^{*\prime} + 2\Phi_j'\Phi_p^{*\prime\prime} = \varphi_{k1}''' - \varphi_{k2}'' + (\lambda - v + u')\varphi_{k1} + u(\varphi_{k1}' - \varphi_{k2}). \tag{I.36}$$

Hence it follows that

$$\Phi_j''\Phi_p^{*\prime} + \Phi_p^{*\prime\prime}\Phi_j' = \frac{1}{3}(\varphi_{k1}'' + u\varphi_{k1})',$$

and consequently

$$\Phi_j'\Phi_p^{*\prime} = \frac{1}{3}(\varphi_{k1}'' + u\varphi_{k1}). \tag{I.37}$$

Then

$$\Phi_j''\Phi_p^{*} = \varphi_{k2}' - \frac{1}{3}(\varphi_{k1}'' + u\varphi_{k1}) \tag{I.38}$$

$$\Phi_j\Phi_p^{*\prime\prime} = \frac{2}{3}\varphi_{k1}'' - \varphi_{k2}' - \frac{1}{3}u\varphi_{k1}. \tag{I.39}$$

Differentiating (I.35) and (I.36) with respect to x, eliminating the products $\Phi_j^{(m)}\Phi_p^{*(r)}$ with help of (I.34)–(I.39), $\ell\Phi_j = \lambda\Phi_j$, $\ell^*\Phi_p^* = \lambda\Phi_p^*$, and replacing $\varphi_{k1} = \psi_{k1}$, $\varphi_{k2} = \psi_{k2}'$, we obtain $L^*\psi_k(x,\lambda) = 0$, $k = \overline{1,3}$. Lemma I.3 is proved.

Let $g^*(x,s,\lambda)$ be the solution of the Cauchy problem

$$L_\lambda^* g^*(x,s,\lambda) = 0, \qquad \frac{\partial^m}{\partial x^m} g^*(x,s,\lambda)\Big|_{x=s} = -\delta_{m4}\begin{bmatrix}1 & 0\\ 0 & 1\end{bmatrix}, \qquad m = \overline{0,4}, \qquad x \le s$$

$$g^*(x,s,\lambda) = 0, \qquad x > s.$$

Denote

$$\mathcal{G}^*(x,s,\lambda) = \sum_{k=1}^{3} D_k(s,\lambda)\psi_k(x,\lambda) + g^*(x,s,\lambda), \tag{I.40}$$

where the functions $D_k(s,\lambda) = [D_{k1}(s,\lambda), D_{k2}(s,\lambda)]^T$ are defined from the relations

$$\sum_{k=1}^{3} D_k(s,\lambda)\psi_k(0,\lambda)\begin{bmatrix}1\\0\end{bmatrix} = -g^*(0,s,\lambda)\begin{bmatrix}1\\0\end{bmatrix},$$

$$\sum_{k=1}^{3} D_k(s,\lambda)\psi_k'(0,\lambda) = -g_x^*(0,s,\lambda).$$

The determinant of this algebraic system is equal to 1. It is clear that $\mathcal{G}^*(0,s,\lambda)[1,0]^T = 0$, $\mathcal{G}_x^*(0,s,\lambda) = 0$.

Let the functions $N_k(x,\lambda) = [N_{k1}(x,\lambda), N_{k2}(x,\lambda)]^T$, $k = \overline{1,10}$, be solutions of the system

$$L_\lambda^0\mathcal{Y}(x) \equiv \mathcal{Y}^{(5)}(x) - \lambda\Omega\mathcal{Y}''(x) = 0, \tag{I.41}$$

with the initial conditions $N_{kj}^{(m)}(0,\lambda) = \delta_{k-j,2m}$, $m = \overline{0,4}$. For every sector $\arg\rho \in \left(\frac{\pi n}{6}, \frac{\pi(n+1)}{6}\right)$ we enumerate the roots ω_k of the equation $w^6 + 27 = 0$ such that $\mathrm{Re}\,(\rho\omega_1) < \ldots < \mathrm{Re}\,(\rho\omega_6)$ and denote $\beta_k = \frac{1}{2} - \frac{1}{18}\omega_k^3$. Then the functions

$$H_{k+4}(x,\lambda) = [-\beta_k \exp(\rho\omega_k x), \exp(\rho\omega_k x)]^T, \qquad k = \overline{1,6},$$

$$H_k(x,\lambda) = N_k(x,\lambda), \qquad k = \overline{1,4},$$

are solutions of (I.41), and

$$N_k(x,\lambda) = \sum_{m=1}^{10} \alpha_{mk}(\lambda)H_m(x,\lambda), \quad k = \overline{1,10}; \quad \alpha_{mk}(\lambda) = \alpha_{mk}^0\rho^{-\sigma_{mk}}, \tag{I.42}$$

where α^0_{mk} is constants, and

$$\sigma_{mk} = 0, \qquad k = \overline{1,4},$$

$$\sigma_{mk} = \left[\frac{k-3}{2}\right], \qquad k = \overline{5,10}, \quad m = 3,4,$$

$$\sigma_{mk} = \left[\frac{k-1}{2}\right], \qquad k = \overline{5,10}, \quad m = 1,2,\overline{5,10}.$$

We consider the function

$$\mathcal{G}^0(x,s,\lambda) = \sum_{k=1}^{3} N_{k+6}(x,\lambda)A_k(s,\lambda) + g^0(x-s,\lambda), \tag{I.43}$$

where

$$g^0(x,\lambda) = \begin{cases} [N_9(x,\lambda), N_{10}(x,\lambda)], & x \geq 0 \\ 0, & x < 0. \end{cases}$$

The functions $A_k(s,\lambda) = [A_{k1}(s,\lambda), A_{k2}(s,\lambda)]^T$ are determined from the relations

$$\sum_{k=1}^{10} \alpha_{m,k+6}(\lambda)A_{kj}(s,\lambda) = -\alpha_{m,8+j}(\lambda)\exp(-\rho\omega_{m-4}s), \qquad m = \overline{8,10}. \tag{I.44}$$

Let us show that $\alpha = \det[\alpha^0_{mk}]_{m=\overline{8,10};k=\overline{7,9}} \neq 0$, i.e. the determinant of the system (I.44) differs from zero. Since the $\alpha_{mk}(\lambda)$ are determined from the relations

$$\sum_{m=1}^{10} \alpha_{mk}(\lambda)H^{(j)}_m(0,\lambda) = N^{(j)}_k(0,\lambda), \qquad j = \overline{0,4}, \qquad k = \overline{1,10}; \qquad \alpha^0_{mk} = \rho^{\sigma_{mk}}\alpha_{mk}(\lambda),$$

it is sufficient to prove that $\Gamma = \det[\gamma_{mk}]_{m=\overline{4,6};k=\overline{3,5}} \neq 0$, where

$$\gamma = \left([-\beta_k, 1, -\beta_k\omega_k, \omega_k, -\beta_k\omega_k^2, \omega_k^2]^T_{k=\overline{1,6}}\right)^{-1}$$

Denote

$$\gamma^0 = \left([\omega_k^3, 1, \omega_k^4, \omega_k, \omega_k^5, \omega_k^2]^T_{k=\overline{1,6}}\right)^{-1}, \qquad \Gamma^0 = \det[\gamma^0_{mk}]_{m=\overline{4,6};k=\overline{3,5}}\,.$$

Evidently $\Gamma^0 \neq 0$ as the Vandermonde determinant. On the other hand, since $\beta_k = \frac{1}{2} - \frac{1}{18}\omega_k^3$, then $\gamma_{m3} = 18\gamma^0_{m3}$, $\gamma_{m4} = 9\gamma^0_{m3} + \gamma^0_{m4}$, $\gamma_{m5} = 18\gamma^0_{ms}$. Consequently $\Gamma \neq 0$, i.e. $\alpha \neq 0$.

Solving the system (I.44), we obtain the estimates

$$\left.\begin{aligned} |A_{kj}(s,\lambda)| &\leq \frac{C}{|\rho|}\,|\exp(-\rho\omega_4 s)|, \qquad k = 1,2, \\ |A_{3j}(s,\lambda)| &\leq C\,|\exp(-\rho\omega_4 s)|. \end{aligned}\right\} \tag{I.45}$$

Using (I.42) and (I.44), we get

$$
\begin{aligned}
\mathcal{G}^0(x,s,\lambda) &= \sum_{m=1}^{4} H_m(x-s,\lambda)[\alpha_{m9}(\lambda),\alpha_{m,10}(\lambda)] \\
&+\sum_{m=1}^{4} H_m(x,\lambda)\sum_{k=1}^{3}\alpha_{m,k+6}(\lambda)A_k(s,\lambda) \\
&+\sum_{m=5}^{7} H_m(x,\lambda)\Big(\sum_{k=1}^{3}\alpha_{m,k+6}(\lambda)A_k(s,\lambda) \\
&+[\alpha_{m9}(\lambda),\alpha_{m,10}(\lambda)]\exp(-\rho\omega_{m-4}s)\Big), \qquad x\geq s
\end{aligned}
\tag{I.46}
$$

$$
\begin{aligned}
\mathcal{G}^0(x,s,\lambda) &= \sum_{m=1}^{7} H_m(x,\lambda)\sum_{k=1}^{3}\alpha_{m,k+6}(\lambda)A_k(s,\lambda) \\
&-\sum_{m=8}^{10} H_m(x,\lambda)[\alpha_{m9}(\lambda),\alpha_{m,10}(\lambda)]\exp(-\rho\omega_{m-4}s), \qquad x\leq s.
\end{aligned}
$$

Let $\mathcal{G}_j^0(x,s,\lambda) = \frac{\partial^j}{\partial x^j}\mathcal{G}^0(x,s,\lambda)$, $j=\overline{0,4}$. From (I.46) we have, for all x, s, λ, according to (I.42) and (I.45)

$$
\mathcal{G}^0(x,s,\lambda)=O((x+1)\rho^{-3}), \qquad \mathcal{G}_j^0(x,s,\lambda)=O(\rho^{j-4}), \qquad j=\overline{1,3}.
$$

Furthermore, from (I.43) with $x=0$ we conclude that

$$
\mathcal{G}_j^0(0,s,\lambda)=0, \qquad j=\overline{0,2}; \qquad \mathcal{G}_4^0(0,s,\lambda)[0,1]^T=0.
$$

Let $f(x)\in\mathcal{L}(0,\infty)$. Then the function

$$
\mathcal{Y}(x,\lambda)=\int_0^\infty \mathcal{G}^0(x,s,\lambda)f(s)\,ds
$$

satisfies the relations

$$
\mathcal{Y}^{(5)}(x,\lambda)-\lambda\Omega\mathcal{Y}''(x,\lambda)=f(x)\mathcal{Y}^{(j)}(0,\lambda)=0, \qquad j=\overline{0,2}
$$

$$
[0,1]\mathcal{Y}^{(4)}(0,\lambda)=0, \qquad |\mathcal{Y}_m(x,\lambda)|\leq C(x+1).
$$

For fixed s, λ, we consider the system

$$\widetilde{\mathcal{G}}_j(x,s,\lambda) = \mathcal{G}_j^0(x,s,\lambda) - \int\limits_0^\infty \mathcal{G}_j^0(x,r,\lambda) \sum_{m=0}^{3} P_m(r)\widetilde{\mathcal{G}}_m(r,s,\lambda)\,dr, \tag{I.47}$$

$$j=\overline{0,3}, \qquad x>0.$$

Since $P_3(x)$, $P_2(x)$, $P_1(x)$, $(1+x)P_0(x) \in \mathcal{L}(0,\infty)$, then it follows from the properties of the functions $\mathcal{G}_j^0(x,s,\lambda)$ that for $|\lambda| \geq \lambda_0$ the system (I.47) has a unique solution $\widetilde{\mathcal{G}}_j(x,s,\lambda)$, $j=\overline{0,3}$, where $\widetilde{\mathcal{G}}_j(x,s,\lambda) = \frac{\partial^j}{\partial x^j}\widetilde{\mathcal{G}}(x,s,\lambda)$, $\widetilde{\mathcal{G}}(x,s,\lambda) = \widetilde{\mathcal{G}}_0(x,s,\lambda)$, and

$$\widetilde{\mathcal{G}}(x,s,\lambda) = O((x+1)\rho^{-3}), \qquad \widetilde{\mathcal{G}}_j(x,s,\lambda) = O(\rho^{j-4}), \qquad j=\overline{1,3}, \tag{I.48}$$

$$\widetilde{\mathcal{G}}(x,s,\lambda) - \mathcal{G}^0(x,s,\lambda) = O((x+1)\rho^{-4}).$$

Lemma I.4. $\mathcal{G}^*(x,s,\lambda) = \widetilde{\mathcal{G}}(s,x,\lambda)$.

Proof. It follows from (I.47) and the properties of the functions $\mathcal{G}_j^0(x,s,\lambda)$ that for a fixed s

$$L_\lambda \widetilde{\mathcal{G}}(x,s,\lambda) = 0, \qquad \widetilde{\mathcal{G}}_j(0,s,\lambda) = 0, \qquad j=\overline{0,2}, \qquad \widetilde{\mathcal{G}}_4(0,s,\lambda)[0,1]^T = 0,$$

where $\widetilde{\mathcal{G}}_4(x,s,\lambda) = \frac{\partial}{\partial x}\widetilde{\mathcal{G}}_3(x,s,\lambda)$. Furthermore, the functions $\widetilde{\mathcal{G}}_j(x,s,\lambda)$, $j=\overline{0,4}$ are continuous for $x \in [0,s)$, (s,∞), and

$$\widetilde{\mathcal{G}}_j(s+0,s) - \widetilde{\mathcal{G}}_j(s-0,s) = \delta_{j4}\begin{bmatrix}1 & 0\\ 0 & 1\end{bmatrix}.$$

Let $f(x) = [f_1(x), f_2(x)]^T \in \mathcal{L}(0,\infty)$, $f^*(x) = [f_1^*(x), f_2^*(x)] \in \mathcal{L}(0,\infty)$, and let $f^*(x)$ be a finite function. For a fixed λ ($\operatorname{Im}\lambda = 0$) we consider

$$\mathcal{Y}(x) = \int\limits_0^\infty \widetilde{\mathcal{G}}(x,s,\lambda) f(s)\,ds, \qquad Z(x) = \int\limits_0^\infty f^*(x)\mathcal{G}^*(x,s,\lambda)\,ds.$$

It is obviously follows from the properties of the functions $\widetilde{\mathcal{G}}(x,s,\lambda)$, $\mathcal{G}^*(x,s,\lambda)$ that

$$L_\lambda \mathcal{Y}(x) = f(x), \qquad \mathcal{Y}^{(j)}(0) = 0, \qquad j=\overline{0,2}$$

$$\mathcal{Y}_2^{(4)}(0) = 0, \qquad L_\lambda^* Z(x) = f^*(x)$$

$$Z_1(0) = Z'(0) = 0, \qquad \lim_{x\to\infty} \mathcal{Y}_i^{(m)}(x) Z_j^{(r)}(x) = 0.$$

Then

$$\int\limits_0^\infty Z(x) L_\lambda \mathcal{Y}(x)\,dx = \int\limits_0^\infty L_\lambda^* Z(x)\mathcal{Y}(x)\,dx,$$

or

$$\int_0^\infty Z(x)f(x)\,dx = \int_0^\infty f^*(x)\mathcal{Y}(x)\,dx.$$

Consequently

$$\int_0^\infty \left(\int_0^\infty f^*(s)\mathcal{G}^*(x,s,\lambda)\,ds\right) f(x)\,dx = \int_0^\infty f^*(x)\left(\int_0^\infty \widetilde{\mathcal{G}}(x,s,\lambda)f(s)\,ds\right)dx.$$

Since $f(x)$ and $f^*(x)$ are arbitrary, we conclude that $\mathcal{G}^*(x,s,\lambda) = \widetilde{\mathcal{G}}(s,x,\lambda)$. Thus, Lemma I.4 is proved.

It follows from (I.40) and Lemma I.4 that

$$\widetilde{\mathcal{G}}(x,s,\lambda) = \sum_{k=1}^{3} D_k(x,\lambda)\psi_k(s,\lambda) + g^*(s,x,\lambda). \tag{I.49}$$

Let (I.7) now hold true. We denote

$$J_n(x) = \frac{1}{2\pi i}\int_{|\lambda|=R_n} \mathcal{Y}(x,\lambda)\,d\lambda, \qquad \mathcal{Y}(x,\lambda) = \int_0^\infty \widetilde{\mathcal{G}}(x,s,\lambda)f(s)\,ds, \qquad R_n \to \infty.$$

Using (I.48) we obtain for any $b > 0$

$$\mathcal{Y}(x,\lambda) = \int_0^\infty \mathcal{G}^0(x,s,\lambda)f(s)\,ds + O(\rho^{-4}), \qquad x \in [0,b].$$

It evidently then follows from (I.46) that

$$J_n(x) = \alpha^* \int_0^x (x-s)f(s)\,ds + \varepsilon_n(x), \qquad \lim_{n\to\infty} \varepsilon_n(x) = 0, \tag{I.50}$$

uniformly for $x \in [0,b]$, where $\alpha^* = [\alpha^0_{mk}]_{m=3,4;k=9,10}$.

Let us show that $\det \alpha^* \neq 0$. Indeed, let $f(x) \in C^3[0,\infty)$ be a finite function, $f^{(j)}(0) = 0$, $j = \overline{0,2}$, and

$$\widetilde{f}(x) = \Omega^{-1}\int_0^x (x-s)f(s)\,ds.$$

Comparing (I.43) with (I.49), we find that the functions $A_k(s,\lambda)$, $g^0(x-s,\lambda)$ satisfy the equations

$$A_k^{(5)}(s,\lambda) + \lambda A_k''(s,\lambda)\Omega = 0, \qquad \frac{\partial^5}{\partial s^5} g^0(x-s,\lambda) + \lambda\frac{\partial^2}{\partial s^2} g^0(x-s,\lambda)\Omega = 0. \tag{I.51}$$

Then using (I.43) and (I.51), we get

$$\int_0^\infty \mathcal{G}^0(x,s,\lambda)f(s)\,ds = \sum_{k=1}^3 N_{k+6}(x,\lambda)\int_0^\infty A_k(s,\lambda)f(s)\,ds + \int_0^x g^0(x-s,\lambda)f(s)\,ds$$

$$= \sum_{k=1}^3 N_{k+6}(x,\lambda)\int_0^\infty A_k''(s,\lambda)\Omega\widetilde{f}(s)\,ds + \int_0^x \left(\frac{\partial^2}{\partial s^2}g^0(x-s,\lambda)\Omega\right)\widetilde{f}(s)\,ds$$

$$= -\frac{1}{\lambda}\sum_{k=1}^3 N_{k+6}(x,\lambda)\int_0^\infty A_k^{(5)}(s,\lambda)\widetilde{f}(s)\,ds - \frac{1}{\lambda}\int_0^x \left(\frac{\partial^5}{\partial s^5}g^0(x-s,\lambda)\right)\widetilde{f}(s)\,ds$$

$$= -\frac{\widetilde{f}(x)}{\lambda} + \frac{1}{\lambda}\int_0^\infty \mathcal{G}^0(x,s,\lambda)\widetilde{f}^{(5)}(s)\,ds.$$

Thus, $J_n(x) = -\widetilde{f}(x) + \varepsilon_n(x)$. Comparing it with (I.50) we obtain $\alpha^* = -\Omega^{-1}$, and consequently $\det\alpha^* \neq 0$.

On the other hand, by virtue of (I.7) and (I.49), the function $\mathcal{Y}(x,\lambda)$ is entire in λ, and consequently $J_n(x) \equiv 0$. Then it follows from (I.50) that

$$\int_0^x (x-s)f(s)\,ds = 0,$$

i.e. $f(x) = 0$ a.e. Theorem I.1 is proved.

APPENDIX II.

INTEGRABLE DYNAMICAL SYSTEMS CONNECTED WITH HIGHER-ORDER DIFFERENCE OPERATORS

1. For a fixed $q \geq 1$ we consider the following Cauchy problem for the nonlinear semi-infinite system:

$$\dot{a}_{nj}(t) = a_{n1}(t)a_{n+1,j-1}(t) - a_{n+j-1,1}(t)a_{n,j-1}(t), \tag{II.1}$$

$$a_{nj}(0) = a^0_{nj}. \tag{II.2}$$

Here $-q+1 \leq j \leq 1$; $n \geq q$ for $j = 1$ and $n \geq q - j$ for $j \leq 0$; $a_{n,-q} = 1$, $a_{q-1,1} = 0$, $a_{q+j-1,-j} = 0$ for $j = \overline{1, q-1}$; a^0_{nj} are complex numbers, and $a^0_{n1} \neq 0$. System (II.1) is a difference analog for equations like the KdV equation, and is equivalent to the Lax equation $\dot{L} = [A, L]$ where $A = [a_{n1}\delta_{n,j-1}]_{n,j\geq q}$, $L = [a_{n,j-n}]_{n,j\geq q}$; $a_{nj} = 0$ for $j > 1$ and $j < -q$. Thus, the solution of the Cauchy problem (II.1)–(II.2) is connected with the study of spectral properties and the solution of the IP for higher-order difference operators

$$(\ell y)_n = \sum_{\mu=0}^{q+1} a_{n,\mu-q} y_{n+\mu-q}, \qquad a_{n1} \neq 0, \qquad a_{n,-q} = 1. \tag{II.3}$$

We note that there are no restrictions on the behaviour of a_{nj} as $n \to \infty$.

For $q = 1$, the system (II.1) is the Toda chain, which has been studied fairly completely (see [43], [44], [111], [116] and references therein). Things are more complicated for $q > 1$, and integrable dynamical systems connected with higher-order difference operators have not been investigated enough. In this direction we mention the articles [117]–[118], in which important integrable systems are pointed out, which are connected with two-term difference operators of the form (II.3) for $q > 1$, when $a_{n,-j} = 0$ for $j = \overline{0, q-1}$.

In this section we study the Cauchy problem (II.1)–(II.2) by the inverse problem method. An algorithm for the solution, along with necessary and sufficient conditions of the solvability of (II.1)–(II.2) in the classes of analytic and meromorphic functions are obtained. For this we use the results of §12 devoted to the IP for higher-order difference operators (II.3).

2. For a fixed $q \geq 1$, we consider the difference equation

$$(\ell y)_n = \sum_{\mu=0}^{q+1} a_{n,\mu-q} y_{n+\mu-q} = \lambda y_n, \qquad n \geq q, \tag{II.4}$$

where $y = [y_n]_{n\geq 0}$, a_{nj} are complex numbers, $a_{n,-q} = 1$, $a_{n1} \neq 0$ for $n \geq q$, and $a_{n,-j} = 0$ for $n - q + 1 \leq j \leq q - 1$, $q \leq n \leq 2q - 2$.

Denote A_R (M_R) by the set of analytic (meromorphic) for $|t| < R$ functions. Let $A = \bigcup_{R>0} A_R$, $M = \bigcup_{R>0} M_R$, and let A' be the set of sequences $\{\alpha_k\}_{k\geq 1}$ such that $\alpha_k = O\left(\left(\frac{k}{\delta}\right)^k\right)$ for a certain $\delta > 0$. We will write $\{f_k(t)\}_{k\geq 1} \in A$ (A^0) if there exists $R > 0$ such that $f_k(t) \in A_R$ ($f_k(t) \neq 0, |t| < R$) for all k.

As in §12, we denote by $\mathcal{F}^+ = \mathcal{F}^+(\mathbb{C})$ the set of GF

$$P(\lambda) = \sum_{k=0}^{\infty} \frac{1}{\lambda^k} P_k, \qquad P_k \in \mathbb{C}.$$

Let $\mathcal{F}_\ell^+$ be the set of GF $P \in \mathcal{F}^+$ such that $P_k = 0$ for $k < \ell$. If $\{P_k\}_{k\geq 0} \in A'$, then we can define the GF $\sigma(\lambda, t) = e^{\lambda t} P(\lambda)$ with the moments

$$\sigma_k(t) = \sum_{j=0}^{\infty} P_{k+j} \frac{t^j}{j!}.$$

Let $\Phi_n(\lambda) = [\Phi_n^i(\lambda)]_{i=1,q}^T$, $n \geq 0$ be the WS for (II.4) under the conditions

$$\Phi_n^i = \delta_{i,n+1} \qquad (n = \overline{0, q-1}); \qquad \Phi_n^i(\lambda) = \sum_{k=0}^{\infty} \frac{1}{\lambda^k} \Phi_{kn}^i \in \mathcal{F}_0^+. \tag{II.5}$$

Substituting (II.5) into (II.4), we obtain

$$\Phi_{k+1,n}^i = \sum_{\mu=0}^{q+1} a_{n,\mu-q} \Phi_{k,n+\mu-q}^i, \qquad n \geq q, \qquad i = \overline{1,q}, \qquad k \geq 0, \tag{II.6}$$

$$\Phi_{0n}^i = 0, \qquad n \geq q \tag{II.7}$$

$$\Phi_{0n}^i = \delta_{i,n+1}, \qquad \Phi_{kn}^i = 0, \qquad k \geq 1, \qquad n = \overline{0, q-1}. \tag{II.8}$$

The WS $\{\Phi_n(\lambda)\}_{n\geq 1}$ can be constructed from (II.6)–(II.8) successively for $k = 0, 1, 2, \ldots$. In particular,

$$\Phi_{kn}^i = 0 \qquad \text{for} \qquad n \geq q(k+1). \tag{II.9}$$

Let $s = [n/q]$, where $[\cdot]$ denotes the greatest integer in the number. Then $n = qs + m$, $0 \leq m \leq q-1$, and (II.6), (II.9) become

$$\Phi_{k+1,qs+m}^i = \sum_{\mu=0}^{q+1} a_{qs+m,\mu-q} \Phi_{k,q(s-1)+m+\mu}^i, \tag{II.10}$$

$$s \geq 1, \qquad k \geq 0, \qquad i = \overline{1,q}, \qquad m = \overline{0, q-1},$$

$$\Phi_{k,qs+m}^i = 0 \qquad \text{for} \qquad k < s, \tag{II.11}$$

and consequently

$$\Phi_{qs+m}^i(\lambda) = \sum_{k=s}^{\infty} \frac{1}{\lambda^k} \Phi_{k,qs+m}^i \in \mathcal{F}_s^+.$$

Using (II.10) for $k = s-1$, we compute

$$\Phi^i_{s,qs+m} = \sum_{\mu=0}^{i-1-m} a_{qs+m,\mu-q}\Phi^i_{s-1,q(s-1)+m+\mu}, \qquad m \le i-1 \tag{II.12}$$

$$\Phi^i_{s,qs+i-1} = 1, \qquad \Phi^i_{s,qs+m} = 0, \qquad i \le m. \tag{II.13}$$

We introduce the WM $\mathfrak{M}(\lambda) = [\mathfrak{M}^i(\lambda)]^T_{i=\overline{1,q}}$ by the formula $\mathfrak{M}^i(\lambda) = \Phi^i_q(\lambda)$. It is clear that

$$\mathfrak{M}^i(\lambda) = \sum_{k=1}^{\infty} \frac{1}{\lambda^k}\mathfrak{M}^i_k \in \mathcal{F}^+_1, \qquad \mathfrak{M}^i_1 = \delta_{i1}, \qquad i = \overline{1,q}.$$

The IP is formulated as follows: given the WM $\mathfrak{M}(\lambda)$ construct the operator ℓ. Such an IP was solved in a more general formulation in §12. Now we briefly reformulate the results of §12 for our particular case.

Let $P_n(\lambda)$, $n \ge 0$ be the solution of (II.4) under the conditions $P_n(\lambda) = \delta_{nq}$, $n = \overline{0,q}$. Then

$$P_{k+q}(\lambda) = \sum_{i=0}^{k} c_{ik}\lambda^i, \qquad k \ge 0, \qquad c_{kk} \ne 0, \qquad c_{00} = 1. \tag{II.14}$$

Substituting (II.4) into (II.4), we calculate

$$\sum_{\mu=0}^{q+1} a_{k+q,\mu-q}c_{i,k+\mu-q} = c_{i-1,k}, \qquad i = \overline{0,k+1},$$

where $c_{ik} = 0$ for $i < 0$ and $i > k$. From this we obtain the recurrence relations for determining a_{nj} from c_{ik}:

$$a_{nj} = (c_{n+j-q,n+j-q})^{-1}\left(c_{n+j-q-1,n-q} - \sum_{\mu=j+1}^{1} a_{n\mu}c_{n+j-q,n+\mu-q}\right), \tag{II.15}$$

where $-q+1 \le j \le 1$; $n \ge q$ for $j = 1$, and $n \ge q-j$ for $j \le 0$.

Denote

$$R_n(\lambda) = [R^i_n(\lambda)]_{i=\overline{1,q}}, \qquad R^i_{qs+m}(\lambda) = \delta_{i,m+1}\lambda^s, \qquad m = \overline{0,q-1},$$

$$\Delta_k = \det\,[\mu_{in}]_{i,n=\overline{0,k}}, \qquad \mu_{in} = (1, \lambda^i\mathfrak{M}(\lambda)R_n(\lambda)), \qquad i,n \ge 0.$$

It is easy to see that $\mu_{in} = 0$ for $0 \le i < n \le q-1$, and $\mu_{00} = 1$ or $\mathfrak{M}^1_1 = 1$, $\mathfrak{M}^i_k = 0$ for $1 \le k < i \le q$. Furthermore, $\Delta_k \ne 0$ for all $k \ge 0$, and

$$c_{ik} = \frac{(-1)^{k-1}}{\Delta_k}\det\,[\mu_{j\nu}]_{j=\overline{0,k}\setminus i;\nu=\overline{0,k-1}}, \qquad i = \overline{0,k}. \tag{II.16}$$

Denote the set of matrices $\mathfrak{M}(\lambda) = [\mathfrak{M}^i(\lambda)]^T_{i=\overline{1,q}}$, $\mathfrak{M}^i(\lambda) \in \mathcal{F}^+_1$ such that $\mu_{00} = 1$, $\mu_{in} = 0$ $(0 \le i < n \le q-1)$ by M^0. The following theorem is the particular case of Theorem 12.5 for $p = 1$.

Theorem II.1. *For a matrix $\mathfrak{M}(\lambda) \in M^0$ to be the WM for ℓ of the form* (II.4), *it is necessary and sufficient that $\Delta_k \neq 0$ for all $k \geq 0$. The operator ℓ can be found by the following algorithm.*

Algorithm II.1. (i) from the given WM $\mathfrak{M}(\lambda)$ construct c_{ik}, $0 \leq i \leq k$ via (II.16); (ii) compute a_{nj} by (II.15).

3. In this section we study the Cauchy problem (II.1)–(II.2).

Denote the WM for the operator ℓ^0 of the form (II.3), constructed from the initial data $\{a_{nj}^0\}$ by $\mathfrak{M}^0(\lambda) = [\overset{o}{\mathfrak{M}}{}^i(\lambda)]^T_{i=\overline{1,q}}$. Let $\{\overset{o}{\mathfrak{M}}{}^i_j\}_{j\geq 1}$ be the moments of $\overset{o}{\mathfrak{M}}{}^i(\lambda)$. Suppose that there exists an solution $\{a_{nj}(t)\} \in A$ which is analytic at $t = 0$ of the Cauchy problem (II.1)–(II.2). We consider the corresponding difference operator $\ell = \ell(t)$ of the form (II.3). Let $\{\Phi_n(t,\lambda)\}$ and $\mathfrak{M}(t,\lambda)$ be the WS and the WM for $\ell(t)$, respectively. In particular, it follows from (II.1) for $j = 1$ that $a_{n1}(t) \neq 0$. Hence, by Theorem II.1, $\Delta_k(t) \neq 0$, $k \geq 0$.

In the following theorem the evolution of the WM with respect to t is obtained.

Theorem II.2.

$$\dot{\mathfrak{M}}^1 = (\lambda - a_{q0})\mathfrak{M}^1 - 1, \tag{II.17}$$

$$\dot{\mathfrak{M}}^i = (\lambda - a_{q0})\mathfrak{M}^i - a_{q+i-2,1}\mathfrak{M}^{i-1}, \qquad i = \overline{2,q}. \tag{II.18}$$

Proof. By construction, the WS's $\Phi^i = [\Phi^i_n]_{n\geq q}$, $i = \overline{1,q}$ satisfy the equations

$$L\Phi^i + \alpha^i = \lambda\Phi^i, \qquad i = \overline{1,q}, \tag{II.19}$$

where $\alpha^i = [\delta_{n,i+q-1}]_{n\geq q}$. Differentiating (II.19) with respect to t, we obtain $\dot{L}\Phi^i + L\dot{\Phi}^i = \lambda\dot{\Phi}^i$. Since $\{a_{nj}(t)\}$ are solutions of (II.1), it follows that $\dot{L} = [A, L]$, and hence

$$L(\dot{\Phi}^i - A\Phi^i) = \lambda(\dot{\Phi}^i - A\Phi^i) + A\alpha^i, \qquad i = \overline{1,q}. \tag{II.20}$$

Denote $\psi^i = \dot{\Phi}^i - A\Phi^i$, $\psi^i = [\psi^i_n]_{n\geq q}$. We compute $\psi^i_n = \dot{\Phi}^i_n - a_{ni}\Phi^i_{n+1}$, $n \geq q$. It is clear that $\psi^i_n \in \mathcal{F}^+_0$. The relations (II.20) become

$$L\psi^i = \lambda\psi^i + A\alpha^i, \qquad i = \overline{1,q}. \tag{II.21}$$

It is convenient to define ψ^i_n for $n = \overline{0, q-1}$ as follows:

$$\psi^1_n = 0, \qquad \psi^i_n = -a_{q+i-2,1}\Phi^{i-1}_n, \qquad i = \overline{2,q}. \tag{II.22}$$

Then, by virtue of (II.21), we have

$$\sum_{\mu=0}^{q+1} a_{n,\mu-q}\psi^i_{n+\mu-q} = \lambda\psi^i_n, \qquad n \geq q, \qquad i = \overline{1,q}.$$

Since $\psi^i_n \in \mathcal{F}^+_0$ and satisfy the initial conditions (II.22) for $n = \overline{0, q-1}$ we get, in view of uniqueness of the WS, that (II.22) are valid also for $n \geq q$. Thus, we arrive at the evolution equations for the WS's.

$$\dot{\Phi}^1_n = a_{n1}\Phi^1_{n+1}, \qquad \dot{\Phi}^i_n = a_{n1}\Phi^i_{n+1} - a_{q+i-2,1}\Phi^{i-1}_n, \qquad i = \overline{2,q}, \qquad n \geq q.$$

In particular, for $n = q$ we have

$$\dot{\mathfrak{M}}^1 = a_{q1}\Phi^1_{q+1}, \qquad \dot{\mathfrak{M}}^i = a_{q1}\Phi^i_{q+1} - a_{q+i-2,1}\mathfrak{M}^{i-1}, \qquad i = \overline{2,q}. \tag{II.23}$$

Since

$$a_{q1}\Phi^1_{q+1} = (\lambda - a_{q0})\Phi^1_q - 1, \qquad a_{q1}\Phi^i_{q+1} = (\lambda - a_{q0})\Phi^i_q, \qquad i = \overline{2,q},$$

then it follows from (II.23) that (II.17) and (II.18) are valid. Theorem II.2 is proved.

Since

$$\mathfrak{M}^i(\lambda) = \sum_{k=1}^{\infty} \frac{\mathfrak{M}^i_k}{\lambda^k} \in \mathcal{F}^+_1,$$

then (II.17) and (II.18) are equivalent to the following relations for the moments $\mathfrak{M}^i_k$:

$$\left.\begin{aligned} \dot{\mathfrak{M}}^1_k &= \mathfrak{M}^1_{k+1} - a_{q0}\mathfrak{M}^1_k, \\ \dot{\mathfrak{M}}^i_k &= \mathfrak{M}^i_{k+1} - a_{q0}\mathfrak{M}^i_k - a_{q+i-2,1}\mathfrak{M}^{i-1}_k, \qquad i = \overline{2,q}. \end{aligned}\right\} \tag{II.24}$$

From this we calculate

$$\mathfrak{M}^1_1 = 1, \qquad \mathfrak{M}^i_k = 0, \qquad (1 \le k < i \le q), \qquad \mathfrak{M}^i_i = \prod_{j=0}^{i-2} a_{q+j,1}, \qquad i = \overline{2,q}. \tag{II.25}$$

Let us now integrate the evolituion equations (II.17), (II.18) with the initial conditions $\mathfrak{M}^i(0,\lambda) = \overset{o}{\mathfrak{M}}{}^i(\lambda)$. We note that except for $\mathfrak{M}^i(t,\lambda)$ the functions $a_{q0}(t)$, $a_{q+i-2,1}(t)$, $i = \overline{2,q}$ in (II.17) and (II.18) are also unknown.

Denote

$$B(t) = \exp\left(\int_0^t a_{q0}(s)\,ds\right).$$

It follows from (II.17) and (II.18) that

$$B(t)\mathfrak{M}^1(t,\lambda) = e^{\lambda t}\overset{o}{\mathfrak{M}}{}^1(\lambda) - \int_0^t e^{\lambda(t-\tau)}B(\tau)\,d\tau, \tag{II.26}$$

$$B(t)\mathfrak{M}^i(t,\lambda) = e^{\lambda t}\overset{o}{\mathfrak{M}}{}^i(\lambda) - \int_0^t e^{\lambda(t-\tau)}B(\tau)a_{q+i-2,1}(\tau)\mathfrak{M}^{i-1}(\tau,\lambda)\,d\tau, \; i = \overline{2,q}. \tag{II.27}$$

The integral in (II.26) is a function which is entire in λ. Therefore $B(t)(1,\mathfrak{M}^1(,\lambda)) = (1, e^{\lambda t}\overset{o}{\mathfrak{M}}{}^1(\lambda))$. In virtue of (II.25), we get $(1,\mathfrak{M}^1(t,\lambda)) = \mathfrak{M}^1_1(t) = 1$, and consequently $B(t) = (1, e^{\lambda t}\overset{o}{\mathfrak{M}}{}^1(\lambda))$ or

$$B(t) = \sum_{j=0}^{\infty} \overset{o}{\mathfrak{M}}{}^1_{j+1}\frac{t^j}{j!}. \tag{II.28}$$

The same result can be obtained from (II.24). Indeed, it follows from (II.24) that $\frac{d}{dt}(\mathfrak{M}_k^1(t)B(t) = \mathfrak{M}_{k+1}^1(t)B(t)$. Hence

$$\mathfrak{M}_k^1(t) = \frac{B^{(k-1)}(t)}{B(t)}, \qquad B^{(k-1)}(0) = \overset{0}{\mathfrak{M}}{}_k^1.$$

In particular, $\{\overset{o}{\mathfrak{M}}{}_k^1\}_{k\geq 1} \in A'$.

Furthermore, in view of (II.25), we have $(\lambda^{i-2}, \mathfrak{M}^i(t,\lambda)) = \mathfrak{M}_{i-1}^i(t) = 0$, $i = \overline{2,q}$. It now follows from (II.27) that

$$(\lambda^{i-2}, e^{\lambda t}\,\overset{o}{\mathfrak{M}}{}^i(\lambda)) = \int\limits_0^t B(\tau)a_{q+i-2,1}(\tau)(\lambda^{i-2}, e^{\lambda(t-\tau)}\mathfrak{M}^{i-1}(\tau,\lambda))\,d\tau.$$

From this, taking (II.25) into account, we obtain

$$\begin{aligned}\frac{d}{dt}(\lambda^{i-2}, e^{\lambda t}\,\overset{o}{\mathfrak{M}}{}^i(\lambda)) &= B(t)a_{q+i-2,1}(t)(\lambda^{i-2}, \mathfrak{M}^{i-1}(t,\lambda)) \\ &= B(t)a_{q+i-2,1}(t)\mathfrak{M}_{i-1}^{i-1}(t) = B(t)a_{q+i-2,1}(t)\prod_{j=0}^{i-3} a_{q+j,1}(t).\end{aligned}$$

Hence

$$a_{q+i-2,1}(t) = \left(B(t)\prod_{j=0}^{i-3} a_{q+j,1}(t)\right)^{-1} \sum_{j=0}^{\infty} \overset{0}{\mathfrak{M}}{}_{j+1}^i \frac{t^j}{j!}, \qquad i = \overline{2,q}, \tag{II.29}$$

and $\{\overset{o}{\mathfrak{M}}{}_{j+1}^i\}_{j\geq 0} \in A'$. Using (II.29), we calculate the functions $a_{q+i-2,1}(t)$ successively for $i = 2, 3, \ldots, q$. Then from (II.26) and (II.27) we find $\mathfrak{M}^i(t,\lambda)$, $i = \overline{1,q}$. Since $\Delta_k(t) \neq 0$, $k \geq 0$ we calculate $\{a_{nj}(t)\}$ by solving the IP. Thus, we obtain the following algorithm for the solution of the Cauchy problem (II.1)–(II.2) by the inverse problem method.

Algorithm II.2. Let $\{a_{nj}^0\}$, $a_{n1}^0 \neq 0$ be given. We then

(1) construct $\{\overset{o}{\mathfrak{M}}{}_j^i\}_{j\geq 1}$, $i = \overline{1,q}$;

(2) compute $B(t)$, $a_{q+i-2}(t)$, $i = \overline{2,q}$ by (II.28) and (II.29);

(3) find $\mathfrak{M}^i(t,\lambda)$, $i = \overline{1,q}$ by (II.26)–(II.27), and calculate $\Delta_k(t)$, $k \geq 0$ by the formulas

$$\Delta_k(t) = \det[\mu_{in}(t)]_{i,n=\overline{0,k}}, \qquad \mu_{in}(t) = (1, \lambda^i \mathfrak{M}(t,\lambda)R_n(\lambda));$$

(4) construct the functions $\{a_{nj}(t)\}$ by solving the IP with the help of Algorithm II.1.

Remark. Algorithm II.2 also works when $a_{nj}(t) \in M$, i.e. in the class of meromorphic functions.

4. Let us seek the solution of the Cauchy problem (II.1)–(II.2) in the class M. It is obvious that if the solution in the class M exists then it is unique. Let us formulate necessary and sufficient conditions for the existence of the solution of (II.1)–(II.2) in the class of meromorphic functions.

Theorem II.3. *For the Cauchy problem* (II.1)–(II.2) *to have a solution* $a_{nj}(t) \in M$, *it is necessary and sufficient that* $\{\overset{0}{\mathfrak{M}}{}^i_j\}_{j\geq 1} \in A'$, $i = \overline{1,q}$. *This solution can be constructed by applying Algorithm II.2. In addition,* $a_{nj}(t) \in M_R$ *if and only if*

$$\mathfrak{N}_i(t) \equiv \sum_{j=0}^{\infty} \overset{0}{\mathfrak{M}}{}^i_{j+i} \frac{t^j}{j!} \in M_R, \qquad i = \overline{1,q}.$$

The following theorem gives necessary and sufficient conditions of solvability in narrower classes A and A_R.

Theorem II.4. *For the Cauchy problem* (II.1)–(II.2) *to have a solution* $\{a_{nj}(t)\} \in A$, *it is necessary and sufficient that* $\{\overset{0}{\mathfrak{M}}{}^i_j\}_{j\geq 1} \in A'$, $i = \overline{1,q}$ *and* $\{\Delta_k(t)\}_{k\geq 0} \in A^0$. *In addition,* $a_{nj}(t) \in A_R$ *if and only if*

$$\mathfrak{N}_i(t) \in A_R, \qquad \mathfrak{N}_i(t) \neq 0, \qquad \Delta_k(t) \neq 0, \qquad |t| < R, \qquad i = \overline{1,q}, \qquad k \geq 0.$$

Let us go on to the proof of Theorem II.3 and II.4. The necessity was proved above. We shall prove the sufficiency. From the given initial data $\{\overset{0}{a}_{nj}\}$, according to Algorithm II.2, we successively construct $\mathfrak{M}^0(\lambda)$, $B(t)$, $a_{q+i-2,1}(t)$, $i = \overline{2,q}$, $\mathfrak{M}(t,\lambda)$, $\Delta_k(t)$ and $\{a_{nj}(t)\}$. By construction the functions $a_{nj}(t)$ belong to the corresponding classes, as was pointed out in the theorems, and $\mathfrak{M}(t,\lambda)$ is the WM for $\ell(t)$. In addition, $a_{nj}(0) = \overset{0}{a}_{nj}$.

Let us show that the functions $\{a_{nj}(t)\}$ which have been constructed are solutions of (II.1). For this we consider the matrix $\Omega = [\Omega_{nj}]_{n,j\geq q}$ defined by the formula $\Omega = \dot{L} - [A, L]$, where $A = [A_{nj}]_{n,j\geq q}$, $L = [L_{nj}]_{n,j\geq q}$, $A_{nj} = a_{n1}\delta_{n,j-1}$, $L_{nj} = a_{n,j-n}$; $a_{nj} = 0$ for $j > 1$ and $j < -q$. It is clear that $\Omega_{nj} = 0$ for $j > n+1$ and $j < n-q+1$. Denote $\Omega_{nj} = 0$ for $j < q$.

Let $\{\Phi^i_n(t,\lambda)\}$ be the WS for $\ell(t)$. Denote $\psi^i = \dot{\Phi}^i - A\Phi^i$, $\psi^i = [\psi^i_n]_{n\geq q}$, $\Phi^i = [\Phi^i_n]_{n\geq q}$. Then $\psi^i_n = \dot{\Phi}^i_n - a_{n1}\Phi^i_{n+1}$, $n \geq q$. Define ψ^i_n for $n = \overline{0, q-1}$ by the formula

$$\psi^1_n = 0, \qquad \psi^i_n = -a_{q+i-2,1}\Phi^{i-1}_n, \qquad i = \overline{2,q}. \tag{II.30}$$

By construction, (II.17) and (II.18) are valid, and hence

$$\dot{\Phi}^1_q = a_{q1}\Phi^1_{q+1}, \qquad \dot{\Phi}^i_q = a_{q1}\Phi^i_{q+1} - a_{q+i-2,1}\Phi^{i-1}_q, \qquad i = \overline{2,q}.$$

Then

$$\psi^1_q = 0, \qquad \psi^i_q = -a_{q+i-2,1}\Phi^{i-1}_q, \qquad i = \overline{2,q}, \tag{II.31}$$

i.e. (II.30) is also valid for $n = q$.

Further, differentiating the equality $L\Phi^i + \alpha^i = \lambda\Phi^i$, $i = \overline{1,q}$ where $\alpha^i = [\delta_{n,j+q-1}]_{n\geq q}$, with respect to t and using the relation $\dot{L} = [A, L] + \Omega$, we obtain

$$L\psi^i - \lambda\psi^i - A\alpha^i + \Omega\Phi^i = 0, \qquad i = \overline{1,q}. \tag{II.32}$$

In view of (II.30), we can rewrite (II.32) in the form

$$\sum_{\mu=0}^{q+1} a_{n,\mu-q}\psi^i_{n+\mu-q} - \lambda\psi^i_n + \sum_{\xi=0}^{q} \Omega_{n,n-q+1+\xi}\Phi^i_{n-q+1+\xi} = 0, \tag{II.33}$$

$$n \geq q, \quad i = \overline{1,q}.$$

Let us show by induction that for $k \geq q$

$$\Omega_{k\xi} = 0, \qquad \xi = \overline{k-q+1, k+1}, \tag{II.34}$$

$$\psi^1_k = 0, \qquad \psi^i_k = -a_{q+i-2,1}\Phi^{i-1}_k, \qquad i = \overline{2,q}. \tag{II.35}$$

Indeed, by virtue of (II.30) and (II.31), relations (II.35) are valid for $k = \overline{0,q}$. Fix $n \geq q$ and assume that (II.35) are valid for $k \leq n$. Then (II.33) becomes

$$\left.\begin{aligned} &a_{n1}\psi^1_{n+1} + \sum_{\xi=0}^{q} \Omega_{n,n-q+1+\xi}\Phi^1_{n-q+1+\xi} = 0, \\ &a_{n1}\psi^i_{n+1} + \sum_{\xi=0}^{q} \Omega_{n,n-q+1+\xi}\Phi^i_{n-q+1+\xi} \\ &\qquad + a_{q+i-2,1}\left(\lambda\Phi^{i-1}_n - \sum_{\mu=0}^{q} a_{n,\mu-q}\Phi^{i-1}_{n+\mu-q}\right) = 0, \quad i = \overline{2,q}. \end{aligned}\right\} \tag{II.36}$$

Denote $s = \left[\frac{n}{q}\right]$, i.e. $n = qs + m$, $0 \leq m \leq q-1$. We rewrite (II.36) as follows

$$a_{qs+m,1}\psi^1_{qs+m+1} + a_{q+i-2,1}J^{i-1}_{m,s} + I^i_{m,s} = 0, \qquad i = \overline{1,q}, \tag{II.37}$$

where

$$J^0_{m,s} = 0, \quad J^i_{m,s} = \lambda\Phi^i_{qs+m} - \sum_{\mu=0}^{q} a_{qs+m,\mu-q}\Phi^i_{qs+m+\mu-q}, \quad i = \overline{1,q-1}, \tag{II.38}$$

$$I^i_{m,s} = \sum_{\xi=0}^{q} \Omega_{qs+m,(s-1)q+m+\xi+1}\Phi^i_{(s-1)q+m+\xi+1}, \qquad i = \overline{1,q}. \tag{II.39}$$

Below one of the same symbols $\kappa_\ell(\lambda)$ will denote various GF from $\mathcal{F}^+_\ell$.

Using (II.38) and (II.10)–(II.13), we transform J^i_{ms}, $i = \overline{1,q-1}$:

$$\begin{aligned} J^i_{m,s} &= \lambda\Phi^i_{qs+m} - \sum_{\mu=q-m}^{q} a_{qs+m,\mu-q}\Phi^i_{qs+m+\mu-q} - \sum_{\mu=0}^{q-m-1} a_{qs+m,\mu-q}\Phi^i_{(s-1)q+\mu+m} \\ &= \frac{1}{\lambda^{s-1}}\left(\Phi^i_{s,qs+m} - \sum_{\mu=0}^{q-m-1} a_{qs+m,\mu-q}\Phi^i_{s-1,(s-1)q+\mu+m}\right) \\ &\quad + \frac{1}{\lambda^s}\left(\Phi^i_{s+1,qs+m} - \sum_{\mu=0}^{q} a_{qs+m,\mu-q}\Phi^i_{s,(s-1)q+\mu+m}\right) + \kappa_{s+1}(\lambda). \end{aligned}$$

By virtue of (II.12) and (II.13), the coefficient for $1/\lambda^{s-1}$ is equal to 0. Indeed,for $i \geq m+1$, we have

$$\Phi^i_{s,qs+m} - \sum_{\mu=0}^{q-m-1} a_{qs+m,\mu-q}\Phi^i_{s-1,(s-1)q+\mu+m}$$

$$= \Phi^i_{s,qs+m} - \sum_{\mu=0}^{i-m-1} a_{qs+m,\mu-q}\Phi^i_{s-1,(s-1)q+\mu+m} = 0,$$

and for $i \leq m$ each term in the sum with $1/\lambda^{s-1}$ is equal to 0.

Furthermore, by virtue of (II.10), the coefficient for $1/\lambda^s$ is equal to $a_{qs+m,1}\Phi^i_{s,qs+m+1}$. Thus, we get

$$J^i_{m,s} = \frac{1}{\lambda^s} a_{qs+m,1}\Phi^i_{s,qs+m+1} + \kappa_{s+1}(\lambda), \tag{II.40}$$

$$i = \overline{1, q-1}, \quad m = \overline{0, q-1}, \quad s \geq 1.$$

Since $\psi^i_n = \dot{\Phi}^i_n - a_{n1}\Phi^i_{n+1}$, it follows from (II.11) and (II.13) that

$$\left.\begin{aligned} &\psi^i_{qs+m+1} \in \mathcal{F}^+_s, && i \geq m+3, \\ &\psi^i_{qs+m+1} \in \mathcal{F}^+_{s+1}, && i \leq m+2, \\ &\psi^i_{qs+m+1} \in \mathcal{F}^+_{s+2}, && i = 1,\ m = q-1. \end{aligned}\right\} \tag{II.41}$$

Now from (II.37) we find $I^i_{m,s}$:

$$I^i_{m,s} = -a_{qs+m,1}\psi^i_{qs+m+1} - a_{q+i-2,1}J^{i-1}_{m,s},$$

and use (II.40), (II.41) and (II.13). This yields

$$\left.\begin{aligned} &I^i_{m,s} \in \mathcal{F}^+_s, && i \geq m+3, \\ &I^i_{m,s} \in \mathcal{F}^+_{s+1}, && i \leq m+2, \\ &I^1_{q-1,s} \in \mathcal{F}^+_{s+2}. \end{aligned}\right\} \tag{II.42}$$

On the other hand, we can use (II.39). Calculations must be carried out separately for $m = \overline{0, q-2}$ and $m = q-1$. At first, let $0 \leq m \leq q-2$. Then, using (II.39), (II.11) and (II.13), we calculate:

(1) for $i = m+2$,

$$\begin{aligned} I^{m+2}_{m,s} &= \Omega_{qs+m,(s-1)q+m+1}\left(\frac{1}{\lambda^{s-1}} + \kappa_s(\lambda)\right) \\ &\quad + \sum_{\xi=1}^{q-1} \Omega_{qs+m,(s-1)q+m+\xi+1}\kappa_s(\lambda) + \Omega_{qs+m,qs+m+1}\left(\frac{1}{\lambda^s} + \kappa_{s+1}(\lambda)\right). \end{aligned}$$

(2) for $i > m+2$,

$$I^i_{m,s} = \sum_{\xi=0}^{i-m-3} \Omega_{qs+m,(s-1)q+m+\xi+1}\kappa_{s-1}(\lambda) + \Omega_{qs+m,(s-1)q+i-1}\left(\frac{1}{\lambda^{s-1}} + \kappa_s(\lambda)\right)$$

$$+ \sum_{\xi=i-m-1}^{q} \Omega_{qs+m,(s-1)q+m+\xi+1}\kappa_s(\lambda).$$

(3) for $1 \le i \le m+1$,

$$I^i_{m,s} = \sum_{\xi=0}^{q+i-m-3} \Omega_{qs+m,(s-1)q+m+\xi+1}\kappa_s(\lambda) + \Omega_{qs+m,qs+i-1}\left(\frac{1}{\lambda^s} + \kappa_{s+1}(\lambda)\right)$$

$$+\sum_{\xi=i}^{m+1} \Omega_{qs+m,qs+\xi}\kappa_{s+1}(\lambda).$$

For $m = q-1$ we calculate analogously:
(1) for $i = 1$,

$$I^1_{q-1,s} = \Omega_{qs+q-1,qs}\left(\frac{1}{\lambda^s} + \kappa_{s+1}(\lambda)\right) + \sum_{\xi=1}^{q-1} \Omega_{qs+q-1,qs+\xi}\kappa_{s+1}(\lambda)$$

$$+\Omega_{qs+q-1,(s+1)q}\left(\frac{1}{\lambda^{s+1}} + \kappa_{s+2}(\lambda)\right).$$

(2) for $i > 1$,

$$I^i_{q-1,s} = \sum_{\xi=0}^{i-2} \Omega_{qs+q-1,qs+\xi}\kappa_s(\lambda) + \Omega_{qs+q-1,qs+i-1}\left(\tfrac{1}{\lambda^s} + \kappa_{s+1}(\lambda)\right)$$

$$+\sum_{\xi=i}^{q} \Omega_{qs+q-1,qs+\xi}\kappa_{s+1}(\lambda).$$

Comparing the relations obtained with (II.42), we arrive at the equalities $\Omega_{n\xi} = 0$ for $\xi = \overline{n-q+1, n+1}$. Thus, (II.34) is valid for $k = n$. But then (II.36) takes the form

$$a_{n1}\psi^1_{n+1} = 0,$$

$$a_{n1}\psi^i_{n+1} = -a_{q+i-2,1}\left(\lambda\Phi^{i-1}_n - \sum_{\mu=0}^{q} a_{n,\mu-q}\Phi^{i-1}_{n+\mu-q}\right)$$

$$= a_{q+i-2,1}a_{n1}\Phi^{i-1}_{n+1}, \qquad i = \overline{2,q},$$

and consequently, (II.35) is valid for $k = n+1$.

Thus, (II.34) and (II.35) are valid for all $k \geq q$. In particular, it means that $\Omega = 0$, i.e. $\dot{L} = [A, L]$. Hence the functions $\{a_{nj}(t)\}$ are solutions of (II.1). Theorems II.3 and II.4 are proved.

REFERENCES

1. Ambarzumian V.A., *Ueber eine Frage der Eigenwerttheorie*, Zs.F.Phys. **53** (1929), 690–695.
2. Borg G., *Eine Umkehrung der Sturm–Liouvilleschen Eigenwertaufgabe*, Acta Math. **78** (1946), 1–96.
3. Levinson N., *The inverse Sturm–Liuoville problem*, Math. Tidsskr. **13** (1949), 25–30.
4. Tikhonov A.N., *On uniqueness of the solution of a electroreconnaissance problem*, Dokl. Akad. Nauk SSSR **69** (1950), 797–800.
5. Marchenko V.A., *Some questions in the theory of a second-order differential operator*, Dokl. Akad. Nauk SSSR **72** (1950), 457–460.
6. Marchenko V.A., *Some questions in the theory of linear differential operators*, Trudy Moskov. Mat. Obshch. **1** (1952), 327–420.
7. Marchenko V.A., *Expansion in eigenfunctions of non-self-adjoint singular second-order differential operators*, Mat. Sb. **52**(94) (1960), 739–788.
8. Marchenko V.A., *The spectral theory of Sturm–Liouville operators*, "Naukova Dumka", Kiev, 1972.
9. Marchenko V.A., *Sturm–Liouville operators and their applications*, "Naukova Dumka", Kiev, 1977; English transl., Birkhauser, 1986.
10. Gel'fand I.M. and Levitan B.M., *On the determination of a differential equation from its spectral function*, Izv. Akad. Nauk SSSR, Ser. Mat. **15** (1951), 309–360; English transl. in Amer. Math. Soc. Transl. (2) **1** (1955).
11. Gasymov M.G. and Levitan B.M., *Determination of a differential equation by two of its spectra*, Usp. Mat. Nauk **19**, No. 2 (1964), 3–63; English transl. in Russian Math. Surveys **19** (1964), 1–64.
12. Levitan B.M., *Inverse Sturm–Liouville problems*, "Nauka", Moscow, 1984; English transl., VNU Sci.Press. Utrecht, 1987.
13. Poschel J. and Trubowitz E., *Inverse spectral theory*, Academic Press, New York, 1987.
14. Krein M.G., *Solution of the inverse Sturm–Liouville problem*, Dokl. Akad. Nauk SSSR **76** (1951), 21–24.
15. Krein M.G., *On a method of effective solution of the inverse problem*, Dokl. Akad. Nauk SSSR **94** (1954), 987–990.
16. Blokh M.S., *On the determination of a differential equation from its spectral matrix-function*, Dokl. Akad. Nauk SSSR **92** (1953), 209–212.

17. Rofe-Beketov F.S., *The spectral matrix and the inverse Sturm-Liouville problem. Theory of Functions, Funct. Analysis and their Applications*, Kharkov **4** (1967), 189–197 (Russian).

18. Hald O.H., *Discontinuous inverse eigenvalue problems*, Comm. Pure Appl. Math. **37** (1984), 539–577.

19. McLaughlin J.R., *Analytical methods for recovering coefficients in differential equations from spectral data*, SIAM Rev. **28** (1986), 53–72.

20. Andersson L.E., *Inverse eigenvalue problems for a Sturm–Liouville equation in impendance form*, Inverse Problems **4** (1988), 929–971.

21. Rundell W. and Sacks P.E, *The reconstruction of Sturm–Liouville operators*, Inverse Problems **8** (1992), 457–482.

22. Pavlov B.S., *On the non-self-adjoint Sturm-Liouville operator on the half-line*, Dokl. Akad. Nauk SSSR **141** (1961), 807–810; English transl. in Soviet Math. Dokl. **2** (1961).

23. Pavlov B.S., *On the spectral theory of nonselfadjoint differential operators*, Dokl. Akad. Nauk SSSR **146** (1962), 1267–70; English transl. in Soviet Math. Dokl. **3** (1962); **4** (1963).

24. Sadovnichii V.A., *On some formulations of inverse spectral problems*, Uspekhi Mat. Nauk **38** (1983), No. 5, p. 132.

25. Agranovich Z.S. and Marchenko V.A., *The inverse problem of scattering theory*, Gordon and Breach, New York, 1963.

26. Faddeev L.D., *On a connection the S-matrix with the potential for the one-dimensional Schrödinger operator*, Dokl. Akad. Nauk SSSR **121** (1958), 63–66.

27. Faddeev L.D., *Properties of the S-matrix of the one-dimensional Schrödinger equation*, Trudy Mat. Inst. Steklov, **73** (1964), 139–166; English transl. in Amer. Math. Soc. Transl. (2) **65** (1964), 314–336.

28. Kay J. and Moses H., *The determination of the scattering potential from the spectral measures function I*, Nuovo Cimento **2** (1955), 917–961.

29. Kay J. and Moses H., *The determination of the scattering potential from the spectral measures function II*, Nuovo Cimento **3** (1956), 56–84.

30. Kay J. and Moses H., *The determination of the scattering potential from the spectral measure function III*, Nuovo Cimento **3** (1956), 276–304.

31. Deift P. and Trubowitz E., *Inverse scattering on the line*, Comm. Pure Appl. Math. **32** (1979), 121–251.

32. Shadan k. and Sabatier P., *Inverse problems in quantum scattering*, Springer-Verlag, New York, 1989.

33. Aktosun T., *Scattering and inverse scattering for a second-order differential equation*, J. Math. Phys. **34** (1993), 1619–1634.

34. Grebern B. and Weder R., *Reconstruction of a potential on the line that is apriory known on the half-line*, SIAM J. Appl. Math. **55** (1995), No. 1, 242–254.

35. Berezanskii Y.M., *On a uniqueness theorem in the inverse spectral problem for the Schrodinger equation*, Trudy Moskov. Mat. Obshch. **7** (1958), 3–51.

36. Nizhnik L.P., *Inverse scattering problems for hyperbolic equations*, "Naukova Dumka", Kiev, 1991.

37. Anikonov Y.E., *Methods of investigating multi-dimensional inverse problems for differential equations*, Novosibirsk, 1978.

38. Lavrent'ev M.M., Romanov V.G. and Shishatskii S.P., *Ill-posed problems in mathematical physics and analysis*, "Nauka", Moscow, 1980.

39. Romanov V.G., *Inverse problems in mathematical physics*, "Nauka", Moscow, 1984.

40. Bukhgeim A.L., *Introduction to the inverse problem theory*, Novosibirsk, 1988.

41. Gardner G., Green J., Kruskal M. and Miura R., *A method for solving the Korteweg-de Vries equation*, Phys. Rev. Lett. **19** (1967), 1095–1098.

42. Lax P., *Integrals of nonlinear equations of evolution and solitary waves*, Comm. Pure Appl. Math. **21** (1968), 467–490.

43. Zakharov V.E., Manakov S.V., Novikov S.P. and Pitaevskii L.P., *The solution theory. The inverse problem method*, "Nauka", Moscow, 1980.

44. Ablowitz M.J. and Segur H., *Solitons and the inverse scattering transform*, SIAM, Philadelphia, 1981.

45. Takhtadjan L.A. and Faddeev L.D., *Hamilton approach in soliton theory*, "Nauka", Moscow, 1996.

46. Sakhnovich L.A., *The inverse problem for differential operators of order $n > 2$ with analytic coefficients*, Mat. Sb. **46**(88) (1958), 61–76.

47. Sakhnovich L.A., *The transformation operator method for higher-order equations*, Mat. Sb. **55**(97) (1961), 347–360.

48. Sakhnovich L.A., *On an inverse problem for fourth-order equations*, Mat. Sb. **56**(98) (1962), 137–146.

49. Khachatryan I.G., *The recovery of a differential equation from its spectrum*, Funktsional. Anal. i Prilozhen., **10** (1976), No. 1, 93–94.

50. Khachatryan I.G., *On some inverse problems for higher-order differential operators on the half-line*, Funktional. Anal. i Prilozhen. **17** (1983), No. 1, 40–52; English transl. in Func. Anal. Appl. **17** (1983).

51. Khachatryan I.G., *Necessary and sufficient conditions of solvability of the inverse scattering problem for higher-order differential equations on the half-line*, Dokl. Akad. Nauk Armyan. SSR **77** (1983), No. 2, 55–58.

52. Khachatryan I.G., *On transformation operators for higher-order differential equations*, Izv. Akad. Nauk Armyan. SSR, Ser. Mat. **13** (1978), No. 3, 215–237.

53. Leibenzon Z.L., *The inverse problem of spectral analysis for higher-order ordinary differential operators*, Trudy Moscov. Mat. Obshch. **15** (1966), 70–144; English transl. in Trans. Moscow Math. Soc. **15** (1966).

54. Leibenson Z.L., *Spectral expansions of transformations of systems of boundary value problems*, Trudy Moskov. Mat. Obshch. **25** (1971), 15–58; English transl. in Trans. Moscow Math. Soc. **25** (1971).

55. Baranova E.A., *On the recovery of higher-order ordinary differential operators from their spectra*, Dokl. Akad. Nauk SSSR **205** (1972), 1271–1273; English transl. in Soviet Math. Dokl. **13** (1972).

56. Yurko V.A., *On an inverse problem for higher-order differential operators*, Manuscript 312–75, deposited at VINITI, 1975, (Russian); R. Zh. Mat. 1975, 9B719.

57. Yurko V.A., *The inverse problem for differential operators of even order*, Diff. Eq. and Numerical Anal., Saratov Univ., Saratov **6** (1976), part 2, 55–66 (Russian).

58. Yurko V.A., *Reconstruction of fourth-order differential operators*, Differentsialnye Uravneniya, **19** (1983), 2016–2017.

59. Yurko V.A., *Uniqueness of the reconstruction of binomial differential operators from two spectra*, Matematicheskie Zametki **43** (1988), No. 3, 356–364; English transl. in Math. Notes., **43** (1988), No. 3–4, 205–210.

60. Yurko V.A., *Reconstruction of higher-order differential operators*, Differentsialnye Uravneniya, **25** (1989), No. 9, 1540–1550; English transl. in Differential Equations, **25** (1989), No. 9, 1082–1091.

61. Strakhov V.A., *On some questions of the inverse problem theory for differential operators*, Mathematicheskie Zametki **21** (1977), No. 2, 151–160.

62. McLaughlin J.R., *An inverse eigenvalue problem of order four*, SIAM J. Math. Anal., **7** (1976), No. 5, 646–661.

63. Kaup D.J., *On the inverse scattering problem for cubic eigenvalue problems*, Stud. Appl. Math., **62** (1980), 189–216.

64. Caudrey P.J., *The inverse problem for the third-order equation*, Phys. Lett., **79**A (1980), 264–268.

65. Deift P., Tomei C. and Trubowitz E., *Inverse scattering and the Boussinesq equation*, Comm. Pure Appl. Math., **35** (1982), 567–628.

66. Beals R., *The inverse problem for ordinary differential operators on the line*, Amer. J. Math., **107** (1985), 281–366.

67. Beals R., Deift P. and Tomei C., *Direct and inverse scattering on the line*, Math. Surveys and Monographs, **28**, Amer. Math. Soc. Providence: RI, 1988.

68. Sukhanov V.V., *The inverse problem for a self-adjoint differential operator on the line*, Math. Sb. **137**(179) (1988), 242–259; English transl. in Math. USSR Sb. **65** (1990).

69. Deift P. and Zhou X., *Direct and inverse scattering on the line with arbitrary singularities*, Comm. Pure Appl. Math. **44** (1991), No.5, 485–533.

70. Niordson F.J., *A method for solving inverse eigenvalue problems. Recent Progress in Applied Mechanics*, The Folk Odquest Volume, Stockholm, 1967, 373–382.

71. Ainola L.Y., *An inverse problem on eigenoscillations of elastic covers*, Appl. Math. Mech. **2** (1971), 358–365.

72. Fage M.K., *Integral representations of operator-valued analytic functions of one independent variable*, Trudy Moskov. Mat. Obshch. **8**(1959), 3–48.

73. Leont'ev A.F., *A growth estimate for the solution of a differential equation for large parameter values and its application to some problems of function theory*, Sibirsk. Mat. Zh. **1**(1960), 456–487.

74. Hromov A.P., *Transformation operators for differential equations of arbitrary orders*, Diff. Eq. and Function Theory, Saratov Univ., Saratov. **3**(1971), 10–24 (Russian).

75. Matsaev V.I., *The existence of operator transformations for higher-order differential operators*, Dokl. Akad. Nauk SSSR **130** (1960), 499–502; English transl. in Soviet Math. Dokl. **1**(1960), 68–71.

76. Fage M.K., *Solution of certain Cauchy problem via enlargement of number of independent variables*, Mat. Sb. **46**(88) (1958), 261–290.

77. Yurko V.A., *An inverse problem for integral operators*, Matematicheskie Zametki **37** (1985), No.5, 690–701; English transl. in Math. Notes, **37** (1985), No.5–6, 378–385.

78. Yurko V.A., *Recovery of differential operators from the Weyl matrix*, Dokl. Akad. Nauk SSSR, **313** (1990), 1368–1372; English transl. in Soviet Math. Dokl., **42** (1991), No.1, 229–233.

79. Yurko V.A., *A problem in elasticity theory*, Prikladnaya matematika i mekhanika **54** (1990), 998–1002; English transl. in J. Appl. Math. Mech. **54** (1990), No.6, 820–824.

80. Yurko V.A., *Recovery of non-self-adjoint differential operators on the half-line from the Weyl matrix*, Math. Sb. **182** (1991), No.3, 431–456; English transl. in Math. USSR Sb. **72** (1992), No.2, 413–438.

81. Yurko V.A., *An inverse problem for integro-differential operators*, Matematicheskie zametki **50** (1991), No.5, 134–146; English transl. in Math. Notes **50** (1991), No.5–6, 1188–1197.

82. Yurko V.A., *An inverse problem for differential operators on the half-line*, Izvestija Vusshikh Uchebnykh Zavedenii, Ser. Math., No.12 (1991), 67–76 (Russian); English transl. in Soviet Math. (Iz. VVZ), 35 (1991), No.12, 67–74.

83. Yurko V.A., *Solution of the Boussinesq equation on the half-line by the inverse problem method*, Inverse Problems, **7** (1991), 727–738.

84. Yurko V.A., *Inverse problem for differential equations with a singularity*, Differentsialnye Uravneniya **28** (1992), 1355–1362; English transl. in Differential Equations **28** (1992), 1100–1107.

85. Yurko V.A., *On higher-order differential operators with a singular point*, Inverse Problems, **9** (1993), 495–502.

86. Yurko V.A., *An inverse problem for self-adjoint differential operators on the half-line*, Doklady RAN, **333** (1993), No.4, 449–451; English transl. in Russian Acad. Sci. Dokl. Math. **48** (1994), No.3, 583–587.

87. Yurko V.A., *On determination of self-adjoint differential operators on the half-line*, Matematicheskie Zametki **57** (1995), No.3, 451–462; English transl. in Math. Notes **57** (1995), No.3–4.

88. Yurko V.A., *On higher-order differential operators with a regular singularity*, Mat. Sb. **186** (1995), No.6, 133–160; English transl. in Sbornik; Mathematics **186** (1995), No.6, 901–928.

89. Yurko V.A., *On integration of nonlinear dynamical systems by the inverse problem method*, Metematicheskie Zametki **57** (1995), No.6, 945–949; English transl. in Math. Notes, **57** (1995), No.5–6.

90. Yurko V.A., *On inverse problems for nonlinear differential equations*, Differentsialnye Uravneniya, **31** (1995), No.10, 1768–1769; Engl. transl. in Differ. Equations **31** (1995), No.10, 1741–1743.

91. Yurko V.A., *On higher-order difference operators*, Journal of Difference Equations and Applications **1** (1995), 347–352.

92. Naimark M.A., *Linear differential operators*, 2nd ed., "Nauka", Moscow, 1969; English transl. of 1st ed., Parts I, II, Ungar. New York, 1967, 1968.

93. Hromov A.P., *Expansion in eigenfunctions of ordinary linear differential operators with irregular separated boundary conditions*, Mat. Sb. **70**(112) (1966), 310–329.

94. Privalov I.I., *Introduction to the theory of functions of a complex variable*, 11th ed., "Nauka", Moscow, 1967.

95. Gakhov F.D., *Boundary value problems*, 3rd ed., "Nauka", Moscow, 1977; English transl. of 2nd ed., Pergamon Press, Oxford and Addison-Wesley, Reading, Mass., 1966.

96. Leont'ev A.F., *Entire functions*, Exponential series, "Nauka", Moscow, 1983.

97. Stashevskaya V.V., *On inverse problems of spectral analysis for a certain class of differential equations*, Dokl. Akad. Nauk SSSR **93** (1953), 409–412.

98. Gasymov M.G., *Determination of Sturm–Liouville equation with a singular point from two spectra*, Dokl. Akad. Nauk SSSR **161** (1965), 274–276.

99. Carlson R., *Inverse spectral theory for some singular Sturm–Liouville problems*, J. Diff. Equations **106** (1993), 121–140.

100. Marchenko V.A. and Maslov K.V., *Stability of recovering the Sturm–Liouville operator from the spectral function*, Math. Sb. **81**(123) (1970), 525–551.

101. Rjabushko T.I., *Stability of recovering the Sturm–Liouville operator from two spectra*, Theory of Functions, Funct. Analysis and Appl., Kharkov **16** (1972), 186–198 (Russian).

102. Yurko V.A., *On stability of recovering the Sturm–Liuoville operators*, Diff. Eq. and Theory of Functions, Saratov Univ., Saratov **3** (1980), 113–124 (Russian).

103. Alekseev A.A., *Stability of the inverse Sturm–Liouville problem for a finite interval*, Dokl. Akad. Nauk SSSR **287** (1986), 11–13.

104. Hochstadt H., *The inverse Sturm–Liouville problem*, Comm. Pure Appl. Math. **26** (1973), 715–729.

105. Hochstadt H., *On the well-posedness of the inverse Sturm–Liouville problems*, J. Different. Equat. **23** (1977), No.3, 402–413.

106. Mizutani A., *On the inverse Sturm–Liouville problem*, J. Fac. Sci. Univ. Tokio. Sect. IA. Math. **31** (1984), 319-350.

107. Berezanskii Y.M., *An eigenfunction expansion for self-adjoint operators*, "Naukova Dumka", Kiev, 1965.

108. Atkinson F.V., *Discrete and continuous boundary problems*, New York: Academic, 1964.

109. Guseinov G.S., *The determination of the infinite non-self-adjoint Jacobi matrix from its generalized spectral function*, Mat. Zametki **23** (1978), 237–248.

110. Nikishin E.M., *The discrete Sturm–Liouville operator and some problems of the theory of functions*, Trudy Seminara I.G. Petrovskogo, Moscow **10** (1984), 3–77.

111. Berezanskii Y.M., *Integration of nonlinear difference equations by the inverse problem method*, Dokl. Akad. Nauk SSSR **281** (1985), 16–19.

112. Sakhnovich L.A., *Evolution of spectral data and nonlinear equations*, Ukrain. Math. Zh. **40** (1988), 533–535.

113. Moses H.E., *A solution of the Korteweg-de Vries equation in a half-space bounded by a wall*, J. Math. Phys. **17** (1976), 73–75.

114. Ablowitz M.J. and Segur H., *The inverse scattering transform: semi-infinite interval*, J. Math. Phys. **16** (1975), 1054–1056.

115. Karachatryan I.G., *On some solutions of the nonlinear string equation*, Dokl. Akad. Nauk Armyan. SSR **78** (1984), 108–112.

116. Toda M., *Theory of nonlinear latties*, Berlin: Springer-Verlag, 1981.

117. Bogoyavlenskii O.I., *Some constructions of integrable dynamic systems*, Isv. Akad. Nauk SSSR. Ser. Mat. **51** (1978), 737–766.

118. Bogoyavlenskii O.I., *Integrable dynamic systems connected with the KdV equation*, Izv. Akad. Nauk SSSR. Ser. Mat. **51** (1987), 1123–1141.

119. Malamud M.M., *On certain inverse problems*, in: Boundary value problems of mathematical Physics, Kiev (1979), 116–124.

120. Yurko V.A., *On reconstruction of integrodifferential operators from their spectra*, in: Theory of Functions and Approximations, Proceedings of Saratov Winter School, Saratov (1983), Part 2, 156–160 (Russian).

121. Yurko V.A., *An inverse problem for integrodifferential operators of the first order*, in: Functional Analysis, Ul'yanovsk (1984), 144–151 (Russian).

122. Eremin M.S., *An inverse problem for a second-order integrodifferential equation with a singularity*, Differen. Uravn. **24** (1988), 350–351.

Subject Index